THERMAL SOLAR DESALINATION

THERMAL SOLAR DESALINATION

Methods and Systems

VASSILIS BELESSIOTIS

SOTERIS KALOGIROU

EMMY DELYANNIS

Amsterdam • Boston • Heidelberg • London
New York • Oxford • Paris • San Diego
San Francisco • Singapore • Sydney • Tokyo
Academic Press is an imprint of Elsevier

Academic Press is an imprint of Elsevier
125 London Wall, London EC2Y 5AS, UK
525 B Street, Suite 1800, San Diego, CA 92101-4495, USA
50 Hampshire Street, 5th Floor, Cambridge, MA 02139, USA
The Boulevard, Langford Lane, Kidlington, Oxford OX5 1GB, UK

Notices
Knowledge and best practice in this field are constantly changing. As new research and experience broaden our understanding, changes in research methods, professional practices, or medical treatment may become necessary.

Practitioners and researchers must always rely on their own experience and knowledge in evaluating and using any information, methods, compounds, or experiments described herein. In using such information or methods they should be mindful of their own safety and the safety of others, including parties for whom they have a professional responsibility.

To the fullest extent of the law, neither the Publisher nor the authors, contributors, or editors, assume any liability for any injury and/or damage to persons or property as a matter of products liability, negligence or otherwise, or from any use or operation of any methods, products, instructions, or ideas contained in the material herein.

British Library Cataloguing-in-Publication Data
A catalogue record for this book is available from the British Library

Library of Congress Cataloging-in-Publication Data
A catalog record for this book is available from the Library of Congress

ISBN: 978-0-12-809656-7

For Information on all Academic Press publications
visit our website at http://www.elsevier.com/

Working together
to grow libraries in
developing countries

www.elsevier.com • www.bookaid.org

Publisher: Joe Hayton
Acquisition Editor: Lisa Reading
Editorial Project Manager: Maria Convey
Production Project Manager: Nicky Carter
Designer: Victoria Pearson

Typeset by MPS Limited, Chennai, India

CONTENTS

AUTHOR BIOGRAPHIES

Dr Vassilis Belessiotis obtained his PhD from Aristotle University of Thessaloniki. He is the Research Director of the Solar and Other Energy Systems Laboratory of NCSR "DEMOKRITOS" (www.solar.demokritos.gr). His main research areas are renewable energy sources and energy conservation. His interests include the areas of thermal engineering, physical processes, and metrology of physical parameters with application to renewable energy and energy conservation systems as well as to the development of methods and simulation models and measurement procedures and characterization of energy (thermal) products. He was project coordinator in a large number of competitive research projects funded from external sources and his scientific work is published in five books, six original contributions in specialized international encyclopedias (*Encyclopedia of Life Support Systems* and *Encyclopedia of Energy*), and more than 250 papers are published in international journals, international and national conferences (with review), as well as many specialized studies. He has approximately 2000 citations on this work. For his scientific activity he has received five honorary awards. He is a permanent member of scientific organizations: ISES, EDS, IDA, ASHRAE, IHT, HellasLab (national member of EUROLAB).

Dr Soteris Kalogirou received his PhD from the University of Glamorgan, Wales. In June 2011 he received, from the same university, the title of D.Sc. He is Visiting Professor at Brunel University, UK, and Adjunct Professor at the Dublin Institute of Technology (DIT), Ireland. For more than 30 years, he has been actively involved in research in the area of solar energy and particularly in flat-plate and concentrating collectors, solar water heating, solar steam generating systems, desalination, and absorption cooling. Additionally, since 1995 he has been involved in pioneering research dealing with the use of artificial intelligence methods, like artificial neural networks, genetic algorithms, and fuzzy logic, for the modeling and performance prediction of energy and solar energy systems. He has 48 books and book contributions and published 306 papers; 140 in international scientific journals and 166 in refereed conference proceedings. Until now, he has received more than 7500 citations on this work. He is Editor-in-Chief of *Renewable Energy* and Deputy Editor-in-Chief of *Energy*, and Editorial Board Member of another 11 journals. He is the Editor of the book *Artificial Intelligence in Energy and Renewable Energy Systems*, published by Nova Science Inc., Co-Editor of the book *Soft Computing in Green and Renewable Energy*

Systems, published by Springer, and author of the book *Solar Energy Engineering: Processes and Systems*, published by Academic Press of Elsevier. He has been a member of World Renewable Energy Network (WREN), American Society of Heating Refrigeration and Air-conditioning Engineers (ASHRAE), Institute of Refrigeration (IoR), and International Solar Energy Society (ISES).

Dr Emmy Delyannis received her PhD from the Athens University of Technology (AUth) and carried out postgraduate studies at the University of Lawrence, Kansas, USA, and Karlsruhe, Germany. She was Assistant Professor of Chemical Engineering and Metallurgy at AUth, where she taught the subjects of heating and chemical technology. For about 20 years she was the General Secretary of the European Working Group on "Fresh Water from the Sea" and she organized six symposiums for the European Federation of Chemical Engineers. She was coordinator with Dr A. El-Nashar in the section on Desalination by Renewable Energy of *Encyclopedia of Live Support Systems* (EOLSS). She was an unpaid Research Associate of the Solar and Other Energy Systems Laboratory of NCSR "DEMOKRITOS". For her contributions in desalination she was honored with the Public Service Award of the Department of the Interior, USA, and with the Certificate of Merit of the International and Environmental Association, USA.

PREFACE

Water, 'Yδωρ, Wasser...: Water is not only a matter of life and of civilization development, but can also be very destructive if its flow is sudden and uncontrolled. If we look into the history of humanity, we shall notice that all the great civilizations were developed and flourished near large water streams of rivers and seas in eras when water was pure, very clean, a natural product, and the benefit of life and development. It is not uniformly distributed around the world with some places having abundant amounts of freshwater and others totally lacking of it.

Nowadays water quality becomes scarcest as good water sources, streams, aquifers, and natural water deposits are depleted and/or contaminated. Nature continues to produce pure usable water but the use of the physical water sources and deposits is incorrect and irrational. Further more the irrational soil contamination with dangerous materials, such as pesticides, artificial fertilizes, and various chemicals which, by rain water, are accumulated in the various physical water deposits. The rapid demographic increase and the increasing industrial wastes result in the lack of *quality water*.

Today quality water is already a problem as it is severely contaminated and this affects life and surviving conditions of many human communities. Water is a natural property for humanity and must be available to all living beings (humans and animals), but at the same time it can be a weapon against weak human groups because of diversion of rivers, desiccation of lakes, etc. Additionally, clean water is unequally distributed and exists in places that are totally arid.

As a consequence of all these water problems, desalination processes were developed in order to purify contaminated and/or salty waters to produce clean water suitable for any use. Desalination is the elimination of the majority of salts from very salty waters, seawaters, and very brackish waters unsuitable for any use.

Desalination is known from the ancient times as a method to purify salty waters, but the applications were very primitive. The first known description was in the Bible (Exodus, chapter 15):

...... So Moses brought Israel from the Red sea, and they went out into the wilderness of Shur; and they went three days in the wilderness, and found no water. And when they came to Marah, they could not drink of the waters of Marah, for they were bitter: therefore the name of it was called Marah. And the people murmured against Moses, saying, what shall we drink?

And he cried unto the LORD; and the LORD shewed him a tree, which when he had cast into the waters, the waters were made sweet....

The large-scale technical applications of desalination started at about 1950, where the United States and other countries constructed and experimented in large-scale installations, mainly on various distillation techniques which quickly turned out to large industrial applications. They used conventional fuels as energy sources and many were operated as dual plants in conjunction with power plants. Later on, for small production amounts of desalinated water, alternative energy sources, mainly solar and wind, were used.

Desalinated water is in fact an industrial product different from the natural water sources, and desalination processes, like all industrial operations, are not totally eco-logical (green) procedures, that is, the huge amounts of brine which are returned to the sea can harm the sea environment and life in the region where the desalination plant is located.

In this book we try to give basic information on solar-driven desalination meth-ods, including small capacity installations, which found practical applications or they are erected as demonstration pilot units, as well as the basic mathematical models that govern these methods. They are described in a way that can be understandable for people who come into contact for the first time with desalination methods.

In whole nature no other material exists which is so malleable and as yielding as it is water. However, if enough time is given to it, it deteriorates even the hardest rocks.
Nothing can beat it because nothing can to avoid it.
Tao Te King

Dr. Emmy Delyannis

INTRODUCTION

*The real journey of discovery
does not demand only search for new spots,
but demand also to have new eyes.*

Marchel Prust

This book has a small history. It started as an extended paper on solar distillation process. During the progress of writing the paper, some questions have been raised about the confusion that exists in some papers about the solar distillation and solar-driven distillation, because both are solar desalination methods. Thus the paper was extended to the solar-driven desalination which is more complicated, that is, two different connected systems. Then it was extended to all possible desalination methods that may be operated or are in operation by solar energy and thus became finally a whole book.

Nowadays water quality becomes more and more scarcer as good water sources are depleted and/or contaminated. But there exist also dry places possessing only very salty water, in large amount, unsuitable for any human use. These places, in the most cases, have intensive solar radiation. Therefore, this book is dedicated to these regions, hoping that it will be a useful guidance to people who are really interested in this field, and also become a basis for further studies on the subject.

It is evident that such a book has a target to introduce this type of desalination method presenting some views of the subject that will help practicing engineers and extend their knowledge. This monogram is addressed to academic people, other than in desalination field, who want to know the procedures and the technology of a, in some way, special field of desalination. But at the same time this is a book that everyone, as students and simple people, can read to extend their knowledge on what has been achieved up today in the desalination technique.

The book is arranged in a way that is an easy reading for beginners who want to understand the desalination and the solar energy conversion systems. Fundamental conceptions and mathematical equations or models are shortly presented. They are not only suitable for students who enter into this field of knowledge but also may be good for scientists already involved in one of these fields and want to know about other desalination system.

A short historical introduction on what the pioneers achieved in this field is given in most of the methods described.

This book has six chapters and three appendices that include:

Chapter One is a general introduction on the overall existing desalination methods, conventional and nonconventional; thus the reader can have an overview of all desalination methods, thermal and nonthermal. A very short description introduces the reader to the availability of coupling of these methods with solar or renewable energy.

Chapter Two is dedicated to the water, the raw material and also the precious product, the heart of desalination. The properties of water—sea, brackish, natural, and/or desalinated water—are of great importance for an optimal economic operation of any desalination plant, independently of the method used. The more important properties of waters are described, which are useful for proper desalination operation. Where necessary, mathematical equations to calculate these properties are presented and tabulated values are also given for easy use. All seawater properties are based on the new, improved, oceanographic properties established by International Association for Properties of Water and Steam (IAPWS, 2008—10), which includes tables of seawater and freshwater properties.

Chapter Three presents solar distillation method and techniques, the older desalination method, known from the antiquity. It is a method that has been studied extensively by researchers mainly on the so-called *sunny belt* of the world, where exist the most poor, small, and remote communities lacking freshwater and in most of the cases electrical grid as well. This is a difficult part of the book as large amount of research is done on solar, but only little work on solar distillation plants exists. The chapter presents the most important work done in this field. Over the last 15—20 years only theoretical and very small-scale experimental works have been traced in the extended bibliography on the subject. No reference on solar distillation plants is found.

Chapter Four is dedicated to a rather new solar-driven and/or conventional fuel system, the "membrane distillation" (MD) method, under commercial development. It is a very promising system for desalination; thus a whole chapter is dedicated to it where the basic mathematical models and laws that govern the operation features of the method and the various techniques of operation are presented. Special attention was given to the hydrophobic microporous membranes that are used in this method. This chapter is written by Dr Panagioti Boutiko, external collaborator of Solar and Other Energy Systems Laboratory of NCSR "DEMOKRITOS".

Chapter Five is dedicated to the rather new method of humidification/dehumidification (HD). It is a method that can operate by any fuel but is mainly addressed to solar energy as an energy source. Solar distillation is in fact an HD method, only the operational technique differs, and in the way of solar thermal energy used. In fact it is a solar-driven nonconventional system. It is, by the time being a promising method, still in the stage of research and development.

Chapter Six includes the most important part of the solar-driven desalination field. The coupling of two different systems, both mature and totally commercialized, is presented: the solar energy conversion system is not part of this book, but thermal solar conventional desalination systems are shortly described and analyzed. The novelties of solar energy conversion systems and what is new in desalination are described shortly, giving the corresponding bibliography for further information if needed.

The last part of this book, the appendices, refers to tables and diagrams concerning water or desalination parameters.

While tracing bibliography in desalination in general we discovered that there is a variety of symbols, although less for units, that create confusion many times. Many authors use SI, others the Anglo-Saxon, a mixture of them or their own symbols. Here we tried to give some tables based on SI that we recommend as basis for all desalination methods. Thermal desalination methods of mathematical analysis depend mainly on heat transfer symbols and units. For electrodialysis and reverse osmosis, the situation is in some extent different as there have been established few new concepts. Thus we recommend in the corresponding tables the most used symbols and SI units. In Tables A.5—A.7, for application in solar energy systems, the units are given as proposed by the International Solar Energy Society (ISES). Overleaping may exist with the other tables which we believed that should be remained as they are. It should be noted that for applied desalination, in many cases, the Anglo-Saxon system is still used.

Appendix A is dedicated to tables of general interest and of symbols and units based on the SI (International System of Units).

Appendix B presents tables of chemical and trace elements in seawater and tables containing properties of physical water. Solubility of gases in seawater, especially oxygen and carbon dioxide, is included as these two gases attack the metallic surfaces of desalination installation creating corrosion problems.

Appendix C contains some diagrams of seawater characteristics and properties especially assembled for desalination purposes.

Finally, it is our pleasant obligation to express our sincere gratitude to Ms. Maria Christodoulidou, MSc, Energy Engineering and Quality Manager at the Solar & Other Energy Systems Laboratory of NCSR "DEMOKRITOS" for her valuable and willingly offered assistance in the preparation of this first edition of the book.

Vassilis Belessiotis
Soteris Kalogirou
Emmy Delyannis

CHAPTER ONE

Desalination Methods and Technologies—Water and Energy

1.1 INTRODUCTION

Although the main scope of this book is to describe and analyze solar desalination methods and related fields, few words about conventional desalination methods are necessary for students or scientists to refresh their knowledge on what in general desalination is. When we refer to conventional desalination methods, we mean these methods that use as a driving force conventional energy sources, ie, fuels and/or grid electricity. Exactly the same methods may use alternative energy sources like solar, wind, etc., but in this case the desalination system is of smaller capacity and adopted to fit the alternative energy collection system.

Desalination is bounded to the quality of fresh water. Nowadays demographic increase, and tremendous industrial consumption increase, which leads to the rapid water resources exhaustion. Natural fresh water resources and simultaneously natural water streams and wells are polluted by industrial, domestic, or community wastes. The biggest polluting section is agriculture. Of all available water in a region, agriculture consumes about 70% the rest remaining for all other needs. Large amounts of chemical fertilizers, pesticides, and insecticides are drawn away by rain water, polluting water streams, aquifers, natural water aqueducts, thus decreasing available fresh water sources.

World wide there exist regions with no fresh water resources and a minimum of rainfall if at all there is any precipitation. Meanwhile the intensive solar radiation evaporates almost the last drops of existing soil humidity.

Worldwide about 33,000 km of coasts surrounding dry regions are situated mainly in the Indian Ocean, the Arabian Peninsula, the North African coasts and Mediterranean, the Caribbean Islands, and around Australia. In many other places, water is more abundant but quality water resources are rapidly exhausting due to intensive exploitation, and water streams became continuously more polluted. For all these regions, desalination becomes inevitable.

Vigotti and Hoffman (2009) refer that clean fresh water demand has been tripled the last 50 years. About 2000 worldwide, fresh water demand was about 4200 km^3, ie, 30% of the available natural fresh water. It is estimated that at about 2025, it will be 70%.

But what do we really mean by desalination methods, described in this chapter?

Thermal Solar Desalination

1.2 WHAT IS DESALINATION?—WHERE DOES IT APPLY?

Desalination is a physical procedure of separating the excess of dissolved salts from waters, brackish and seawater, or any aqueous salt solution in order to collect low-salt content water for any suitable use, such as drinking, industrial, pharmaceutical, municipal, or household water. Desalination is a pure industrial procedure and an intensive energy system independently of the process, rending energy cost as the major economic problem. Energy consumption of distillation process is higher than any other in the chemical industry and as a consequence desalinated water is still costly despite the dramatic decline in cost during the recent years. Its price is higher than transportation of the same amount of natural water from a short distance. However, transportation or pumping of water from longer distances may be as expensive or more expensive as desalination systems, depending on the distance and the morphology of the site.

Cost of natural water pumping is due mainly to capital cost of the installation. At the beginning of the operation, consumption may be lower than real capacity installed and cost relatively high. Consumption usually increases rapidly and cost decreases until the water source is almost exhausted. Then another source must be found further away from the site to start again. A desalination plant is more flexible to demand fluctuations and capacity may be increased according to demand by installing new units. Furthermore is the only solution for places blessed with abundant brackish water wells or are surrounded by seawater but have no fresh water resources.

1.2.1 The Desalination Processes

The progress in and the development of desalination technology and methods resulted from the activities of Office of Saline Water (OSW) of the US Department of the Interior, from 1952 up to about 1972. During these years, the intensive OSW activities in R&D and the construction of five demonstration plants, of 1 mgd each, developed and promoted desalination to its large industrial application. Since 1972, industry undertook R&D and technical applications.

Many desalination methods have been studied intensively but few of them have found wide industrial application. These methods are divided into two general groups: thermal and nonthermal processes or they may be divided according to the element separation mode, ie, if water or salt is the element that is separated. Both classifications are technical in order to facilitate analysis of the methods. The main desalination processes include, according to the energy type applied:

1. *Thermally driven desalination methods (or distillation processes)*
   - Solar distillation (SD)
   - Humidification—dehumidification (H/D)

- Multiple humidification—dehumidification (MHD)
- Membrane distillation (MD)
- Multiple effect distillation (MED)
- Multistage flash distillation (MSF)
- Thermal vapor compression (TVC).

All are distillation methods applying different operation and equipment techniques, independently of the type of energy used. SD, MD, and H/D do not belong yet to the conventional desalination methods. They are of small capacity procedures and still under investigation not totally commercialized. Conventional desalination thermal methods are fired by steam produced by conventional fuels.

2. *Electrically driven desalination methods*
- Reverse osmosis (RO)
- Electrodialysis (ED), electrodialysis reversal (EDR)
- Mechanical vapor compression (MVC).

Fig. 1.1 presents a flow sheet of the existing conventional desalination processes, either still under investigation or totally commercialized. High capacity thermal processes are driven by conventional energy sources, as fuels, in the form of vapor

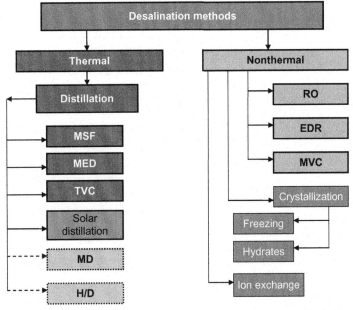

Figure 1.1 Flow sheet of the existing desalination methods according to type of energy used. Thermal conventional desalination methods are presented in red and conventional desalination methods fired by electricity in green. Blue color refers to methods that were applied but found no wide application. (For interpretation of the references to color in this figure legend, the reader is referred to the web version of this book.)

pressure and nonthermal by grid electrical energy. From the above methods, MSF, MED, TVC, RO, MVC, and ED/EDR are technologically mature. Most of their operational problems have been solved and main effort today is to reduce energy consumption to increase efficiency and to construct more compact units in order to reduce transportation weight, installation cost, and consequently cost of produced fresh water. The rest of the methods, except for MD and H/D which are still in the stage of industrial development, have found no wide practical application due to operational and cost problems, which are described later.

Today there exist more than 15,000 conventional desalination units worldwide, producing ~ 70 million m^3 fresh water per day. 50% of this desalinated water is produced in Middle East countries. Of these units, 60% are RO installations having energy consumption $\sim 5\,kWh\,m^{-3}$ and about 27% are MSF plants consuming $25\,kWh\,m^{-3}$. MED installations contribute with about 8% of worldwide desalinated water production and share 15% of the thermally desalination methods (IDA, 2010−2011).

Another way of classifying the flow sheet of the existing conventional desalination processes is according to the separation mode of water or of salts as shown in Fig. 1.2.

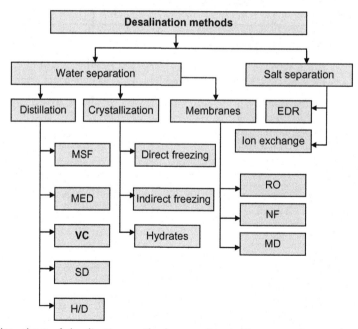

Figure 1.2 Flow sheet of desalination methods according to the separation mode of water or of salts.

1.2.1.1 Distillation Processes

Conventional distillation techniques started the application of large-scale water production in about 1950. They are thermally driven methods. Today there are mature methods commercially available in large capacities. Distillation processes are phase changing procedures separating dissolved salts from salty waters or aqueous solutions by changing water phase to vapor, which condenses into pure water. For each kilogram of evaporated water, an amount of about 2230 kJ is needed. This amount of heat is partly recovered from the condensates as sensible heat. The remaining liquid is the concentrated brine which in most of the cases is rejected back to the sea. Some trials have been made to recover a few of the chemical compounds in the concentrated brine, as magnesium, etc., but turned out to be uneconomic procedures. Desalination by distillation is a pure evaporation procedure.

Solar distillation. This is a procedure that uses solar radiation directly. Incident solar rays are captured in devices called "*solar stills*" where solar radiation is absorbed by the water in the basin of the still and is converted into thermal energy, which evaporates the water in the still. The created vapor condenses on the inside transparent colder cover surface of the still and collected by gutters. It is an H/D procedure taking place simultaneously in the same device, the solar still. It is a very low to low capacity procedure.

Humidification–dehumidification. This is an alternative of SD, a method that uses separate compartments, called towers or chambers, for the evaporation (humidification) and condensation (dehumidification) procedures. Solar energy is collected separately mainly by solar flat plate collectors. It is a low to medium capacity method.

Membrane distillation. In this method, hot seawater feed flows across a membrane surface which is permeable only to water vapor. Vapor permeates through the membrane pores and condenses on the other side of the membrane. This method uses microporous hydrophobic membranes permeable only to water vapor. It is a low to medium capacity method.

Multiple effect distillation. This is a middle to high capacity desalination method. The feed seawater is heated by pressurized steam in the first effect and then forwarded in a battery of consecutive chambers each having lower pressure from the previous one. Created vapors are condensed giving the condensation enthalpy to incoming feed seawater.

Multistage flash distillation. The hot feed water heated by steam in the first stage enters a battery of compartments, each at lower pressure than the previous one. The hot water expands into each chamber releasing vapor which condenses giving fresh water and simultaneously its condensation enthalpy to the incoming seawater feed.

Thermal vapor compression. By this distillation technique, part of the released vapor which remains in vapor phase during operation is compressed to high pressure by means of a steam thermocompressor. The compressed vapor is used to heat the feed seawater entering the distillation compartment.

1.2.1.2 Reverse Osmosis

RO separates water from salts by applying mechanical energy, in the form of pressure, to a semipermeable membrane. Pressure is higher than the osmotic pressure of the salt solution and its level depends on the salt concentration in the solution. RO is a one phase procedure that operates at low temperature, having very low energy requirements for operating the corresponding pump.

RO was applied for the first time in around 1962, in the beginning only for brackish waters (BWRO) due to the short life of membranes. Extended research on semipermeable membranes achieved a longer membrane life and better performance of the procedure and since about the middle of 1973 seawater reverse osmosis desalination (SWRO) entered the water industry. Since that time RO has been totally commercialized, it found a wide application for brackish and seawater desalination and is competitive to distillation methods, now used in about 50% of plants around the world. There exist various configurations of membrane modules such as tubular, spiral wound, and hollow fiber (Figs. 1.3 and 1.4).

Typically an RO module is about 1.0 m long and the pressure vessel has a diameter of 10−20 cm, and total membrane surface ~ 33.9 m^2, depending on the manufacture and the kind of solution to be treated. A new RO module was presented by Koch Membrane Systems, USA (von Gottberg, 2004; Voutchkov, 2004; Bergman and Lozier, 2010). The new SWRO module, called "*MegaMagnum*," has dimensions: pressure vessel diameter = 45.7 cm (18″) and 1.55 m (61″) length, and total membrane surface of ~ 241.5 m^2. This new RO module incorporates into the pressure vessel only five membrane elements. The manufacturers claim economy in O-rings and some other features. Fig. 1.5 gives a view of the usual way the RO modules are installed in stacks.

1.2.1.3 Electrodialysis and Electrodialysis Reversal

Both variations use ion exchange membranes, which separate salt ions from a solution giving pure water. The principle of operation is shown in Fig. 1.6. Simple ED was

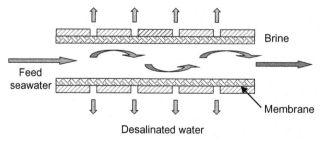

Figure 1.3 Principle of tubular membrane element.

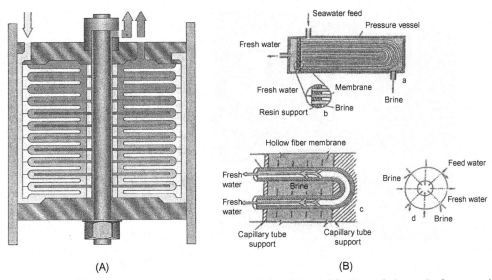

Figure 1.4 Hollow fiber reverse osmosis (RO) modules. (A) Modification of the early frame and plate RO of ROCHEM Company, Germany, the so-called "disk-tube membrane module" (www.rochemindia.com/technology/pt_module.html). (B) (a and b) RO permeators with hollow fiber capillary tubes arrangement. (c) Enlargement of a U-shaped hollow fiber tube with the circulation of the seawater and fresh water streams. (d) Cross-section of the enlarged hollow fiber.

Figure 1.5 Stacks of conventional reverse osmosis modules (16′ × 40′) (Morris, 1998).

applied first in about 1960, in small industrial size installation only for brackish waters having salt content <5000 mg L^{-1}. EDR was first applied at about 1973. The difference of the two procedures depends on the operation mode. The ED cell technology is exactly the same. In common ED electrode, poles remain stable continuously. The EDR procedure reverses the polarity in about every 20′. This mode of operation

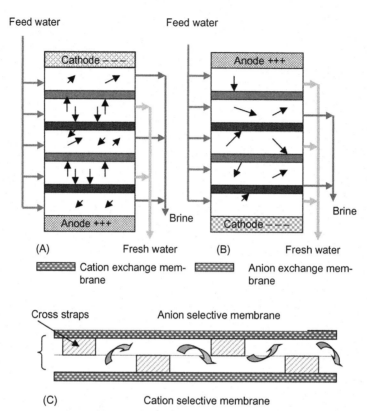

Figure 1.6 Principle of electrodialysis and electrodialysis reversal: (A) electrodialysis unit with positive polarity (Ionics Inc., 2002; von Gottberg and Siwak, 1998); (B) the unit after polarity is reversed having negative polarity; and (C) brackish water flowing through anion and cation membranes.

decreases considerably scale formation and fouling onto the membrane surface. Energy consumption in ED is proportional to salt concentration in the solution and thus it is applied mainly to brackish waters.

A membrane stack having anionic and cationic membranes separated by a narrow channel, the spacer. The spacers represent the flow path of product pure water and the concentrated brine streams. They are made of low-density polyethylene material and are placed in the membrane stack so that each stream is manifolded together (Fig. 1.6C).

Two ED installations are shown in Fig. 1.7.

ED produces very high purity water which is suitable for pharmaceuticals, food industry, and any industry that uses high purity water.

1.2.1.4 Mechanical Vapor Compression
Part of vapor from the last stage, which does not condenses during operation, in a distillation technique is compressed by mechanical means and used as heating energy for the seawater evaporation.

Figure 1.7 Two electrodialysis installations: (A) The Ionics Inc., Watertown, MA, USA (Ionics Inc., 2002), and (B) electrodialysis cells designed by Ashahi Chemical Industry Com, Japan (Hamada, 1992).

1.2.1.5 Crystallization Processes

Crystallization processes comprise freezing of seawater and hydrate formation. Both despite their simplicity have found no wide industrial applications.

The Freezing Process. This is simple in application, as during freezing seawater crystallize almost to pure water ice leaving the dissolved salts in the solution, which is continuously concentrated. Despite its simplicity and despite intensive research, freezing has not found a wide industrial application, mainly due to small crystals formed during the freezing procedure. Small crystals are not easily separated from the ice brine slurry and almost half of the fresh water produced is used to wash out the salts from the ice surface, considerably reducing the efficiency of the method, increasing product cost.

Hydrate Formation. This is an alternative to produce pure crystals. Water combines with other substances to form hydrate crystals. For desalination purposes, hydrocarbons like propane or butane have been studied. During crystal formation, all impurities like the dissolved salts in seawater are excluded and the crystals formed are pure hydrates. After hydrate formation, the gas is released giving pure water. The process found no large-scale commercial applications due to many problems arising during operation.

1.2.1.6 Ion Exchange Processes

Ion exchange is a physicochemical method using anion and cation exchange resins to desalinate waters. It is mainly applied to natural hard water or low-grade brackish water. Studies have been performed for seawater as well, but the procedure has not found wide industrial application, mainly due to quick saturation of resins with ions, which made the procedure costly. Today it is used for seawater pretreatment to avoid hard scale formation.

1.3 OPERATION STEPS OF A DESALINATION PLANT

Independent of the desalination method, water must be treated, before entering the desalination unit, by filtration to remove coarse-suspended matter, organic material and microorganisms, such as sand and plankton, and then should be pretreated by a chemical method to remove or to prevent calcium and magnesium salts from forming hard scale on the warm metallic surfaces of the thermal system or onto the membrane surfaces. Posttreatment is also necessary for making desalinated water palatable. Disinfection is necessary only for low temperature desalination methods. The flow sheet of Fig. 1.8 shows the general steps to be followed in a desalination installation, independently of the desalination technique used. For desalination by MSF or MED, the desalination system may be combined with a power station. Fresh water from the desalination unit is used to produce steam for the electricity system and the station's waste heat is used as thermal energy for the desalination plant thus decreasing total cost of operation (dual-purpose plant).

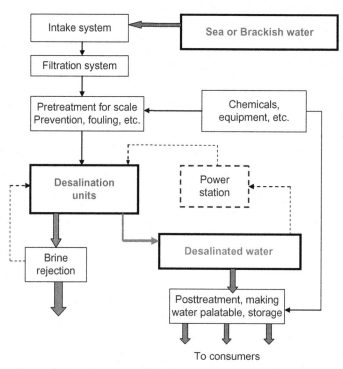

Figure 1.8 Flow sheet of the main steps of desalination systems operation, independently of the desalination method applied.

1.4 WATER AND ENERGY

As referred previously, desalination methods, especially distillation, are intensive energy enterprises which render energy as the basic factor to an economic operation of the system and consequently a mean for low price desalinated water.

Water and energy are two physical sources, the basis of technological development and consequently of the society in general. Conventional energy sources are continuously depleted and simultaneously water sources are continuously contaminated or wasted.

Today the primary energy consumption amounts, worldwide, to about 500 EJ y^{-1} ($500 \times 10^{18} \text{ J y}^{-1}$). The last 130 years the energy consumption increased about five times per capita increasing considerably carbon dioxide emission. Before the industrial revolution, carbon dioxide emission was stable (by only physical activations) and about 280 ppm. Between 2001 and 2007, mean emission increased to 18×19^9 tons per year (Schindewolf and Böddeker, 2010), mainly by human activities, closely approaching a value of 400 ppm.

Renewable energies are available in huge amounts and can be used without or at least with minor carbon dioxide emissions. Solar energy is available worldwide but in unequal intensity. Its intensity is a function of the hour of the day, the day of the season, and of the local region, ie, of the latitude of the site. In most cases, high solar radiation intensity and lack of natural fresh water rend solar energy as an inexpensive energy source for seawater desalination.

The highest solar energy intensity on earth's surface is 1 kWh m^{-2} and global radiation on earth's surface amounts in $4 \times 10^{24} \text{ J y}^{-1}$, which is 10^4 times the primary energy consumption ($5 \times 10^{20} \text{ J y}^{-1}$) worldwide.

The % renewable energies contribution to the global energy production (left column, in EJ per year (per annum)) is presented in Fig. 1.9. In Fig. 1.10, the diagram depicts the renewable energies (solar energy is not included) that are potential energy sources for desalination methods.

It should be noted that nuclear energy is an alternative energy source for large capacity desalination plants. For desalination, it is a renewable energy source as long as we take in consideration the huge amounts of energy, in the form of vapor waste heat, but sure is not an environmental friendly source.

May be we have to refer here that conventional distillation methods can be operated easily by using the huge amounts of waste vapor delivered from nuclear plants producing electricity. It is a safe, ie, without any radiation, waste heat thermal energy material. There exist many studies on coupling nuclear power plants with large capacity distillation plants (Belessiotis et al., 2010), but in fact only one dual-purpose, nuclear desalination plant has been erected worldwide, not any more in operation.

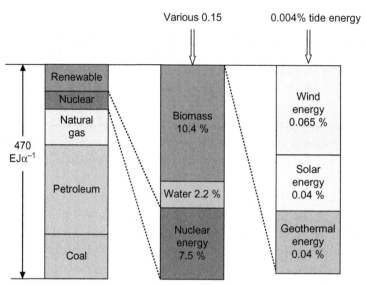

Figure 1.9 Contribution of renewable energy sources (%) (right and middle columns) to the global production of energy (left column) (Schindewolf and Böddeker, 2010).

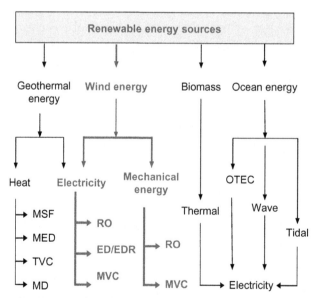

Figure 1.10 Potential utilization of renewable energies in desalination systems. Green color refers to wind energy which has been already applied for small desalination installations, mainly RO. (For interpretation of the references to color in this figure legend, the reader is referred to the web version of this book.)

Table 1.1 Suggestive Amount of Renewable Energies that Fires Various Desalination Methods (ProDes, 2009)

Method	Energy	Amount	Method	Energy	Amount
RO	PV	31%	MEH	Solar	9%
RO	Wind	12%	MD	Solar	11%
ED/EDR	PV	3%	Hybrid		3%
MSF	Solar	7%	Various[a]	Solar	15%

[a]Majority refers to solar energy application.

We refer to the Sevschenko nuclear-desalination plant in ex USSR (Novikov et al., 1969; Baranov et al., 1983; Egorov et al., 1980).

The suggestive amount of renewable energies which are used in desalination methods (ProDes, 2009) is presented in Table 1.1.

1.4.1 Energy Necessary for Desalination

In general, every source of energy is suitable for desalination after the proper conversion to thermal energy or electricity. Large capacity conventional distillation plants ($<< 10,000$ m^3 per day) use pressurized steam as energy source, generated by conventional fuels, and electrical driven desalination uses electricity from grid to ensure continuous and efficient operation of the corresponding desalination system.

Alternative energy sources are the renewable energies, thermal or nonthermal, such as solar, wind, wave, geothermal, and biomass, although have not been applied practically in larger scale they are suitable for small capacity desalination installations. They are still today under investigation and the total capacity of existing plants is roughly about 0.1% of the global conventional desalination installations.

From all renewable energy sources, solar energy found wider application. The advantages of solar energy against other renewable energy sources are:
- It is available worldwide on the earth's surface, delivers about 89,000 TW per day when global earth's consumption is only about 15 TW (Hamada, 1992)
- It is converted easily to thermal or electrical energy and up to day is the most studied and advanced renewable energy source.

There exist places on the earth, the so-called "*Sunny Zones*" where sunshine last 8−10 hours per day, having high solar radiation intensity. Richter (2009) refers that in 1 km^2 of desert can be collected, by concentrating collectors, about 250×10^6 kWh per year electricity and generate 60×10^6 m^3 per year fresh water from it.

Nowadays the collection and conversion technology of solar radiation to useful energy has application to small (<10 m^3 per day) and medium (10−100 m^3 per day) installations. There exist some theoretical studies for large plants up to 100,000 m^3 per day but yet found no wider applications (Mangano and Fabi, 2010; Hofmann, 1992; Sagie et al., 2005).

For solar energy utilization, there exist some restrictions in collection and use that affect the operation of solar desalination systems such as:

- Solar radiation is available only during day hours directly or if there is high diurnal intensity the surplus energy can be stored for night operation
- It is a spread out and fluctuating energy having low flow density. In high rates incident solar radiation can be collected by an amount of about 50—68%, highest.

The low density can be overcome by the use of solar concentrators in order to achieve higher temperatures suitable for MES or MSF units. Typically, systems converting solar radiation to thermal or electrical energy have low efficiencies and need large surface areas to be installed.

Table 1.2 presents indicative combinations of the main desalination methods with the corresponding suitable solar energy collection systems (ProDes, 2009).

The selection of a desalination system fired by solar energy, especially for remote and arid regions, must present the following characteristics (Abdul-Fattah, 1986):

- Simplicity
- Easy handling
- Systems characteristics must fit to the site of installation
- Possibility of future capacity extension and efficiency optimization.

1.5 THERMAL SOLAR DESALINATION

Solar desalination refers to the extraction of pure water from an aqueous solution, which contains dissolved salts, by using the energy from the sun. An extension may be also the separation of water from various dissolved or suspended matters from waste waters. In this case, dissolved matter may also be organic compounds. There exist two main methods of solar desalination depending on the way by which solar energy is captured and utilized.

Solar distillation. In this method, the solar radiation is used directly to distill the water into solar stills. Solar stills are simple tools that convert solar radiation into heat and simultaneously evaporate the water. Solar stills may be combined, as auxiliary tools, with flat plate collectors and/or solar ponds to increase the system's efficiency.

Solar desalination. It is also called "solar driven, solar powered, and/or solar assisted desalination." In this method, solar radiation is used indirectly by capturing solar energy in a solar collector field. Solar driven desalination plants are dual plant

Table 1.2 Typical, Indicative, Characteristics of the Various Desalination Technologies and the Corresponding Suitable Energy Sources

Desalination Method	Capacity (m³ per day)	Collector Technology	Temperature Region (°C)	Necessary Energy
SD	<0.1	Flat plate	40–80	$4400{-}5800\ kJ\ kg^{-1}$ $1166{-}2188\ kWh_{th}\ kg^{-1}$
MEH	1–100	Flat plate or vacuum tube	65–80	$100\ kWh_{th}\ m^{-3} + 1.5\ kWh_e\ m^{-3}$
MD	0.15–10	Flat plate or vacuum tube	70	$150{-}299\ kWh\ m^{-3}$
MED	>5000	CPC	110–160	$60{-}70\ kWh_e\ m^{-3} + 1.5{-}2.0\ kWh_e\ m^{-3}$
MED$_{LT}$	>5000	Flat plate or waste steam	70–80	$60{-}70\ kWh_e\ m^{-3} + 1.5{-}2.0\ kWh_e\ m^{-3}$
MSF	>>10^4	CPC/CDC	90–120	$60{-}70\ kWh_e\ m^{-3} + 1.5{-}2.0\ kWh_e\ m^{-3}$
TVC	>5000	Vacuum tube	Up to 110	$8{-}10\ kWh_{th}\ m^{-3}$
RO	<100	PV	Up to 35	$SW = 4{-}5\ kWh_e\ m^{-3}$ $BW = 0.5{-}1.5\ kWh_e\ m^{-3}$
EDR	<100	PV	Up to 35	$3{-}4\ kWh_e\ m^{-3}$

MED$_{LT}$, low temperature MED; SW, seawater; BW, brackish water.

techniques consisting of two totally different independent systems: the solar radiation conversion system and the conventional desalination units. Capture and conversion systems may be:

Solar collectors that convert directly the solar radiation into heat which is used either as heat to power conventional distillation units such as MSF, MED, and combined TVC–MED or is converted to electricity (for high temperature concentrators) mainly of medium capacity to fire ED, RO, and MVC units.

Photovoltaic cells, which convert solar energy directly into electricity, or wind energy, an indirect solar energy system, which may also be used to produce electricity for desalination purposes.

The collectors that can be used are: flat plate (FPC), evacuated tubes (ETC) for low or medium temperatures or concentrating collectors as parabolic trough (PTC), compound parabolic collectors (CPC) depending on the desalination method and the corresponding capacity. Central receivers may provide electricity and high temperature steam or low temperature waste heat from the turbine for desalination purposes. Thus a solar collector field may be used either by heating feed seawater or by generating steam to drive the conventional distillation plant.

Solar driven desalination plants, in general, are of small to medium capacity installations, up to 5000 m³ per day fresh water production. SD plants are even smaller. The largest SD plant ever built was of a maximum peak output of 40 m³ per day and normally of about 20 m³ per day. Fig. 1.11 presents the utilization of solar energy in various desalination fields.

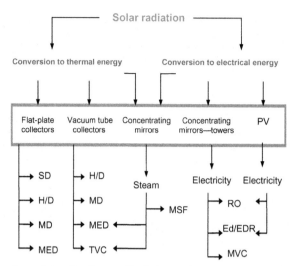

Figure 1.11 The coupling of desalination methods to a solar energy converting system.

1.6 NEW TRENDS IN DESALINATION

During the recent years, the efforts of researches and engineers are to achieve more energy-efficient desalination technology. In this search, combinations of desalination methods and/or combined desalination and power generation plants result in performance and economic improvements. These new approaches comprise also revived interest on old desalination methods that found not practical application, ie, freezing. The economic advantages of desalination/power generation dual-purpose plants are already known. The new trends nowadays are described below.

1.6.1 Hybridization

This refers to combination of desalination methods with or without power generation systems. The main scope of hybridization is the overall plant economic and operation optimization. The first ideas about this innovation were given by Awerbuch et al. (1987). Papers describing the benefits of hybridization are: in addition to the one mentioned above by Awerbuch et al. (1987), a detail economic study on this novel trend presented by Hellal (2009), other suitable combinations by Awerbuch et al. (1989), various combinations of these novel desalination technology by Mahbub et al. (2009), and a paper by Agasichen (2004) who presents a detail economic study and concludes that hybridization of thermally driven plants can provide a number of advantages to the new systems.

1.6.2 Forward Osmosis

This is an innovative procedure, a new emerging technology that gains increasing attention in seawater desalination. It was proposed first by Kessler and Moody (1976) and Moody and Kessler (1976). It is an osmotic process that separates water from dissolved salts by means of a semipermeable membrane by creating an osmotic gradient. The basis of the procedure is the use of a draw solution such as liquid ammonia-carbon dioxide mixture which poses a high osmotic coefficient and is separated easily from potable water through a suitable semipermeable membrane. It is a low energy consumption procedure still under investigation (Low, 2009).

In the new trends the vessel-based approach may be added, not a desalination method but the concept to use large vessels, carriers, etc. with desalination installation, to provide up to 9×10^6 m^3 per day desalinated water to regions lacking water. All desalination methods are applied comprising those fired by renewable energy according to demand and the size of the vessel.

REFERENCES

Abdul-Fattah, A.F., 1986. Selection of a solar desalination system for supply water in arid zones. Desalination. 60, 165–189.

Agashichev, S.P., 2004. Analysis of integrated co-generative schemes present value of examples and levelized cost of water. Desalination. 164, 281–302.

Awerbuch, L., Van der Mast, V., Soo Hoo, R., 1987. Hybrid desalting systems—a new alternative. Desalination. 64, 51–63.

Awerbuch, L., May, S., Soo Hoo, R., Van der Mast, V., 1989. Hybrid desalting systems. In: Proceedings of 4th World Congress on Desalination and Water Reuse, Kuwait, November 4–8, 1989, Vol. IV, pp. 189–197.

Baranov, Ju. S., Kabluchko, N.A., Lebedev, P.K., Samarkin, A.A., Egorov, A.P., Podberezni, V.L., 1983. Operating experience of desalination units in Shevchenko. Desalination. 45, 167–174.

Belessiotis, V., Papanikolaou, E., Delyannis, E., 2010. Nuclear desalination: a review on past and present. Desalin. Water Treat. 20 (1–3), 45–50.

Bergman, R.A., Lozier, J.C., 2010. Large diameter membrane elements and their increasing global use. IDA Desalin. Water Reuse. 2 (1), 16–23.

Egorov, A.P., Shatsillo, V.G., Borissov, B.M., Lazarev, I.P., Vakhin, I.G., 1980. Questions in improvement in thermal desalination technology at operating plants in the town of Shevchenko. In: Proceedings of 7th International Symposium on "Fresh Water from the Sea", Vol. 1, pp. 205–218.

Hamada, M., 1992. Brackish water desalination by electrodialysis. Int. Desalin. Water Reuse. 2 (4), 8–16.

Hellal, A.M., 2009. Hybridization—a new trend in desalination. Desalin. Water Treat. 3, 120–135.

Hoffman, D., 1992. The application of solar energy for large scale desalination. Desalination. 89, 115–184.

IDA Desalination Yearbook, 2010–2011. Water desalination Report. Media Analytic Ltd., Oxford, UK (2010–2011).

Ionics Inc., 2002. Bright future forecast for EDR. Desalin. Water Reuse. 12 (3), 38–41.

Kessler, J.O., Moody, C.D., 1976. Drinking water from sea water by forward osmosis. Desalination. 18, 297–306.

Low, S.C., 2009. Preliminary studies of seawater desalination using forward osmosis. Desalin. Water Treat. 7, 41–46.

Mahbub, F., Hawlander, M.N.A., Mudjumdar, A.S., 2009. Combined water and power plant (CWPP)—
    a novel desalination technology. Desalin. Water Technol. 5, 172—177.
Mangano, R., Fabri, C., 2010. Desalination must pay more attention to solar energy. Desalin. Water
    Reuse. 19 (4), 34—38.
Moody, C.D., Kessler, J.O., 1976. Forward osmosis extractors. Desalination. 18, 283—295.
Morris, A., Curaçao, K.A.E., 1998. Desalin. Water Reuse. 8, 40—44.
Novikov, E.R., Chernozubov, V.B., Golub, S.I., Shatillo, V.G., Tkach, V.I., Sobolev, E.A., et al., 1969.
    Nuclear industrial desalination plant with fast nuclear reactor at Shevschenko. Desalination. 6, 34,
    349—367.
PRODES, 2009. Roadmap for development of desalination powered by renewable energy. In:
    Paprapetrou, M., Wieghause, M., Biercamp, C. (Eds.). Available from: www.prode-project.org.
Richter, C., 2009. The role of solar thermodynamic system in delivering clean water. IEA, Working
    Party on RE Technologies, Paris. Available from: www.prodes-project.org.
Sagie, D., Magdenberg, E., Weinberg, J., 2005 Commercial scale solar-powered desalination, Orlando.
    ISES, August 8—12, 2005. Paper No. 155.
Schindewolf, U., Böddeker, K.W., 2010. Renewable energies. Desalin. Water Treat. 13, 1—12.
Vigotti, R., Hoffman, A. The water scarcity and water security: political and social implications. In:
    Workshop on Renewable Energy, IEA Working Party on Renewable Energy, Technology and Water,
    Rome, March 23, 2009.
Von Gottberg, A.J.M., 2004. Introducing world largest RO element. Int. Desalin. Water Reuse. 14,
    22—26.
von Gottberg, A.J.M., Siwak, L.R., 1998. Re-engineering of the electrodialysis reversal process. Int.
    Desalin. Water Reuse. 7, 33—37.
Voutchkov, N., 2004. Reducing SWRO costs and impact by power plant co-location. Desalin. Water
    Reuse. 14 (3), 32—36.

GENERAL INTEREST BOOKS

The first seven are old books that may be useful as theoretical method basis but not as a contemporary
    technical desalination development. They are presented here just for the reader's knowledge.

Spiegler, A.S., 1966. Principles on Desalination. Academic Press, New York, 566 p.
Merten, U., 1966. Desalination by Reverse Osmosis. The MIT Press, Cambridge, MA, 300 p.
Lakshminarayanaiah, N., 1969. Transport Phenomena in Membranes. Academic Press, New York, 516 p.
Lacey, R.E., Loeb, S., 1972. Industrial Processing with Membranes. Willey Interscience, New York, 348 p.
Khan, A.H., 1986. Desalination Processes and Multistage Flash Distillation Practice. Elsevier, Amsterdam, 596 p.
Heitmenn, H.-G., 1990. Saline Water Processing. VCH Verlagsgesellschaft, Weinheim, 332 p.
Staude, E., 1992. Membranen und Membranprozesse. VCH Verlagsgesellschaft, Weinheim, 224 p.
Spiegler, K.S., El-Sayed, Y.M.A., 1994. A Desalination Primer. Balaban Desalination Publications, 216 p.
Faller, A.K., 1999. Reverse Osmosis and Nanofiltration. American Water Works Association.
El-Dessouky, H.T., Ettouney, H.M., 2002. Fundamentals of Salt Water Desalination. Elsevier, 670 p.
IAEA, 2000. Introduction to Nuclear Desalination: A Guidebook. International Atomic Energy Agency.
Lattemann, S., Höpner, T., 2003. Seawater Desalination, Impacts of Brine and Chemical Discharge on
    the Marine Environment. Balaban Desalination Publications and DEStech Publications, 142 p.
Schafer, A.I., Fane, A.G., Waite, T.D., 2004. Nanofiltration Principles and Applications. Elsevier
    Publishing Co.
Van Rijn, C.J.M., 2004. Nano and Micro Engineered Membrane Technology. Elsevier Publishing, Co.
El-Nashar, M., 2008. Multiple-Effect Distillation of Seawater Using Solar Energy. Nova Science
    Publishers, 112 p.
Somariva, C., 2005. Desalination Management and Economics. Faversham House Group, South
    Croydon, Surrey, 80 p.
Sheikholeslami, R., 2007. Fouling in Membranes and Thermal Units. Balaban Desalination Publications.
Wilf, M., 2007. The Guidebook to Membrane Desalination Technology. Balaban Desalination
    Publications.

Rizzuti, L., Etouney, H.M., Chipollina, A., 2007. Solar Desalination for the 21st Century. Springer Verlag, 416 p.

Voutchkov, N., 2008. Pre-Treatment Technologies for Membrane Seawater Desalination. Australian Water Association.

Chipollina, A., Micale, G., Rizzuti, L., 2009. Seawater *Desalination*. Springer Verlag, Berlin, New York.

Somariva, C., 2010. Desalination and Advanced Water Treatment. Economics and Financing, Balaban Desalination Publication.

Voutchkov, N., 2010. Seawater Pretreatment. Water Treatment Academy (TechnoBiz Communications Co., Ltd), 174 p.

Wilf, M., 2010. The Guide Book to Membrane Technology for Waste Water Reclamation. Balaban Desalination Publication.

Pearce, G.L., 2011. Water Treatment Academy (TechnoBiz Communications Co., Ltd). http://www. watertreatment-academy.org/books.html.

Pearce, G.L., 2011. UF/MF Membrane Water Treatment: Principles and Design, 387 pages. Water Treatment Academy, enquiry@watertreatment-academy.org.

Voutchkov, N., 2011. Desalination Plant Concentrate Management. Water Treatment Academy (TechnoBiz Communications Co., Ltd), 182 p.

Voutchkov, N., 2011. Desalination Cost Assessment and Management. Water Treatment Academy (TechnoBiz Communications Co., Ltd), 182 p.

Lior, N., 2012. Advances in Water Desalination. Wiley Online Library.

Khayet, M., Matsuura, T., 2011. Membrane Distillation. Elsevier B.V.

Schorr, M., 2011. Desalination, Trends and Technologies. Publ. In Tech.

Hillal, N., Khayet, M., Wright, C., 2012. Membrane Modification. CPC Press.

Kucera, J., 2014. Editor, Desalination Water for Water. Scrivener Publishing, Wiley, p. 664.

N. Voutvhkov, Desalination Engineering: Operation and Maintenance, 2014, 330 p.

Bundschbuch, J., Hoinkis, J., 2012. Renewable Energy Applications for Water Production. CRC Press, Taylor Francis Group.

Raza, E., 2013. Solar Humidification, Dehumidification, Desalination System. Lambert Academic Publishing, p. 76.

Voutchkov, N., 2013. Desalination Engineering: Planning and Design. We press, McGraw Hill, p. 688.

GENERAL INTEREST JOURNALS AND WEBSITES

Advances in Water Desalination. John Willey, Vol. 1, 2007.

Desalination Directory on Line. http://www.desline.com.

Desalination Journal. www.sciencedirect.com.

EOLSS. Desalination Part (DESWARE). http://www.deware.net.

IDA Journal (International Desalination & Water Reuse). www.idadesal.org; www.awwa.org.

Desalination and Water Treatment. www.deswater.com.

CHAPTER TWO

Water, the Raw Material for Desalination

2.1 INTRODUCTION

Water is a matter of life and development but can also be very destructive if its flow is sudden and uncontrolled. It is not uniformly distributed around the world, with some places having abundant amounts of fresh water and others totally lacking in it. Today there is a crisis in availability of good-quality fresh water, which is increasing dramatically with unnecessary waste and contamination due to human activity. In many places worldwide the only water source that remains is the immense water of the seas and oceans. It is the raw material to be desalinated for the production of fresh water.

As desalination is energy intensive consumption enterprise water produced by desalination methods is high-cost water, a totally industrial product and is not yet compatible to physical fresh water resources. Thus water and energy are two of the most important parameters for social and economic development linked inextricably and reciprocally, as for example, the production of fuel needs large amounts of water and vice versa, treatment of water depends on low-cost energy.

To operate properly a desalination plant, independent of the method used, the properties of the various waters and their behavior according to the salt content and eventually some contaminants, are essential for functional parameters, such as corrosion, scale formation prevention, fouling, and especially for the proper and economic operation of the whole system. Here we try to give some of the more important water features that can help to understand the proper water characteristics of desalinated, brackish, seawater, and natural waters.

The total surface of the earth is $510 \times 10^{12} \, m^2$. Out of this surface only about $150 \times 10^{12} \, m^2$ is solid soil. The rest, 360×10^{12} consists of the water surfaces of lakes, seas, oceans, that is, mostly of high salinity waters. The total sea water volume is about 97.2%, and only the remaining of $\sim 2.80\%$ consists of the fresh potable waters from which again only a small portion, about 0.75%, is directly available to us for use. This 2.8% of total fresh water is distributed as follows (Frenkel, 2008):

- Ice in the poles and glaciers $\sim 30 \times 10^{15}$ m^3
- Waters up to a depth of 750 m $\sim 4.4 \times 10^{15}$ m^3
- Waters in depths from 750 up to 3500 m $\sim 5.6 \times 10^{15}$ m^3
- Lake waters $\sim 12 \times 10^4$ m^3
- Rivers and creeks $\sim 1.2 \times 10^4$ m^3
- Soil humidity $\sim 2.4 \times 10^4$ m^3
- Atmospheric humidity $\sim 1.3 \times 10^4$ m^3

Fresh water comes mainly from atmospheric vapor condensation, that is, as rain water. Water vapors are formed on the ocean, seas, lakes, and soil-free surfaces by the activity of solar radiation, by plant transpiration activities, the clouds, and atmospheric humidity. Pure water exists only in the state of vapor. As condensation of water vapor starts impurities are accumulated from the surrounding atmosphere, such as oxygen (O_2), and carbon dioxide (CO_2), that exist naturally in the air but also various other gases, created by human activities, such as sulfur dioxide (SO_2), nitrogen dioxide (NO_2), etc. The gases dissolve into condensate droplets and then form acids increasing hydrogen content. This is the reason for acid rain.

During the rainy season condensed water returns to the earth's surface, penetrates into the soil forming aqueducts and surface creeks or torrents. In both cases the final place is the ocean or a lake. Flowing through the soil the water enriches in various soluble salts forming hard or brackish waters. The salts accumulate into the seas and oceans forming their salinity. This procedure is called the *"water cycle or hydrological cycle"* which may be considered as a natural desalination operation, a huge, open solar distillation procedure going on continuously through the millennia.

2.1.1 What is Really Water?

Water as a pure chemical compound consists of two hydrogen molecules and one oxygen molecule (H_2O). As all dipolar molecules, it is a good solvent for ions. Compared to other solvents, water and its aqueous solutions show unique properties that are very sensitive to temperature and pressure. Water vapor is really the only pure state of water and has no structure. Liquid pure water is something of an enigma as it possesses unusual properties and is characterized in general as anomalous. It is characterized by the following extreme properties (Drost-Hansen, 1965):

1. *It is very stable* though easily changes phases from solid—to liquid to vapor and vice versa
2. *It has a comparatively high boiling point*, due to strong intermolecular forces, making it difficult for molecules to escape into pure vapor phase
3. *Its melting enthalpy* is the highest except for ammonia
4. *It has a high melting point* due to a kind of quasicrystalline structure
5. *Its evaporation enthalpy* is the highest of all substances. This property is extremely important in heat transfer procedures

6. *Its heat capacity* is the highest of all liquids and solids except for ammonia. This is a property that prevents extreme temperature ranges

7. *Its specific heat* is almost the highest of all liquids

8. *It has the highest heat conductivity* of all liquids

9. *Its dielectric constant* is high compared to most liquids (78.30 at 25°C). This results in high dissociation of dissolved substances

10. *It has the highest surface tension* of all liquids. This is an important characteristic because it controls droplet formation and behavior

11. *Its maximum density* is at a temperature about 4°C above freezing point. With increasing salinity temperature freezing point decreases

12. *It dissolves easily*, in large amounts, solids, liquids, and gases of more substances than any other liquid. Dissolved solids and liquids are in ionic form and gases are in a molecular. Thus it is a strong electrolyte solution

13. *Its boiling point* varies significantly with small changes of pressure

Because it is a strong solvent it cannot be found in nature in a very pure condition. It is pure only at the moment of condensation in the atmosphere and can be prepared only by physical or chemical methods in very pure form. Even the most purified water has a small electrical conductance (about $0.75 \times 10^{-6} \ \Omega^{-1} \ cm^{-1}$ at 18°C). This is due to dissolved carbon dioxide in equilibrium with atmospheric carbon dioxide and to a slight dissociation of the water molecules:

$$H_2O \rightarrow H^+ + OH^- \quad \text{or} \quad 2H_2O \rightarrow H_3O^+ + OH^- \qquad (2.1)$$

In nature it exists only as an electrolyte, as a salt solution, of various salt concentrations, the so-called *natural water*. In the vapor phase it may be as pure water but always has a small amount of dissolved gases especially carbon dioxide (CO_2). These properties lead to the conclusion that there exist some particularly strong forces of attraction between molecules in the liquid state.

2.1.1.1 The Molecular Structure of Water

The molecular structure of water is H_2O. In the gaseous state water molecules have established that the H−O−H bond angle is 104.31° and the O−H intermolecular distance is 0.97 Å. Really the bond is not linear but has the following molecular structure (Bernal, 1965):

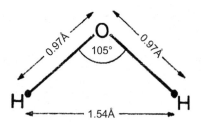

One cubic meter of water contains about 3×10^{22} molecules which have a diameter of 3×10^{-8} cm (3.0 Å). The strong intermolecular forces of liquid water are caused by molecular electrical polarity and make the water molecules mutually attractive. This explains some of the above extreme or anomalous properties as to why it is a very good solvent. Most of the anomalies occur near ambient temperature up to $100°C$.

2.1.2 Natural Waters

Natural or ordinary waters are formed during the hydrological cycle and fall as precipitation on the earth's crust. Running on the earth's surface or percolating into the soil the water dissolves various chemical materials from the soil or disintegrates rocks and is enriched continuously with new ingredients.

The various components that are dissolved from the soil or various rocks, either during the flow onto the earth's surface or in the underground regions, comprise various inorganic and organic substances which give the water its physical and chemical characteristics.

The main materials that dissolve are chemical compounds of calcium, magnesium, sodium, and potassium. Minor chemical compounds are lead, copper, etc., and various other minerals such as microelements. Besides inorganic material included are suspended solid matter and organic material. Organic compounds exist generally in very small amounts and include humic acid that originates from the decay of plant and animal matter, detergents, pesticides from the agricultural runoffs, and any other hazardous organic or inorganic material resulting from human activities and from untreated hazardous waste discharges. Most of the natural organic materials are carbon compounds of hydrogen, oxygen, nitrogen, phosphorus, and sulfur, which render water unsuitable for drinking purposes and domestic or municipal utilization.

The chemical compounds, suspended matter, and organic material and their concentration determine its quality which may be classified in several ways. For desalination purposes the most convenient classification is to group its physical, chemical, and in some extend biological property characteristics. For drinking water odor, hardness, alkalinity, biochemical oxygen demand, and toxicity, etc., are important. Inorganic substances comprise various salts whose concentration determines the suitability for use as drinking water or any other use. From a physical point of view natural water includes a small amount of hydrogen and oxygen isotopes, that is, for 99.985 moles of hydrogen exist 0.015 moles ^{1}H and for 99.959 moles of oxygen exist 0.027 ^{17}O and 0.0202 ^{18}O, which are not of any importance for desalination.

The qualities of the various types of waters can be identified by the simple procedure, of the logarithm of concentration (logC) of total dissolved solids (TDS). The same expression may be used for the partial substances, for example, Cl, Na or other ions and molecules. Table 2.1 presents the values of logarithmic impurities concentration. In this pure salt and the Dead Sea water are given as the extreme cases of salt concentration. The last column refers to impurities in the waters expressed by the

Table 2.1 Quality Prediction of Natural Waters

Natural Water	Log(C)	Units	PHI/F
Pure salt	0	%	12
Dead Sea water	−1	%	11
Sea water	−2	%	10
Well water	−3	$g\,kg^{-1}$	9
Lake water	−4	ppm	8
Spring water	−5	ppm	7
River water	−6	ppm	6
Dew (humidity concentrate)	−7	ppb	5
Rain water	−8	ppb	4
Snow water	−9	ppb	3
Hail water	−10	ppt	2
Ice water	−11	ppt	1
Pure water	< −12	ppt	<0

ppm, parts per million; ppb, parts per billion; ppt, parts per trillion.

parameter, PHI (Power Hydro-Impurities or F) which is a parameter similar to pH for hydrogen concentration where the impurities are given in values of negative logarithms (Delyannis and Belessiotis, 1995).

$$PHI = F = 12 + \log(C) \qquad (2.2)$$

Water and energy are two of the most important, the basic parameters which are mutually influenced and govern our life and our technical development. All goods that improve our standard of life are based primarily on energy and water use. Water is itself a dynamically potential energy source that produces energy, that is, hydropower, wave power, and of the tides. Thus water resources have become of high importance.

2.1.3 Water Resources

As mentioned previously less than 1% of fresh water is available to us for use. This availability has an uneven distribution that makes some places blessed with abundant water flow and other regions totally dry. Fig. 2.1, presents the water volume distribution, per km^3 of various types of water.

The rapid growth of population, by now more than 7 billion, which is expected to increase by the year 2025 to about 8.5 billion leads to a tremendous waste of water mainly in countries having rich water resources, exhausting rapidly the existing quality water resources. There are countries that consume fresh water more than their natural resources can afford. On the other hand simultaneously, the overload with pollutants of the water streams and aquifers makes the water resources unsuitable for any use. As a consequence pure fresh water especially drinking water becomes continuously more precious. Consequently as quality decreases, water regulations become more stringent.

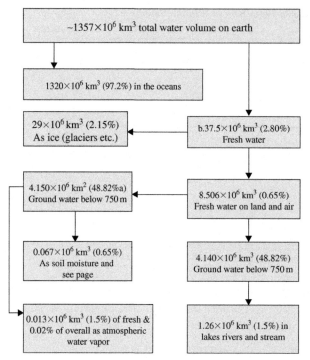

Figure 2.1 Global volume distribution of various types of waters worldwide (not in scale).

Besides the regions with plenty of water resources, but to some extent already polluted, many places, including North Africa, the Caribbean Islands, the Arabic Peninsula, some places in Australia, etc., suffer from soil dryness and severe lack of natural water. They are regions in the so-called *solar zones* that have abundant and intensive solar radiation. For these conditions what is the right solution? The answer depends on a variety of parameters and the policy is to find the optimum conditions between transportation from natural resources, if any exist, or desalination, which must be the solution only if all other possibilities are excluded.

2.1.4 The Water Problem

Water is a natural product, one of the most precious goods that nature provides to humanity. It is absolutely necessary for the survival of all species on earth and for the evolution of mankind. Looking at the history of mankind, civilization and water were always bound together from the prehistoric era until today. Rivers, seas, lakes, oceans, and even oases providing plenty of clean natural water were the attractive poles for creation of cities and installation sites, for water is a source of life. It is not just a coincidence that all great civilizations and the economic power of many nations were developed and flourished by people living along great water streams.

Today the situation is very different. Nowadays water demand exceeds available sources, in many cases, due to tremendous expansion of industry which consumes and pollutes large amounts of water. For example, the production of 1 kg aluminum needs 100 m³ of water. The rapid increase of population in the large cities has led to disproportional water use and unnecessary waste. Simultaneously pollution of fresh water streams and reservoirs increases rapidly with dangerous pesticides and other chemical ingredients, such as fertilizers, etc.

Nature donates its products in inexhaustible amounts but the irrational exploitation of natural reserves, such as water has led to physical unbalanced distribution, because earth's physical operations are very slow in replacing missing reserves. This is a great problem that must be solved by rational exploitation of water resources.

2.1.5 Water Demand

Water demand differs considerably according to geographic, social, economic, and cultural standard of life and to its availability.

The World Health Organization (WHO) developed and issued, guidelines in which regulations are included for desalinated water. The guideline has five sections (Abouzaid and Cotruvo, 2004; Feldman, 2007; Cotruvo and Lattemann, 2008).
1. Desalination technology and technical issues associated with desalination
2. Chemical aspects of desalinated water
3. Sanitary microbiology of production and distribution of desalinated water
4. Monitoring surveillance and regulations
5. Environmental impact assessment (EIA) of desalination projects.

Water demand can be classified into four main groups: Domestic, community, industrial use, and water for agriculture.

Domestic use comprises drinking water, household cleaning, cooking, small household industries, garden irrigation, and water use in shops.

Community use comprises irrigation water of public gardens; local water distribution networks, fire-fighting water, and all water used for transportation, as canals, rivers, etc.

Industrial water includes all water used in chemical industries, in power plants for vapor production or other industries, water for hydroelectric power stations, cooling water for industries, etc.

Agriculture, in general, consumes the larger water amounts for farming, raising cattle, and fisheries. All are very sensitive parameters requiring quality waters, having specialized water specification.

For all these purposes the water sources are mainly ground waters, water from springs and surface water streams including river waters and lakes. The quality for each group may differ considerably and sometimes according to the WHO specifications, which may be also more or less flexible according to the local national laws.

2.1.6 Recycling and Reuse of Water

Specifications for other uses are not so stringent, except for the food industry and pharmaceuticals. As water is regarded a raw material the problem of the future recycling or reuse technology of wastewaters becomes important. Reuse of water is an indirect way to increase water resources. Modern technological developments of wastewater treatment made feasible the recycling of all kinds of industrial wastewaters decreasing environmental impacts and increasing resources.

Community wastewaters can be used again as drinking waters, as household water, or better for agriculture purposes. In most cases conventional purification methods are used or combined conventional and desalination methods for higher purity of the reused water.

2.2 WATER AND SEAWATER PROPERTIES—DEFINITIONS

Seawater properties are the most important factors that dominate in the development, design, and operation of desalination systems. Of these salinity (g kg^{-1} or ‰), temperature (K or °C), and pressure (Pa, bar), are the most important as they usually determine the other thermophysical properties. This differs from pure water, where only pressure and temperature determine its physical properties. Seawater is an aqueous electrolyte solution following almost the laws of dilute solutions.

2.2.1 General Definitions

The following definitions are directly connected to desalination operations concerning various characteristics of natural and salty waters. They are given for the readers who are not familiar with chemistry, chemical compounds, and water chemistry and/or technology.

Anion is a positive charged particle of a salt in aqueous solution, for example Na Cl $\leftrightarrows$ Na$^+$ + Cl$^-$. Sodium chloride dissociates into positively charged sodium anion and negatively charged chloride cation.

Atom is the smallest substance or particle of an element that has all the chemical characteristics of the reference element.

Atomic mass is the mass of an atom expressed in g. Atoms combine, separate, and change positions always proportionally to their atomic mass or their multiples.

Cation is a negatively charged salt particle resulting from the dissociation of a salt to anions and cations, for example Na Cl $\leftrightarrows$ Na$^+$ + Cl$^-$, where Cl$^-$ is the cation.

Compounds are complex substances that can split into more elements or can be formed from one or more elements, for example sodium chloride (NaCl), which is formed from two elements, sodium and chlorine.

Electrical charge is the positive or negative electrical charge of particles, that is, of an anion or cation, respectively.

Electrolyte is a salt that when dissolved in a solvent (in water in our case) is dissociated into anions and cations.

Element is a chemical substance, which cannot be divided, by the application of a chemical treatment in simpler form, for example elements are: iron, sodium, chlorine, calcium, etc.

Mole is the combination of two or more atoms of the same element.

Molecular mass is the mass of a molecule expressed in g (g mol^{-1}).

Valence z, is the ability of an element to combine with one or more other elements. Valence is calculated from the number of hydrogen atoms, which by definition equals unity, for example water (H_2O) contains two hydrogen atoms and one oxygen atom, defining that oxygen is bivalent, Sodium chloride (NaCl) contains one sodium atom and one chloride atom, defining both as monovalent. Calcium chloride ($CaCl_2$) has two monovalent atoms of chloride defining calcium as divalent. Metallic elements have positive and nonmetallic elements negative valence. Valence and electrical charge refer as follow:

Ca^{2+} = Calcium, bivalent—positive charge

Cl^- = Chlorine, monovalent—negative charge

Roots result from the combination of elements with oxygen or other nonmetallic elements. Roots have always-negative charge and combine with positive metallic elements forming salts. Roots are: for example, carbonates (CO_3^{2-}), nitrates (NO^{3-}), sulfates (SO_4^{2-}), etc.

Salts result from the combination of a metallic to a nonmetallic element or a root.

$$Ca^{2+} + CO_3^{2-} \leftrightarrows CaCO_3, \quad Ca^{2+} + 2Cl^- \leftrightarrows CaCl_2 \qquad (2.3a)$$

$$Mg^{2+} + SO_4^{2-} \leftrightarrows MgSO_4, \quad Mg^{2+} + 2Cl^- \leftrightarrows MgCl_2 \qquad (2.3b)$$

Acids are chemical compounds that dissolve in water forming hydrogen ions H^+ making the water acid.

Bases are chemical compounds that when dissolve in water form hydroxyls that turn the water alkaline.

When acids and bases dissolve simultaneously in water they form salts while hydrogen and hydroxyl ions combine to form water.

$$HCl + NaOH \rightarrow NaCl + H_2O \qquad (2.4a)$$

$$H_2SO_4 + 2NaOH \rightarrow Na_2SO_4 + 2H_2O \qquad (2.4b)$$

Salts, acids, and bases dissolve easily in water and dissociate into electrical charged particles, the ions. Negative charged are called cations and positive ions anions. Chemical compounds that dissociate into ions by dissolving in water are called "*electrolytes*" because they are conductive to electric current. Dissociation of salts is not total and its degree depends on the type of chemical compound. If into water containing electrolytes two electrodes are inserted and through them electrical current is transmitted, anions will conduit to the anode and cations to the cathode.

$$CaCO_3 \leftrightarrows Ca^{2+} + CO_3^{2-} \tag{2.5a}$$

$$NaCl \leftrightarrows Na^+ + Cl^- \tag{2.5b}$$

Solution is an amount of water (or of another liquid) that contains electrolyte ions. The solution is neutral as cations and anions are in equal quantities:

$$\Sigma_{anions} = \Sigma_{cations} \tag{2.6}$$

In a solution anions and cations are expressed as equivalent per m^3, "$eq\ m^{-3}$."

Solutes are the dissolved substances (the salts) in an amount of water.

Solvent is pure water (or another pure liquid) where various substances are dissolved forming an aqueous (or in general a solution) solution.

2.2.2 Basic Chemical Relations

For this section material from Tsobanoglou and Schroeder (1985) is used.

Molarity (M) is the amount of molecules of a solute A, in one m^3 of solvent:

$$M = \frac{\text{moles of } A}{m^3 \text{ of water}} \quad mol\ m^{-3} \tag{2.7}$$

A mole of solute refers to its molecular mass. For natural waters may be that instead of m^3, moles are referred per liter.

Molality (m) refers to the moles of a solute A, per kg of solvent:

$$m = \frac{\text{moles of } A}{\text{kg of water}} \quad mol\ kg^{-1} \tag{2.8}$$

In Table 2.2 is presented the ionic form and the molality m, of the major constituents in seawater at various salinities.

Chemical equivalent (eq) of a root or element is the amount of mass that can rebound or be displaced by the mass of one hydrogen atom, for example:

$$CaCO_3 + H_2SO_4 \rightarrow CaSO_4 + CO_2 + H_2O \tag{2.9a}$$

$$Na_2CO_3 + 2HCI \rightarrow 2NaCI + CO_2 + H_2O \tag{2.9b}$$

Because calcium, in Eq. (2.9a), displaces two hydrogen atoms, has the chemical equivalent 2. In Eq. (2.9b) one atom of sodium displaces one hydrogen atom. Its equivalent is 1. Hydrogen has an atomic mass of 1.00784 g. Equivalent mass of any other element is calculated from its atomic mass divided by its valence, for example calcium has atomic mass 40.078 and is bivalent, thus its equivalent

Table 2.2 Molality m, of Major Constituents in Seawater (mol kg^{-1})

Constituent	Salinity 30.0‰	Salinity 34.80‰[a]	Salinity 40.0‰
Cl$^-$	0.48243	0.56241	0.064997
Na$^+$	0.41417	0.48284	0.55801
Mg^{2+}	0.04666	0.05440	0.06287
SO$_4^{2-}$	0.02495	0.02909	0.03362
Ca^{2+}	0.00909	0.01059	0.01224
K$^+$	0.00902	0.01052	0.01215
HCO$_3^-$	0.00211	0.00245	0.00284
Br$^-$	0.00074	0.00087	0.00100
B(OH)$_3$	0.00038	0.00044	0.00051
Sr^{2+}	0.00008	0.00009	0.00011

[a] ~Normal seawater.

is: $= 40.78/2 = 20.039$. This means that 20.039 g of calcium displaces 1.00784 g of hydrogen; sodium (Na) is monovalent and has atomic mass 22.98976 displacing 1.00784 g hydrogen. Reducing of element or root mass to their equivalent value facilitates the comparison of the mass amounts and as well as facilitates control of water analysis. In Appendix, Table B.4 the equivalents of the main constituents in seawater are presented.

Chemical equivalent is expressed in "$eq\ m^{-3}$" or in the case of very dilute solutions in meq L^{-1} (milli-equivalent per liter). The equivalent mass is expressed as:

$$\text{Mass of solute } A,\ \text{g eq}^{-1} = \text{Atomic (molecular) mass}/z,\ \text{g eq}^{-1} \qquad (2.10)$$

Normality (N) is the equivalent (eq) mass of a solute A, in 1 m^3 of solution.

Mass concentration (c_m) is mass quantity of solute A, in gm^{-3} of solution:

$$c_A = \frac{\text{grams of solute A}}{\text{m}^3 \text{ of solution}} \qquad (2.11)$$

Usually the expression c_i is used to denote a solute i. In some books mass concentration is referred in mg L^{-1} especially for natural waters or waters with very low salt content. For seawater mass concentration is given by salinity (in g kg^{-1} or ‰. See below in chemical composition of seawater).

Salt concentration (c_s) In dilute solutions, as natural waters, brackish or desalinated waters, salt concentrations are expressed in ppm (parts per million):

$$\text{ppm} = \frac{\text{dissolved mass of solute } A,\ \text{g}}{10^6 \text{ g of solution}} \qquad (2.12)$$

Molar proportionality or mole fraction (x_A) of a solute A is the ratio of number of moles n_A, of the solute A to the water molecules n_w or to the *sum* of all moles in the solution, n_{sm}, in case of many dissolved salts:

$$x_A = \frac{n_A}{n_A + n_w}$$

$$x_w = \frac{n_w}{n_A + n_w}$$

$$x_A = \frac{n_A}{n_W + n_B + n_C + \cdots + n_N}$$

(2.13)

where

n_A the moles of solute A

n_w the moles of the solvent, that is, here water

n_B, n_C ... n_N ... the moles of each component B ... N, dissolved in water

It is: $n_A + n_w = 1.0$ and $n_W + n_A + n_B + n_C + \cdots + n_N = 1.0$

The sum of dissolved salt moles is very small in comparison to the water moles in the solution. For routine calculations the following simple expression may be used:

$$x_A \approx n_A/n_w \approx n_A/55.56$$

(2.14)

where n_w is approximated as: $1000/18$ g $(g/mol)^{-1} \approx 55.56$ mol. For normal seawater of salinity $\sim 35\text{‰}$ the moles of water are: $(1000-35)/18 = 965/18 = 53.61$ mol.

Chemical equilibrium or equilibrium point refers to the law of mass action. Chemical reactions are related to the constituent's concentrations and the concentration of the reaction products. If we consider two constituents A and B reacting to produce C and D products according to the equation:

$$bA + cB \leftrightarrows dC + eD$$

(2.15)

then according to the law of mass action in equilibrium it applies:

$$K = [C]^c[D]^d/[B]^b[A]^a$$

(2.16)

where b, c, d, and e are the number of moles of each constituent. The parameter $\leftrightarrows$ denotes that the reaction of the constituents is in equilibrium. For example for sodium chloride or water we can write:

$$NaCl \leftrightarrows Na^+ + Cl^-$$

(2.17a)

$$H_2O \leftrightarrows H^+ + OH^-$$

(2.17b)

This means that neutral sodium chloride is in kinetic equilibrium with sodium positive charged ions and the negative charged ions of chlorine. The equilibrium is expressed as:

$$K_{NaCl} = \frac{[Na^+][Cl^-]}{[NaCl]} \qquad (2.18a)$$

$$K_{H_2O} = \frac{[H^+][OH^-]}{[H_2O]} \qquad (2.18b)$$

where K is the equilibrium constant. Parameter K is affected by temperature and ionic strength of the solution. The brackets denote the molarity, for example M_{Na+}, M_{NaCl}, etc., and are expressed in mol m^{-3}.

Calcium carbonate equivalent/mass concentration expresses the quantity of an element in equivalent of calcium carbonate ($CaCO_3$). It is used in water hardness calculation, even today in some cases.

$$\text{Equivalent mass of } CaCO_3 = \frac{100.089 \text{ mgmol}^{-1}}{2 \text{ meqmol}^{-1}} = 50.045 \text{ gmeq}^{-1} \qquad (2.19)$$

where 100.089 is the molecular mass (in g) of calcium carbonate.

Chemical potential (μ) is a parameter in a solution that indicates the mass flow rate, at constant temperature and pressure, from one point to another in direction where concentration difference is decreased. It is an analogous phenomenon of heat flow rate. The larger the concentration differences the highest the flow rate. Flow rate decreases until kinetic equilibrium is achieved.

Solubility product. When a solution is saturated to a solute, for example NaCl, the equivalent constant K of Eq. (2.18a) has a value over which solute starts to reject from the solution in solid phase. For example for sodium chloride (NaCl), at saturation point Eq. (2.18a) becomes:

$$[Na^+][Cl^-] = K_{st} \qquad (2.20)$$

where K_{st} is equivalent constant at saturation.

Henry's law refers to the solubility of a gas in water. It is a function between the solubility of the gas and the partial pressure of the gas in the ambient onto the surface of the solution. It is expressed by Henry's law:

$$x_i = K_H p_p \qquad (2.21)$$

where x_i is the fraction of gas i at equilibrium, p_p the partial gas pressure (Pa), and K_H Henry's constant (1/Pa), that is, the equivalent constant at saturation point. Henry's constant is a function of gas type, temperature, and concentration of the salts in the solution. Dissolved gases, especially oxygen and carbon dioxide are of importance in desalination systems for corrosion handling.

2.3 THE CHEMICAL COMPOSITION OF SEAWATER

The term seawater is used to describe a solution where the solute in seawater is referred to as sea salt, the solvent as water. Seawater is a complex, dynamic mixture of dissolved minerals, salts, and organic material. The fundamental equation for Gibbs energy, as a function of salinity, temperature, and pressure is the state equation for seawater (see below sub-chapter 2.4.1.1).

As already mentioned previously, the salt content of various world seas and oceans and large lakes varies considerably. Table 2.3 gives the total salt content of the seas and oceans around the world.

2.3.1 The Chemistry of Seawater

Water has two very important characteristics which distinguish it from all other materials:
• Its stability is very strong though it easily changes phases from solid → to liquid → to vapor and vice versa
• It has strong solvent characteristics. It easily dissolves solids, liquids, and gases.

Dissolved solid and liquid constituents in seawater are in ionic form and gases in molecular form.

In large water basins water is uniformly distributed due to strong currents, but the various seas and oceans have very large differences in the concentration of elements (in total dissolved salts [TDS]), as it can be seen also in Table 2.3. Seawater contains almost all chemical elements in various combinations, some of them in very low

Table 2.3 Mean Value of Total Dissolved Salts in Various Seas and Oceans

Sea or Ocean	TDS (ppm)	Sea or Ocean	TDS (ppm)
Baltic Sea	7000	Indian Ocean	33,800
Caspian Sea	13,500	Mediterranean near Crete	37,900
Black Sea	20,000	Mediterranean	39,000
White Sea	28,000	Red Sea	43,000
North Atlantic	29,000	Arabian Gulf	50,000
Pacific Ocean	33,600	Australian Shark Bay	70,000
Atlantic South	36,000	Kara Bogas Gulf	164,000
North Atlantic	36,200	(in Caspian Sea)	
Adriatic Sea	31,400	Dead Sea	220,000

concentrations. In Appendix B, Tables B.1—B.5 give the composition of ocean standard seawater, the main compounds, the ions, and the trace elements in seawater.

Natural waters such as fresh well waters, river, and lake waters contain small amounts of dissolved salts in various concentrations. Seawater contains larger amounts of dissolved salts, mainly sodium chloride. Dissolved salts of natural waters differ extremely in composition and in amount depending on the type of water, the site of origin, and in some cases from the time or of the season and other parameters such as the rate of evaporation. Natural water sources may be altered by heavy rains, severe floods, and intrusions in wells and aquifers.

The waters may be classified according to the TDS as follows (WHO, 2004):

Up to 500 ppm TDS	*Natural water*: suitable for drinking
500—1000 ppm TDS	*Light brackish water*: in some special cases is permitted as drinking water
1000—2000 ppm TDS	*Brackish water*: in some cases, it is used as domestic water. In general creates stomach troubles by drinking it
2000—10,000 ppm TDS	*Middle brackish water*: it is unsuitable for any domestic, community or agriculture use
>10,000 ppm TDS	*Very brackish waters*: not usable
~19,000 to >45,000 ppm TDS	*Sea and ocean waters*

For desalination systems more important than dissolved salts are the chemical elements present in seawater. Seawater contains larger amounts of dissolved salts, mainly sodium chloride. The main cations Na, Ca, and K are present as hydrated cations but magnesium and the nonchloride ions are more complex. In seawater concentrates, depending from the concentration, Na and K may exist in a more complex form. Although total concentration of seawater salts varies depending on the sea or on the ocean (Table 2.3), *the relative ratio of the major constituents remains almost constant*. This is due mainly to the intrusion and the smooth mixing of the world's ocean waters.

Fundamental determination of physical and chemical properties of seawater has been previously studied from synthetic seawater, called "Standard or Normal Seawater." Synthetic seawater was prepared from very pure distilled water and especially pure chemical reagents. Standard seawater contains salts only in concentrations >0.1 ppm. Trace elements were not added. The first in use was the Normal Seawater prepared by the Hydrographic Laboratories of Copenhagen, Denmark, the "Copenhagen Prototype" which has the composition presented in Table 2.4. Many other prototypes have been proposed. The most recent standard seawater now in use in oceanography, which replaced the Copenhagen Prototype, is based on the North Atlantic surface waters, established by IAPSO (International Association for Physical Science of the Ocean) as the Standard Seawater (Millero et al., 2008). In Table 2.5 the Standard Seawater composition with total TDS 35 ppm is presented.

Table 2.4 The Copenhagen Prototype Seawater (ppm)

Na	1.07678	HCO$_3$	0.01425
Mg	0.12975	ρ (20°C)	1024.8 (kg m^3)
Ca	0.04081	Salinity	35.01 (g kg^{-1})
K	0.03876	Chlorinity	19.381 (g kg^{-1})
Cl	1.93605	Chlorosity	19.862 (g L^{-1})
SO$_4$	0.27017		

Table 2.5 Standard Seawater Composition (35 ppm, pH = 8.1)

Constituent	ppm	Constituent	ppm	Constituent	Ppm
Calcium	410	Silica	0.04−0.08	Magnesium	1310
Chloride	19,700	Sodium	1900	Sulfate	2740
Potassium	390	Fluoride	1.4	Barium	0.05
Bromide	65	Strontium	13	Nitrate	<0.7
Iron	<0.2	Bicarbonate	152	Manganese	<0.01

A lot of improvements and new scientific formulations were introduced in the past years for the thermodynamic properties of water, water vapor, and ice according to the International Association for Properties of Water and Steam (IAPWS (2006) and IAPWS (2008)). The new formulations were released on International Temperature Scale of 1990 (ITS-90) and on the new atomic weights (Wieser and Coplen, 2011) and on IAPWS Release 2008 for the Properties of Water and Steam. The new formulations give more accurate equations and expended ranges of validity. The new formulations comprise:

a. The adoption of IAPSO's (International Association for the Physical Sciences of the Oceans) Standard Seawater
b. A composition model for standard seawater, the "reference composition salinity scale," in short "reference salinity." Reference salinity represents the absolute IAPSO's (Millero et al., 2008) absolute salinity of standard seawater
c. The equation state of seawater
d. The adoption of the seawater modified Gibbs function (Feistel, 1993, 2003), which permits the computation of all thermodynamic properties of seawater.

2.3.2 What Is Salinity(S)—Chlorinity (Cl)

Seawater is an electrolyte solution and its chemistry is dominated by the presence of six main ions, that is, Na^+, K^+, Ca^{2+}, Mg^{2+}, $Cl,^-$ and SO_4^{2-}, which constitute more than 99.5% of dissolved elements. These electrolytes have a significant influence on the physical, chemical, and biological properties and processes that control the chemistry of the seas and oceans. Of importance is the total ion concentration called "Salinity, S." It is defined as: "The total amount of solid material (in g), that is contained in 1.0 kg of sea water after all carbonates are converted to oxides and all bromides and iodides are converted to equivalent chlorides." The particular properties of the ions have no significant impact on the properties of seawater.

Various salinity scales have been established. In use are the practical salinity scale, S_p (Lewis and Perkin, 1978), absolute salinity S_A, and the reference composition salinity S_R (Millero et al., 2008). The new terms and definitions of salinity are based on 1990 International Temperature scale (ITS-90), and the atomic weight of elements revised by IUPAC-2009 (Wieser and Coplen, 2011).

The term KCl-normalized seawater (for short "normalized seawater" is used to describe a seawater sample that has the same specific conductivity (at $t_{90} = 14.996°C$ and normal pressure $p_o = 101\ 324$ Pa), as a solution of KCl in water. Temperature t_{90} corresponds to IPTS-68 scale, $t_{68} = 15.0°C$.

The mass fraction of standard KCl solution is 32.4356 g kg^{-1} of solution. Table 2.6 presents selected properties of the KCl-normalized seawater sample (Millero et al., 2008).

Current definitions of salinity based on the KCl-normalized seawater are:

$$S_R = 35.16504/35 \text{ g kg}^{-1} \quad \text{or} \tag{2.22}$$

$$S_R = (35.16504/35) = 1.00472 S_p \tag{2.23}$$

Absolute salinity (S_A), which is defined as the mass fraction of dissolved material in seawater to the total mass of seawater. Absolute salinity must be expressed in kg kg^1. No established method is available for practical determination of S_A.

Normal salinity (S_n), represents the salinity of KCl-normalized seawater.

Normalized seawater is any seawater sample that has been adjusted to have a practical salinity 35 g kg^{-1}. Normalized seawater provides a convenient way to refer to any seawater sample that has a practical salinity S_p of 35‰.

Table 2.6 Selected Properties of the KCl-Normalized Reference Seawater

Symbol	Value	Uncertainty	Unit	Comment
M_S	31.4038218	0.001	g mol^{-1}	Reference salinity molar mass $M_S = \sum_j x_j M_j$
z	1.2452898	Exact	–	Reference salinity valence factor $z^2 = \sum_j x_j z_j^2$
N_S				1.9176461×10^{22}
	6×10^{17}		g^{-1}	Reference salinity particle number, $N_S = N_A/M_S$
u_{ps}	1.004715...	Exact	g kg^{-1}	Unit conversion factor, 35.16504 g kg^{-1}/35
S_S	35.16504	Exact	g kg^{-1}	Standard reference salinity, 35 u_{ps}
p_o	101.325	Exact	Pa	Standard ocean surface pressure
T_S	273.15	Exact	K	Standard ocean temperature
t_o	0	Exact	°C	Standard ocean temperature

Practical scale salinity (PSS-78) Sp, is by definition 35 g kg^{-1} and is defined only in the range $2 < S_p < 42$. It is based on terms of conductivity ratio K_{15}, at temperature $t_{78} = 15°C$ which is electrical conductivity of the sample at 15°C and pressure $p_o = 101325$ Pa, divided by the conductivity of a standard KCl solution at the same temperature and pressure. It is globally accepted as standard salinity.

Reference-composition salinity scale (or reference salinity for short) S_R was established by Millero et al. (2008) and refers to the salinity of 35.16504 g kg^{-1}. Reference salinity of pure water equals zero. Reference composition salinity can be related to practical salinity by:

Reference-composition salinity is computed from the equation (IAPWS, 2008):

$$S_R = \frac{m_{sw} + m_w}{m_{sw}} S_n \qquad (2.24)$$

where m_{sw} is the amount of the KCl-normalized seawater sample, m_w is the mass of water added to the solution ($m_w \geq 0$), or removed from it ($0 \leq -m_w < m_{sw}$), and S_n is normal salinity = 35.16504 g kg^{-1}.

Martin Knudsen (1901) refers to composition of major substances dissolved in seawater of North Atlantic surface waters.

Reference seawater is defined as any seawater that has the reference composition of solute dissolved in pure water at equilibrium state. Normalized reference seawater to a practical salinity of 35 has a reference composition salinity of exactly $S_R = 35.16504$ g kg^{-1}.

Standard seawater is a seawater sample that, by definition, has the composition of North Atlantic surface waters. It is assumed that seawater pH has a value of 8.1.

The sea salt composition definition for reference salinity of the standard ocean at $t = 25°C$ and $p_o = 101325$ Pa is shown in Table 2.7.

Brackish and seawater are estimated to be solutions and all their properties are calculated according to the laws of dilute solutions. Due to various impurities, especially organic matter present in seawater, analytical determination of salinity is complicated and not accurate. To overcome this difficulty Martin Knudsen in 1900 (1901) and Forch et al. (1902) developed an approximate relation between salinity S (in ‰) and chlorinity, Cl (‰), known as the Knudsen's formula:

$$S‰ = 0.03 + 1805 \, Cl \qquad (2.25)$$

Chlorinity referred as the total amount of chlorine in sea water (g kg^{-1} sea water) after all bromides and iodides are converted to equivalent chlorides.

Various salinity formulae were developed in between (Millero et al., 2008). In 1978 the use of practical salinity S_p was introduced (Lewis and Perkin, 1978). This new salinity approach defined to be equal to 35 g kg^{-1} when electrical conductivity ratio of seawater sample to a KCl solution is 1.000 at $t_{68} = 15.0°C$, as referred previously.

Table 2.7 The Sea Salt Composition Definition for Reference Salinity of the Standard Ocean at $t = 25°$ and $p_o = 101325$ Pa (Feistel, 2008; McDougall et al., 2009)

Solute j	Valence, z_j	M_j, g mol^{-1}	x_j, mole fraction 10^{-7}	$x_j \times z_j\ 10^{-7}$	x_m (w_j), mass fraction
Na^+	$+1$	22.98976928	4188071	418871	0.3065958
Mg^{2+}	$+2$	24.3050	471678	943356	0.0365055
Ca^{2+}	$+2$	40.078	91823	183646	0.0117186
K^+	$+1$	39.0983	91150	91159	0.0113495
Sr^{2+}	$+2$	87.62	810	1620	0.0002260
Cl^-	-1	35.453	4874839	-4874839	0.5503396
SO_4^{2-}	-2	96.0626	252152	-504304	0.7711319
HCO_3^-	-1	61.01684	15340	-15340	0.0029805
Br^-	-1	79.904	7520	-7520	0.0019134
CO_3^-	-2	60.0089	2134	-4268	0.0004078
$B(OH)_4^-$	-1	78.8404	900	-900	0.0002259
F^-	-1	18.998403	610	-610	0.0000369
OH^-	-1	17.00733	71	-71	0.0000038
$B(OH)_3$	0	61.8330	2807	0	0.0005527
CO_2	0	44.0095	86	0	0.0000121
Sum			10000000	0	1.0

Chlorinity of KCl-normalized seawater with composition equal to the reference composition is $35/1.80655$ g kg^{-1} and differs from salinity by a constant factor.

$$S_R = 1.815068 Cl \text{ g kg}^{-1} \tag{2.26}$$

The International Association for the Properties of Water and Steam (IAPWS, 2008) recently defined chlorinity as "0.328523 times the ratio of the mass of pure silver required to precipitate all dissolved chloride, bromide and iodide in seawater to the mass of seawater" ($Cl = 0.3285234$ Ag g kg^{-1}).

These terms are established mainly for oceanographic purposes and are used for accurate calculations of seawater properties which are of importance also for design of desalination systems. Practical values may be used for routine calculations of the desalination systems.

During evaporation of sea water the brine is concentrated progressively, and thus salinity is increasing. The concentration ratio R_s is given as the salinity of the brine S_{br} to the salinity of the initial sea water feed S_{in}, both in g kg^{-1}:

$$R_s = S_{br}/S_{in} \tag{2.27}$$

The parameter R_s is called "concentration ratio" and is of importance for the smooth operation of desalination plants. A normal concentration ratio prevents scale formation on heat transfer surfaces and helps for an economical exploitation of feed water and a smooth disposal of brine.

The salinity profile at a given location may exhibit diurnal and seasonal fluctuations. This is due to evaporation of the water surface. Therefore salinity increases during daytime falling to initial conditions during darkness. Occasionally seasonal or other conditions as heavy rains, melting of snow and ice may also affect the salinity of seawater at a fixed location.

2.3.3 Hardness—Alkalinity

Hardness, (H) is defined as the polyvalent cation concentrations dissolved in the water. Calcium (Ca^{++}) and magnesium (Mg^{++}) are, after sodium, the most common and in larger concentration elements in seawater. In normal sea water calcium has a concentration of 0.01 M and magnesium of 0.05 M. In addition to these cations iron (Fe^{++}), Manganese (Mn^{++}), and Strontium (Sr^{++}) may also contribute to hardness. However, their contribution is negligible or may be accounted as equivalent Ca^{++}. Hardness is expressed as equivalent of calcium and magnesium per m^3 of solution:

$$Hardness = (Ca^{2+}) + (Mg^{2+}) \text{ eq m}^{-3} \tag{2.28}$$

Hardness is divided in two categories; carbonate hardness and noncarbonate hardness. Total hardness equals the sum of noncarbonated and carbonates hardness. Calcium and magnesium carbonate with calcium and magnesium bicarbonate form the carbonate hardness. Noncarbonate hardness is due to calcium and magnesium chloride, sulfate, and nitrate salts. Bicarbonate salts of calcium and magnesium are soluble in water but decompose by heating forming carbonates. Carbonate salts are insoluble and may form scale deposits on the hot metallic surfaces of the distillation plants or on the RO and ED, etc., membranes. The bicarbonate type of hardness is called "temporary or transient hardness":

$$Ca^{2+} + 2HCO_3 \rightarrow CaCO_3 + CO_2 + H_2O \tag{2.29}$$

$$Mg^{2+} + 2HCO_3 \rightarrow MgCO_3 + CO_2 + H2O \tag{2.30}$$

The remaining compounds of magnesium and calcium form the permanent hardness. They form scale, especially the sulfates, only if solubility product is high and the salts are rejected as solids, at their saturation conditions, in the brine solution.

Hardness is reported as equivalent calcium carbonate ($CaCO_3$). The concentration of $CaCO_3$ classifies the water as soft, hard, very hard, etc. Usually water hardness is classified as:

Soft	$CaCO_3$ eq, mg L^{-1}	<75
Moderate hard	>>	75–150
Hard	>>	150–300
Very hard	>>	<300

Alkalinity (*A*) is the ability of water to neutralize a certain amount of acidity and vice versa. Alkalinity is due to the presence of hydroxyl (OH), carbonate (CO₃), and bicarbonate (HCO₃) anions. However borates, silicates, etc., contribute to alkalinity. It is expressed as equivalents of the above roots per m³ of solution:

$$A = (HCO^{3-}) + (CO_3^{2-}) + (OH^-) \ \ eq \ m^{-3} \tag{2.31}$$

More accurate are the calculations of alkalinity taking into consideration the hydrogen ions as well:

$$A = (HCO^{3-}) + (CO_3^{2-}) + (OH^-) - (H^+) \ \ eq \ m^{-3} \tag{2.32}$$

The brackets denote eq m³.

Alkalinity is usually reported as equivalent calcium carbonate (CaCO₃). Any hardness greater than the alkalinity (eq. CaCO₃ hardness > eq CaCO₃ alkalinity) represents noncarbonated hardness.

In Appendix C, Figs. C.1−C.4, give, indicatively, the stability diagrams of calcium carbonate, for 60 and 100°C, calcium sulfate at 100°C, as well the magnesium hydroxide stability diagram (McKetta, 1994).

2.4 PROPERTIES OF SEAWATER

2.4.1 General

The design and operation of desalination plants are directly related to chemical, thermophysical and thermodynamic properties of seawater (Delyannis and Delyannis, 1974). These properties may be grouped into three main categories:

1. *Thermophysical properties* related to heat flow and fluid flow that refer to density, viscosity, thermal conductivity, specific volume, activity coefficient, vapor pressure, etc.
2. *Thermodynamic properties* as enthalpy, entropy, relation between pressure−temperature−volume, heat capacity, boiling point temperature, boiling point elevation temperature, and water freezing/melting point
3. *Various other properties* as electrical conductivity, surface tension, etc.

The new equation of state of seawater and the thermophysical properties derived from the equation of state were developed for accurate oceanographic studies. Dealing with desalination systems the accurate thermophysical properties are of importance for the design, analysis, and operation of the desalination techniques.

A big variety of equations, mathematical formulations, and algorithms for the thermophysical properties of seawater have been developed based on the new terms of salinity, reference composition, and equation of state. For oceanographic purposes measurements have been carried mainly in the range of temperature t_{90} 0−40°C and salinities $S = 0-40$ g kg^{-1}. Developing desalination systems at around 1970 a lot of experimental studies were performed on the seawater properties ranging for temperatures t_{68} 20$-$180°C and salinities $S_p = 10-160$ g kg^{-1}. These old experimental data have been adapted to the new terms of temperature, atomic weights, and salinities. All seawater properties given in the tables of this section are experimental data corrected according to the new terms of the state equation.

A detail State-of-the-Art on the thermodynamic properties of seawater for design and performance for desalination systems is presented by Sharqawy et al. (2010). The properties include density, specific capacity, thermal conductivity, dynamic viscosity, boiling point elevation, specific enthalpy, and osmotic coefficient. The existing correlations for the thermophysical properties of seawater were adopted to the new terms of salinity and reported over ranges of interest for both thermal and membrane desalination processes. Temperature ranges are 0$-$120°C and salinity ranges between 0 and 120 g kg^{-1}. Kretzschmar et al. (2015) describe the IAPWS industrial formulation for the thermodynamic properties of seawater.

Numerical values of thermal properties of pure water and water vapor are given in Steam Tables. The most reliable and complete Steam Tables are up today those of the American Society of Mechanical Engineers (ASME) and the most recent and complete, as already referred, are the Steam Tables of the International Association for properties of Water and Steam (IAPWS-2008, see Wagner and Kretzschmar (2008)). In the Appendix and in some cases in this chapter tables are included indicating the main properties of seawater and pure water. Most of these properties were compiled from experimental tests, as seawater is not considered in practice, a really dilute solution and the corresponding theoretical equations concerning ideal conditions are not very accurate for calculating these properties. The experimental values and the derived equations were developed mainly for use in desalination systems.

The new approach of "Thermodynamic and Equation of State of Seawater" is based on Gibbs function of seawater (Feistel, 2008) which is in the form of the specific Gibbs energy, g (J kg^{-1}), as a function of absolute salinity, temperature, and pressure. The Scientific Committee on Oceanic Research (SCOR) and the International Association of Physical Sciences of Oceans (IASPO) established a Working Group on the thermodynamics and equation of state of seawater (TEOS-10, 2010) which developed a Gibbs function from which all the thermodynamic properties of seawater can be derived in a physically consistent manner (IOC, SCOR and IAPSO, 2010).

2.4.1.1 The Gibbs Function—Equation of State

The Gibbs function of seawater is a function of absolute salinity, temperature, and pressure. The reason for preferring absolute salinity over practical salinity is because the thermodynamic properties are directly influenced by the mass of dissolved chemical constituents. The Gibbs function of seawater $g(S_A, T, p)$ is related to the specific enthalpy h, and entropy s, $g(273.15\text{ K} + t)s$ and is defined as the sum of a pure water part and the salinity part (Feistel, 2008), preferably expressed in Celsius degree. The Gibbs function of seawater can be expressed as the sum of Gibbs function of pure water IAPWS-95 (1995) formulation and a salinity correction $g^S(S_A, t, p)$:

$$g(S_A, t, p) = g^W(t, p) + g^S(S_A, t, p) \qquad (2.33)$$

The Gibbs energy G (J) of seawater is expressed by the mass of water m_w, and the mass of salt m_S, with the chemical potentials of water μ^w and of salt μ^S. Specific Gibbs energy is expressed in terms of mass and chemical potential as:

$$g(S_A, t, p) = \frac{G}{m_W + m_S} = \mu^W + S_A\left(\mu^S - \mu^W\right) \qquad (2.34)$$

where the water and salt potentials are calculated as:

$$\mu^W = \left(\frac{\partial G}{\partial m_w}\right)_{T,p,m_S}, \quad \mu^S = \left(\frac{\partial G}{\partial m_S}\right)_{T,p,m_W} \qquad (2.35)$$

where

G	Specific Gibbs energy for seawater	J kg^{-1}
g^W	Water part of the specific Gibbs energy of seawater	J kg^{-1}
g^S	Saline part of the specific Gibbs energy of seawater	J kg^{-1}
m_s	Is the mass of salt in seawater	kg
m_w	Is the mass of water in seawater	kg
P	Absolute pressure = 101325	Pa
μ^S	Chemical potential of salt in seawater	J kg^{-1}
μ^w	Chemical potential of water in seawater	J kg^{-1}
S_A	Absolute salinity	kg kg^{-1}
T	Absolute temperature (ITS-90)	K
T	Celsius temperature	°C

The saline part of the Gibbs function of seawater is valid over a range $0 < S_A < 120\text{ g kg}^{-1}$, $0 < p < 10^4$ dbar and $-6.0°\text{C} < t < 80°\text{C}$. The Gibbs function (2.33) contains four arbitrary constants that cannot be determined by any thermodynamic measurement. The water part $g^W(T, p)$, is computed from IAPWS-95 for Thermodynamic Properties of Ordinary Water Substances and for the saline part is given as, $g^S(S^A, T, p)$, by a dimensionless polynomial-like function (McDougall et al., 2009; IOC et al., 2010) comprising tabulated coefficients of the function.

The equation of state is presented in the form of specific Gibbs energy as function of independed variables of absolute S_A salinity, Celsius ITS-90 temperature t, and sea pressure p, corresponding to Gibbs potential.

Table 2.8 Pure Water Coefficients of Eq. (2.36)

j	k	g_{jk}	j	K	g_{jk}
0	0	$0.101342743139674 \times 10^3$	3	2	$0.499360390819152 \times 10^3$
0	1	$0.100015695367145 \times 10^6$	3	3	$-0.239545330654412 \times 10^3$
0	2	$-0.254457654203630 \times 10^4$	3	4	$0.488012518593872 \times 10^2$
0	3	$0.284517778446287 \times 10^3$	3	5	$-0.166307106208905 \times 10$
0	4	$-0.333146754253611 \times 10^2$	4	0	$-0.148185936433658 \times 10^3$
0	5	$0.420263108803084 \times 10$	4	1	$0.397968445406972 \times 10^3$
0	6	-0.546428511471039	4	2	$-0.301815380621876 \times 10^3$
1	0	$0.590578347909402 \times 10$	4	3	$0.152196371733841 \times 10^3$
1	1	$-0.270983805184062 \times 10^3$	4	4	$-0.263748377232802 \times 10^2$
1	2	$0.776153611613101 \times 10^3$	5	0	$0.580259125842571 \times 10^2$
1	3	$-0.196512550881220 \times 10^3$	5	1	$-0.194618310617595 \times 10^3$
1	4	$0.289796526294175 \times 10^2$	5	2	$0.120520654902025 \times 10^3$
1	5	$-0.213290083518327 \times 10$	5	3	$-0.552723052340152 \times 10^2$
2	0	$-0.123577859330390 \times 10^5$	5	4	$0.648190668077221 \times 10$
2	1	$0.145503645404680 \times 10^4$	6	0	$-0.189843846514172 \times 10^2$
2	2	$-0.756558385769359 \times 10^3$	6	1	$0.635113936641785 \times 10^2$
2	3	$0.273479662323528 \times 10^3$	6	2	$-0.222897317140459 \times 10^2$
2	4	$-0.555604063817218 \times 10^2$	6	3	$0.817060541818112 \times 10$
2	5	$0.434420671917197 \times 10$	7	0	$0.305081646487967 \times 10$
3	0	$0.736741204151612 \times 10^3$	7	1	$-0.963108119393062 \times 10$
3	1	$-0.672507783145070 \times 10^3$			

1. For pure water (Feistel, 2003; McDougall et al., 2009; IOC et al., 2010):

$$g^{W}(t,p) = g_u \sum_{j=0}^{7} \sum_{k=o}^{6} g_{jk} y^j z^k \tag{2.36}$$

2. For salt in seawater (Feistel, 2008; McDougall et al., 2009):

$$g^{S}(S_A, t, p) = g_u \sum_{j,k} \left\{ g_{1jk} x^2 \ln x + \sum_{i>1} g_{ijk} x^i \right\} y^j z^k \tag{2.37}$$

The terms y and z represent reduced temperature ($y = t/t_u$) and reduced pressure ($z = p/p_u$). The reduced quantities vary from zero to one. The constants t_u, p_u, and g_u

have the following values: $t_u, = 40°C$, $p_u = 10^8$ Pa, and $g_u = 1$ J kg^{-1}. The coefficients of Eqs. (2.36) and (2.37) are presented in Tables 2.8 and 2.9 respectively.

where

g_u	Unit depended energy constant, $g_u = 1$	J kg^{-1}
g_{ijk}	Coefficient of the Gibbs function for the salt in seawater	−
g_{jk}	Coefficient of the Gibbs function for the water in seawater	−
S_u	Unit-dependent salinity constant	g kg^{-1}
X	Dimensionless absolute salinity root = $x^2 = S_A/S_u$	−
p_u	Unit-independent pressure constant = 10^8	Pa
P	Gauge pressure	Pa
p_o	Standard atmospheric pressure (normal pressure)	Pa
t_u	Unit-independent temperature constant = 40	°C
T_o	Celsius zero point temperature	K
Y	Dimensionless Celsius temperature, $t_u xy = t = T - T_o$	−
Z	Dimensionless gauge pressure, $z x p_u = p = p - p_o$	−

Table 2.9 Salt Coefficients Corresponding to Eq. (2.37)

i	j	k	g_{ijk}	I	J	k	g_{ijk}	i	J	k	g_{ijk}
1	0	0	5812.81456626732	2	5	0	−21.6603240875311	3	2	2	−54.1917262517112
1	1	0	851.226734946706	4	5	0	2.49697009569508	2	3	2	−204.889641964903
2	0	0	1416.27648484197	2	6	0	2.13016970847183	2	4	2	74.7261411387560
3	0	0	−2432.14662381794	2	0	1	−3310.49154044839	2	0	3	−96.5324320107458
4	0	0	2025.80115603697	3	0	1	199.459603073901	3	0	3	68.0444942726459
5	0	0	−1091.66841042967	4	0	1	−54.7919133532887	4	0	3	−30.1755111971161
6	0	0	374.601237877840	5	0	1	36.0284195611086	2	1	3	124.687671116248
7	0	0	−48.5891069025409	2	1	1	729.116529735046	3	1	3	−29.4830643494290
2	1	0	168.072408311545	3	1	1	−175.292041186547	2	2	3	−178.314556207638
3	1	0	−493.407510141682	4	1	1	−22.6683558512829	3	2	3	25.6398487389914
4	1	0	543.835333000098	2	2	1	−860.764303783977	2	3	3	113.561697840594
5	1	0	−196.028306689776	3	2	1	383.058066002476	2	4	3	−36.4872919001588
6	1	0	36.7571622995805	2	3	1	694.244814133268	2	0	4	15.8408172766824
2	2	0	880.031352997204	3	3	1	−460.319931801257	3	0	4	−3.41251932441282
3	2	0	−43.0664675978042	2	4	1	−297.728741987187	2	1	4	−31.6569643860730
4	2	0	−68.5572509204491	3	4	1	234.565187611355	2	2	4	44.2040358308000
2	3	0	−225.267649263401	2	0	2	384.794152978599	2	3	4	−11.1282734326413
3	3	0	−10.0227370861875	3	0	2	−52.2940909281335	2	0	5	−2.62480156590992
4	3	0	49.3667694856254	4	0	2	−4.08193978912261	2	1	5	7.04658803315449
2	4	0	91.4260447751259	2	1	2	−343.956902961561	2	2	5	−7.92001547211682
3	4	0	0.875600661808945	3	1	2	83.1923927801819				
4	4	0	−17.1397577419788	2	2	2	337.409530269367				

The Gibbs functions for water and salt permit an accurate evaluation of sea properties. Details on these new terms for seawater properties are presented in UNESCO's Manual and Guide No 56, IAPWS Release 09 (2009), IAPWS (2008) Releases and Feistel (2008).

Sea water has medium salt concentration, that is, of about $0.5\ mol\ L^{-1}$ and its boiling point is $>100°C$ in normal atmospheric pressure. Pure water has salt concentration of $0.02\ mol\ L^{-1}$ and its properties follow the laws of very dilute solutions.

In desalination methods all physical properties are of importance, especially in cases where feed water must be treated to prevent scale formation. Product water has also to be controlled for its suitability according to the final application, that is, if it is for drinking purposes, for irrigation, etc. As has already been referred previously, sea water diluents or concentrates have the same relative ratios of constituent ions. Relative ratios are of importance for the pretreatment of feed water but its properties depend on its concentration.

2.4.1.2 Seawater as an Electrolyte Solution

An ideal solution is defined as one that obeys Raoult's law. Very dilute solutions, as fresh water, approach ideality. On the other hand, seawater of salinity ~35‰ (~0.5 M), is in fact not a dilute, but a moderately concentrated solution, containing very complex electrolytes.

In terms of Raoult's law, both 0.5 M seawater and NaCl solutions have a low deviation from ideality, about 0.4%, in most of their properties, except for electrical conductivity and freezing point depression for which both solutions behave relatively as concentrated systems. Seawater contains other electrolytes as well (in lower concentrations), thus if the total mass of salts in seawater were accounted as NaCl, the concentration of NaCl would be about 0.6 M. As the concentration of electrolytes in a solution increases, is obvious that the divergence of concentration from linearity is evident. This can be seen in the curves of Fig. 2.2 where concentration of NaCl, in molality m is plotted versus involved vapor pressure of water.

This is due to a phenomenon by which the progressive concentration is not proportional to the amount of solute accumulating in the solution. This means that as concentration increases, dissociation decreases and the solutes become progressively less effective. This phenomenon applies especially to seawater concentrates, for example the effluents from the desalination plants. Thus to describe aqueous electrolyte solutions, such as seawater and its concentrates, the effective concentrations of the elements are needed, which are described by their activity (see 2.4.2.5).

The new thermodynamic and equation state of seawater and the thermophysical properties derived from these state equations were developed for oceanographic purposes and are valid in the range $S = 0-40\ g\ kg^{-1}$ and temperatures $t = 0-40°C$.

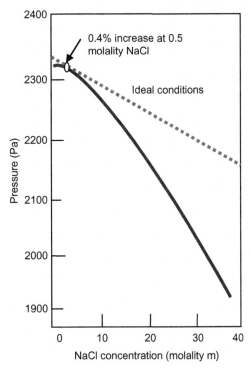

Figure 2.2 Indicative water vapor pressure as a function of NaCl concentration at 20°C at various molalities (m) of NaCl (Walton, 1974).

For desalination systems the properties now are extended in the range of salinity $S = 0$ $(20)-120$ g kg^{-1} and temperatures $t = 0(20) - 120°C$.

2.4.1.3 pH

The concept of pH was introduced by the Danish Søren P.L. Sørensen in 1909 and revised later to accommodate definitions and measurements in terms of electrochemical cells.

The parameter pH is a symbol for hydrogen ions concentration in a solution and defines its intensity of alkalinity or acidity. H stands for hydrogen and p for decimal logarithm. Pure water is diverged into hydrogen anions and hydroxyl cations which are in kinetic equilibrium with water molecules:

$$H_2O \leftrightarrows H^+ + OH^- \tag{2.38}$$

In equilibrium point the following relationship of Eq. (2.18a) is applied:

$$K_w = [H^+][OH^-]/[H_2O] \qquad (2.39)$$

where K_w is equilibrium constant of water and brackets define molar concentration of ions and compounds. Pure distilled water has constant molar water $[H_2O]$ concentration comprised in the equilibrium constant K_w. Thus for pure water the Eq. (2.39) has the simple form:

$$[H^+] \cdot [OH^-] = K_{pw} \qquad (2.40)$$

For atmospheric pressure, $p_o = 1.01325$ bar, and $t = 25°C$ the value of K_{pw} is 10^{-14} and for same temperature and pressure according to Eq. (2.41) the value is 10^{-7}:

$$[H^+] = [OH^-] = 10^{-7} \qquad (2.41)$$

where, 10^{-7} means neutral water. When various salts are added in pure water, equilibrium changes according to the hydrogen or hydroxyl ion concentration in the solution. For example, dissolving $NaOH$ and H_2SO_4 in water we have:

$$NaOH \leftrightarrows Na^+ + OH^-, \quad H_2SO_4 \leftrightarrows 2H^+ + SO_4^- \qquad (2.42)$$

The water will be acidic as two hydrogen ions exist in the solution versus one hydroxyl. If $[H^+] >> [OH^-]$ water becomes acidic, and has $10^{-1} < K_w < 10^{-7}$. If $[OH^-] >> [H^+]$ water is alkalic and has values $10^{-14} > K_w > 10^{-7}$. Sørensen established an exponential expression for hydrogen, deriving from Eqs. (2.40) and (2.41):

$$[H^+]^2 = K_{pw} \qquad (2.43)$$

pH really has no thermodynamic meaning and according to the above Sørensen Eq. (2.43) increase by 1.0 unit of pH means decrease by 10-fold of hydrogen ions. Further Sørensen gave another equation that converts the exponential type of pH into plane numbers:

$$pH = -\log[H^+] = \frac{1}{2}\log K_{pw} = \log\frac{1}{[H^+]} = -\frac{1}{2}\log 10^{-14} = 7.0 \qquad (2.44)$$

Table 2.10 gives the equivalent constant, K_{pw}, of distilled water and its pH at various temperatures.

2.4.1.4 Density (ρ)
Seawater density is a function of temperature and salinity. It is expressed by the Gibbs energy of seawater as:

Table 2.10 Equivalent Constant and pH for Pure Water

°C	K_{pw} (mol^2 L^{-2})	pH	°C	K_{pw} (mol^2 L^{-2})	pH
0	1.13×10^{-15}	7.45	25	1.00×10^{-14}	6.92
5	1.83×10^{-15}	7.37	30	1.45×10^{-14}	6.94
10	2.89×10^{-15}	7.27	35	2.07×10^{-14}	6.84
15	4.46×10^{-15}	7.18	40	2.91×10^{-14}	6.77
20	6.75×10^{-15}	7.09			

Table 2.11 Chebyshef Coefficients for Practical Values (Eq. 2.39)

i	j = 0	j = 1	j = 2	j = 3
0	4.032219	−0.108199	−0.012247	0.000692
1	0.115313	0.001571	0.001740	−0.000087
2	0.000326	−0.000423	−0.000009	−0.000053

$$\rho(S, T, p) - v^{-1} = (\partial g/\partial p)_{S,T}^{-1} = g_p^{-1} \tag{2.45}$$

The validity ranges are, for temperature $0 < t_{90} < 40°C$ and for salinity $0 < S < 40$ g kg^{-1}.

Corresponding pressure is in the range of $0-100$ MPa.

Grunberg (1970), Isdale and Morris (1972), and Isdale et al. (1972) performed experimental measurements of seawater densities in the range of practical salinity $10 < S_P < 160$ g kg^{-1} and temperature range $20 < t_{68} < 180°C$ by using the Chebyshef polynomial which gives smooth density values. It is specially fitted to seawater density calculation. The corrected polynomial according to the new state equation is:

$$\rho_{sw} = 10^3 \sum_{i=0}^{3} \sum_{j=o}^{2} C_{ij} \left(\frac{2S_p - 150}{150} \right) t_{i-68} \left(\frac{2t - 200}{160} \right) \tag{2.46}$$

where C_{ij} is the Chebyshef coefficient, prices of which are given in Table 2.11, T is temperature in K and t_{68} is temperature in °C. By setting $B = (2S_P - 150)/150$ and $A = (2t_{68} - 200)/160$ derive the empirical Chebyshef equation for routine calculations:

$$\rho_{sw} = 0.5a + b \cdot A + c(2A^2 - 1) + d(4A^3 - 3A) \tag{2.47}$$

where the parameters a, b, c, and d are calculated from salinity and the parameter B:

$a = 2.016110 + 0.115313B + 0.000326 (2B^2 - 1)$
$b = -0.0541 + 0.001571B - 0.000432(2B^2 - 1)$
$c = -0.006124 + 0.001740B - 0.9 \times 10^{-5}(2B^2 - 1)$
$d = 0.000346 + 0.87 \times 10^{-4}B - 0.53 \times 10^{-4}(2B^2 - 1)$

Thus density is directly dependent from salt concentration, that is, from the salinity. The calculations accuracy is $\pm 0.1\%$.

An accurate knowledge of density-concentration relationship may determine salt content in a seawater sample or its concentrates without use of chemical analysis. A very practical and simplified expression is to divide seawater solution into 1 part of total dissolved salts and 32 parts of seawater. This is the so-called 1/32th system which applies to seawater and its concentrates, for example if seawater of 1/32th density is concentrated 50% of its original volume then its density becomes 2/32. By plotting chlorinity against 32th derives a straight line (Hampel, 1950).

In Fig. 2.3 the curves of seawater and its concentrates as function of temperature are presented. In Appendix, C Fig. C.5 presents the curves of seawater density as function of temperature chlorosity and salinity (Spiegler and El-Sayed, 1994).

Millero and Poison (1981), proposed the following expression for seawater density which presents an accuracy of 0.01%.

$$\rho_{\sigma\varsigma} = \rho_{pw} + AS_p + BS_p^{2/3} + CS_p \tag{2.48}$$

where

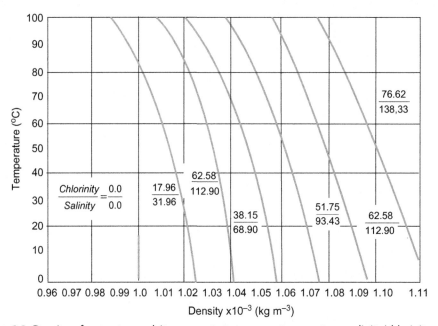

Figure 2.3 Density of seawater and its concentrates versus temperature, salinity/chlorinity ratio. 00/00 refers to pure water and the last curve (S/Cl = 76.62/138.33) refers to the saturation state (Hampel, 1950).

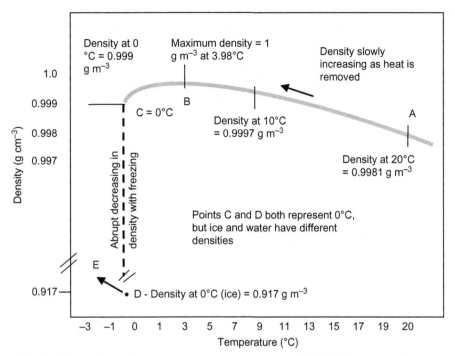

Figure 2.4 Variation of density over temperature changes (OCN-623-Chemical Oceanography 2009).

$$A = 0.824493 - 4.0899 \times 10^{-3} t_{68} + 7.6438 \times 10^{-5} t_{68}^2 + 8.2467 \times 10^{-7} t_{68}^3 + 5.3875 \times 10^{-9} t_{68}^4$$
$$B = 5.72466 \times 10^{-3} + 1.0227 \times 10^{-4} t_{68} - 9.0529 \times 10^{-6} t_{68}^2, \quad C = 4.8314 \times 10^{-4}$$

$$(2.49)$$

Seawater has its highest density in salinities 24.7 g kg^{-1}. Normal seawater freezes to ice at 271.8 K ($-1.332°C$). Fig. 2.4 presents the curve of density variation over temperature changes (OCN 623, Salinity 2012 Web) for water. The density of seawater, for salinities $0-120 \text{ g kg}^{-1}$ and temperatures $0-120°C$ is given in Table 2.12. The values in Table 2.12 are the corrected experimental data by Isdale and Morris (1972), expressed in terms of ITS-90 and the reference absolute-composition salinity scale IARWS-2008 (Sharqawy et al., 2010). Fig. 2.5 presents the relationships between density and salinity for various temperatures at saturation pressure.

Regression coefficients R^2 for the linear fits of salinity versus density are very close to one (Sun et al., 2008). Detailed seawater densities, as function of salinity and temperature are presented by Millero and Huang (2009).

Table 2.12 Density of Seawater, kg m^{-3} at Various Salinities

°C	0[a]	10	30	50	70	90	110	120
				Salinity (g kg^{-1})				
0	999.8	1007.9	1024.0	1040.0	1056.1	1072.1	1088.1	1096.2
10	999.7	1007.4	1023.0	1038.7	1054.4	1070.1	1085.7	1093.6
20	998.2	1005.7	1021.1	1036.5	1051.8	1067.2	1082.6	1090.3
30	996.7	1003.1	1018.2	1033.4	1048.5	1063.6	1078.7	1086.3
40	992.2	999.7	1014.6	1029.5	1044.5	1059.4	1074.4	1081.8
50	988.0	995.6	1010.3	1025.1	1039.9	1054.7	1069.5	1076.9
60	983.2	990.6	1005.3	1020.0	1034.7	1049.5	1064.2	1071.5
70	977.8	985.1	999.8	1014.5	1029.1	10438	1058.5	1065.8
80	971.8	979.1	993.8	1008.5	1023.1	1037.8	1052.5	1059.8
90	965.3	972.6	987.3	1002.0	1016.8	1031.5	1046.2	1053.5
100	958.4	965.7	980.5	995.2	1010.0	1024.8	1039.6	1047.0
110	950.9	958.3	973.2	988.1	1003.0	1017.8	1032.7	1040.2
120	943.1	950.6	966.6	980.6	995.6	1010.6	1025.6	1033.1

[a]Pure water, Accuracy ±0.1%.

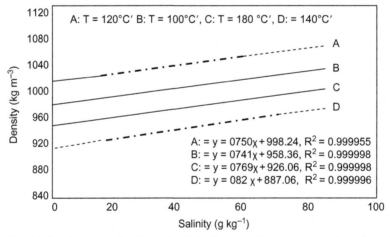

Figure 2.5 The relation between density and salinity for several temperatures based on the experimental data of Grunberg (1970) (Sun et al., 2008).

2.4.1.5 Activity (α) and Activity Coefficients (γ)—Chemical Potential (μ)

The activity of the species introduced first by G.N. Lewis in 1907 during elaboration of the thermodynamic work of J.W. Gibbs.

When pure water is added in a solute the physicochemical properties of water are altered, but not proportionally to the amount of solutes added. As concentration increases the solute becomes less effective and the solutions properties change. This occurs in both weak and strong electrolytes. This phenomenon is described for seawater and for aqueous electrolyte solutions not by the partial dissociation of the solute but by the effective concentration of the solutes, which is called "activity."

Electrolyte species interact with solvent species and to each another. Thus even for quite dilute solutions there is a departure from linear divergence as it is very clearly presented in Fig. 2.2, where the concentration of sodium chloride is plotted against involved vapor pressure of water.

The consequences of these interactions between the various components in the electrolyte solution are described in terms of equilibrium thermodynamics, that is, with the energetic of chemical reactions. When dealing with chemistry of seawater it is assumed that the system is of defined temperature and pressure. To study electrolyte solutions partial free energy is an important thermodynamic property. It is referred to as "Gibbs free energy" presented by (G), or as "chemical potential" (μ).

In such systems the fundamental thermodynamic function of Gibbs free energy G (J) is applied. The change of free energy accompanying a chemical reaction is given by the following equation:

$$\Delta G = \Delta H - T \cdot \Delta S \tag{2.50}$$

where ΔH (J) is the change in enthalpy. It is the measure of heat change that takes place during a chemical reaction and depends on the relative strength of interaction between the various components. ΔS (J K^{-1}) is the change in entropy giving the degree of disorder due to rearrangement of the molecules during reaction.

Gibbs energy G of a seawater sample containing the mass of water m_w and the mass of salt M_s at temperature t and pressure p can be written in the form (Feistel, 2008):

$$G(m_w, m_s, t, p) = m_w \mu^W + m_s \mu^S \tag{2.51}$$

where, μ^W and μ^S are calculated from Eq. (2.35). The specific Gibbs energy is given (Feistel, 2008) as:

$$G(S_A, t, p) = \frac{G}{m_w + m_s} = (1 - S_A)\mu^W + S_A\mu_s = \mu^W + S_A(\mu^S - \mu^W) \tag{2.52}$$

Partial molar quantity. The thermodynamic functions, as free energy, etc., depend on temperature, pressure, and volume, that is, the state of the system and the amount of the various constituents in the solution. For such systems the partial molar value for a constituent i is defined as:

$$\bar{x} = (\partial x / \partial n_i)_{T, p, n1, n2} \tag{2.53}$$

For small changes in the system, that is, if moles of a constituent increase or decrease, at constant temperature and pressure, then the total change dx, temperature and pressure dependent, is given as:

$$d(x)_{T,p} = \bar{x}_1 dn_1 + (-)\bar{x}_2 dn_2 + \ldots + \bar{x}_n dn_i \qquad (2.54)$$

where

$n_1, n_2, \ldots n_n$ Are moles of constituents 1, 2, 3 ... n —
$\bar{x}_1, \bar{x}_2 \ldots$ Are partial molar values of constituents 1, 2, 3, ... n —

The partial molar-free energy-chemical potential. Total Gibbs free energy of a fixed quantity of electrolyte solution and of given composition depends only on pressure and temperature independently of adopted conventions to express partial molar quantities such as activities. In general for a solvent A and a solute B (ie, the anhydrous solute) the partial molar Gibbs free energies are respectively (Robinson & Stokes, 1959):

$$\overline{G_A} = (\partial G/\partial n_A)_{n_B, T, p} \quad or \quad \overline{G_w} = (\partial G/\partial n_w)_{n_S, T, p} \qquad (2.55)$$

$$\overline{G_B} = (\partial G/\partial n_B)_{n_A, T, p} \quad or \quad \overline{G_s} = (\partial G/\partial n_s)_{n_w, T, p} \qquad (2.56)$$

where n_A and n_B are the moles of solvent A and solute B in the system respectively, which will be referred here as n_w and n_s because the solvent in this case is water and the solute a salt. In general, in electrolyte systems we deal more with the variation of chemical potential as composition changes than with absolute values. Thus these quantities are expressed as differences between the absolute value and a value which holds in a specific standard state. This is indicated as $\overline{G_w^o}, \overline{G_s^o}$ or μ_w^o, μ_s^o, and Gibbs free energy is then called "chemical potential." Standard state for electrolyte solutions to which free energy is referred is the pure solvent at the same temperature and pressure, that is, in our case the pure water.

In ideal systems, as is fresh water, solute and solvent do not interact but simply mix. The chemical potential (μ) (ie, the Gibbs free energy) of a component i is related to its molality concentration, m_i, by the following relation:

$$\mu_i - \mu_i^o = RT \ln(m_i/m_o)S \qquad (2.57)$$

When electrolyte solutions are in equilibrium with their vapor, the chemical potential of any component in the solution is equal to its partial vapor pressure. According to Raoult's law, in an ideal solution partial vapor pressure of any constituent is proportional to its mol fraction (x). Thus we can write:

$$\mu_i - \mu_i^o = RT \ln(m_i/m_o) \quad or \quad \mu_i - \mu_i^o = RT \ln(p_i) \qquad (2.58)$$

where p_i is the partial vapor pressure of constituent i, and m_o is a reference molality. For nonideal solutions as is seawater vapor pressure of a component is proportional to its molar ratio multiplied by a correction factor γ_i which denotes the deviation of the ideal state; thus it is $p_i = x_1 \cdot \gamma_i$. The factor γ_i is known as "activity coefficient" being the ratio of the activity (α) of a component i to its molarity (M) or molality (m) of the same component.

$$\gamma_i \equiv \alpha_i / m_i \qquad (2.59)$$

Thus the above Eq. (2.58), for ideal electrolyte solutions becomes:

$$\mu_i = -\mu_o + RT \ln(m_i \cdot \gamma_i) = \mu_o + RT \ln(\alpha_i) \qquad (2.60)$$

where

α_i	Activity of the component i	—
γ_i	Activity coefficient of the component i	$kg\ mol^{-1}$
μ_i	Normal chemical potential of pure water at ideal conditions and pressure of 101.325 kPa	$J\ kg\ mol^{-1}$
μ_o	Reference state chemical potential	$J\ kg\ mol^{-1}$
m_i	Molality of component i	$mol\ kg^{-1}$
m_o	Reference molality state	$mol\ kg^{-1}$
R	Gas constant = 8.31439	$J\ mol^{-1}\ K^{-1}$
T	Thermodynamic temperature	K
x_i	Mole fraction of component i	—

Thus activity is defined with respect to a "standard state," that is, in a state where the solution behaves ideally, so that can be assigned a value to μ_o. When a solution is in standard state, as it approaches ideal conditions, activity coefficient approaches unity. Thus the solute is chosen such that (Horne, 1969):

$$lim\ a_\pm = 1.0 \qquad (2.61)$$

where $a_\pm$ is mean activity. The above expression is valid for all temperatures and pressures. By practical point of view only mean activities of electrolytes can be measured experimentally, that is, both cations and anions. Thus in general, only mean values of activity, $\alpha_\pm$, activity coefficient $\gamma_\pm$, and molality $m_\pm$ are applied, based on the dissociation of a generalized electrolyte. If an electrolyte solution contains n cations (C), and m anions (A), it applies:

$$C_n \cdot A_m \leftrightarrows n \cdot C^{ZC} + m \cdot A^{ZA} \qquad (2.62)$$

where, ZC and ZA are changes of cation C and anion A respectively. According to Eq. (2.60) mean values of activity, activity coefficient, and ionic molality are:

$$\begin{aligned} a_\pm^{n+m} &\equiv \alpha_C^n \cdot a_A^m, \quad \gamma_\pm^{n+m} \equiv \gamma_C^n\ g_A^m \quad \text{and} \\ m_\pm^{n+m} &\equiv m_C^n\ m_A^m \end{aligned} \qquad (2.63)$$

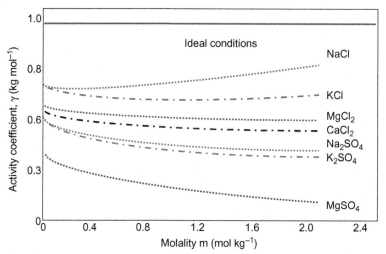

Figure 2.6 Indicative curves of concentration dependence of mean ionic activity coefficient $(\gamma_{\pm})$ for some components in seawater (Horne, 1969).

Table 2.13 Activity + Activity Coefficient of Main Components in Seawater for Salinity $S = 34.3$ g kg^{-1}, $T = 298.15$ K, $p = 101.325$ kPa, and pH = 8.1

Component	Component	Molarity (M)	Activity Coefic. (γ)	Activity (α)
H_2O	Water	55	0.98	54
Cl^-	Chloride	0.55	0.64	0.35
Na^+	Sodium	0.47	0.76	0.36
Mg^{2+}	Magnesium	0.047	0.36	0.017
SO_4^{2-}	Sulfates	0.015	0.12	0.0016
K^+	Potassium	0.0099	0.64	0.0063
Ca^{2+}	Calcium	0.0095	0.28	0.0027
$MgSO_4$	Magnesium sulfate	0.0060	1.13	0.0068
$CaSO_4$	Calcium sulfate	0.00084	1.13	0.00095
HCO_3^-	Hydrogen carbonate	0.0016	0.68	0.0011
$MgCO_3$	Magnesium carbonate	0.00017	1.13	0.00022
$NaHCO_3$	Sodium hydrogen carbonate	0.0001	1.13	0.00011
CO_3^{2-}	Carbonates	0.00002	0.20	0.000004
$CaCO_3$	Calcium carbonate	0.00002	1.13	0.00002

Note: The reference units used must be clearly described. They can be expressed as γ_m, γ_i^C, $\gamma_{\pm}$, etc., referring to molar, ionic, or mean values respectively. Normally activity coefficient is expressed in molality or ionic units.

In Fig. 2.6, some mean ionic activity coefficients of salts in sea water as function of their molality m are given indicatively. In Table 2.13 activity and activity coefficients of main ions in seawater are presented.

Chou and Rowe (1969) give an expression for the logarithmic activity $\ln a_w$, variation of water as:

$$\ln a_w = \ln a_w(p_o) + \frac{1}{RT \int_{po}^{P} \left(\overline{v}_1 - \overline{v}_1^o\right) dp} \tag{2.64}$$

which is related to the Eq. (2.65). In Eq. (2.64) p_o is a fixed pressure (Pa), $\overline{v}_1, \overline{v}_1^o$ are partial volume at the concentration to which the activity α_w refers and partial volume at infinite dilution, respectively. Partial volumes are evaluated from specific volumes of the solution by the equation:

$$\overline{v}_1 = v - x(\partial v/\partial x)_{R,p} \tag{2.65}$$

where x is mole fraction and v is the specific volume of the solution.

2.4.1.6 Thermal Conductivity (λ, k)

Thermal conductivity is expressed by the thermal conductivity coefficient λ (or k, $W\,m^{-1}\,K^{-1}$). It is defined as the amount of heat that a solid material or a thin layer of a fluid membrane permits to pass through per time unit, per surface area unit, and per unit of temperature gradient.

Thermal conductivity of water, or of aqueous solutions containing electrolytes, decreases with increasing salt concentration. For seawater is proportional to its salinity. The following empirical equation, based on experimental data, is applied for the routine calculation of thermal conductivity coefficient in seawater (Jamieson and Tudhope, 1970; Jamieson, 1986):

$$\ln\lambda_{sw} = \ln(\lambda_c + a) + \left(2.3 - \frac{b}{T}\right) \cdot \left(1 - \frac{T}{T_c + c}\right)^{0.333} \tag{2.66}$$

a, b, c are salinity parameters calculated from the following relationships:

$$a = 0.0002S_p, \quad b = 343.5 + 0.375S_p, \quad c = 0.03S_p \; g\,kg^{-1} \tag{2.67}$$

The following equation from experimental data is reformed according to IAPWS formulations for seawater properties (Sharqawy et al. (2010)):

$$\log_{10}(k_{sw}) = \log_{10}(240 = 0.0002S_p) + \left(2.3 - \frac{343.5 + 0.037S_p}{t_{68} + 273.15}\right)\left(1 - \frac{t_{68} + 273.15}{647 + 0.03S_p}\right)^{0.33} \tag{2.68}$$

where

λ_{sw}, k_{sw}	Thermal conductivity of the seawater probe	$10^{-3}\,Wm^{-1}\,K^{-1}$
λ_c	Thermal conductivity of distilled water in critical point, $\lambda_c = 240 \times 10^{-3}$	$Wm^{-1}\,K^{-1}$
S_p	Practical salinity	$g\,kg^{-1}$

t	Temperature of the seawater probe	°C
T	Seawater probe temperature	K
T_c	Critical temperature of distilled water = 647.3	K

The equation is valid in the temperature range $0 < t_{68} < 180°C$ and salinity range of $0 < S_p < 160$ g kg^{-1}. The accuracy is $\pm 3\%$.

A simplified empirical equation for salinities S, $0-100$ g kg^{-1} and temperatures $10-150°C$ is given as (Khan, 1986):

$$\lambda = a + bt_{68} + ct_{68}^2 10^{-3} \ \text{W m}^{-1} \ \text{K}^{-1} \tag{2.69a}$$

where the parameters a, b, and c have the following values:

$$a = 576.6 - 34.64 \cdot d + 7.286 \cdot d^2 \tag{2.69b}$$

$$b = 1.526 + 0.4662 \cdot d + 0.2268 \cdot d^2 + 0.02867 \cdot d^3 \tag{2.69c}$$

$$c = -[0.582 + 2.055 \cdot d - 0.9916 \cdot d^2 + 0.1464 \cdot d^3] \times 10^{-2} \tag{2.69d}$$

The coefficient d is a correcting parameter of practical salinity and is calculated as:

$$d = 28.17 S_p/(1000 - S_p) \tag{2.70}$$

Castelli et al. (1974) present an equation for thermal conductivity based on measurements of standard seawater:

$$0.55286 + 3.4025 \times 10^{-4} p + 1.8364 \times 10^{-3} t_{68} - 3.3058 \times 10^{-7} t_{68}^3 \tag{2.71}$$

The equation is valid for temperatures $0 < t_{68} < 30°C$, for pressures $0 < p < 140$ MPa and salinities $S_p = 35$ g kg^{-1}. Its accuracy is $\pm 0.4\%$.

In Table 2.14, the values for thermal conductivity are based on Eq. (2.68) and are valid for temperatures $0-180°C$ and $S_p = 0-160$ g kg^{-1}, that is, distilled water, up to concentrates of 4.57-fold according to experimental data of Jamieson and Tudhope (1970) corrected for IAPWS-08 terms.

Fresh water thermal conductivity is presented as (Sharqawy et al., 2010):

$$k_{pw} = \sum_{i=1}^{4} a_i (T/300)^{bi} \tag{2.72}$$

Table 2.14 Thermal Conductivity of Seawater, $Wm^{-1} K^{-1}$ at Various Salinities

°C				Salinity (g kg^{-1})				
	0[a]	10	30	50	70	90	110	120
0	0.572	0.571	0.570	0.569	0.568	0.566	0.565	0.565
10	0.588	0.588	0.587	0.585	0.584	0.583	0.582	0.582
20	0.604	0.603	0.602	0.601	0.600	0.599	0.598	0.597
30	0.617	0.617	0.616	0.615	0.614	0.613	0.612	0.611
40	0.630	0.629	0.628	0.627	0.626	0.625	0.624	0.624
50	0.641	0.640	0.639	0.638	0.637	0.636	0.635	0.635
60	0.650	0.650	0.649	0.648	0.647	0.646	0.645	0.645
70	0.658	0.658	0.657	0.656	0.655	0.656	0.654	0.654
80	0.665	0.665	0.664	0.663	0.663	0.662	0.661	0.661
90	0.671	0.671	0.670	0.669	0.669	0.668	0.667	0.667
100	0.676	0.675	0.675	0.674	0.673	0.673	0.672	0.672
110	0.679	0.679	0.678	0.678	0.677	0.676	0.676	0.675
120	0.681	0.681	0.681	0.680	0.679	0.679	0.678	0.678

[a]Pure water, Accuracy ± 3%.

The parameters a_i and b_i have the following values:

$$a_1 = 0.80201, \quad a_2 = -0.25992, \quad a_3 = 0.10024, \quad a_4 = -0.032005$$
$$b_1 = -0.32, \quad b_2 = -5.7, \quad b_3 = -12.0, \quad b_4 = -15.0$$

The equation is valid for temperatures $273.15 \leq T \leq 383.15$ and its accuracy is ±2%.

2.4.1.7 Diffusion—Diffusion Coefficient (D)

By pouring into pure water a small amount of a salt, for example sodium chloride, without stirring the water, sodium chloride will dissolve and partially hydrolyze into ions Na^+ and Cl^-. There will be mutual sodium chloride molecules and ions penetration into the pure water and simultaneously pure water to the more concentrated part of the salt solution. This very slow mutual physical flow of molecules and ions in an aquatic solution is called diffusion. Flow will continue until the phases reach to a kinetic equilibrium. Thus diffusion of a component i in seawater is defined as the amount of moles that flow through the unit surface perpendicular to the flow direction of the solution, in unit time.

Table 2.15 Indicative Diffusion Coefficients (D, $m^2\ s^{-1}$) of Some Water Electrolytes at 25°C and the Corresponding Molarities, M

M	NaCl	KCl	CaCl$_2$	SrCl$_2$	Na$_2$SO$_4$	MgSO$_4$	CsCl
0.005	15.60	19.34	11.79	11.29	11.23	7.10	19.78
0.01	15.45	19.17	–	–	–	–	19.58
0.05	15.07	18.64	11.21	–	–	–	–
0.10	14.83	18.44	11.10	–	–	–	18.71
0.50	14.74	18.50	11.40	–	–	–	18.60
1.0	14.84	18.92	12.03	–	–	–	19.02
1.50	14.95	19.43	12.63	–	–	–	–

$$J_i = -D_i \frac{dc_i}{dx} \quad mol\ m^{-2}\ s^{-1} \tag{2.73}$$

where

J_i	Diffusion of component i	$mol\ m^{-2}\ s^{-1}$
D_i	Diffusion coefficient of component i	$m^2\ s^{-1}$
c_i	Mole concentration of component i	$mol\ m^{-3}$
dc_i/dx	Concentration gradient to the direction x	$mol\ m^{-3}\ m^{-1}$

Diffusion of a component i is affected by concentration gradient of ions in the solution, from the friction resistance in the direction of diffusion and from the electrical gradient of the electrolytes in the solution. For ions of a certain component i, diffusion alters with temperature changes, the concentration gradient of the component, and the concentration of other ions in the solution.

Table 2.15 presents indicative diffusion coefficients of some seawater species at temperature of 25°C.

Diffusion and electrical conductance in electrolyte solutions involve motion of ions. Nevertheless there exist differences between the two processes:

1. During diffusion positive and negative ions move in the same direction while in electrical conduction they move in the opposite direction
2. In very extreme dilution during diffusion the ions move at equal speed while in electrical conduction the ions move independently of one another.

2.4.1.8 Viscosity (η, ν)

Viscosity is defined as the property of a liquid which causes resistance to flow. It is the internal friction (drag forces) which molecular and ionic constituents exert on one another. The adhesion forces of the fluid molecules determine it, that is, viscosity arises primarily from the movement of solvent molecules and provides a useful insight into solvent structure. Viscosity is especially sensitive to

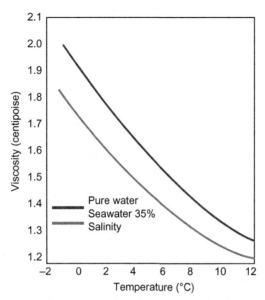

Figure 2.7 Pure and seawater temperature indicative dependence of viscosity (Horne, 1969).

liquid structure and to temperature changes. It is of importance for desalination as it affects heat and fluid flow and is defined by the following equation as the dynamic viscosity.

$$\eta = F/A = dw/ds \qquad (2.74)$$

where F is the shear force created tangentially on a plane of area A and the velocity gradient dw/ds is perpendicular to the plane at the point of application. Fig. 2.7 presents the viscosity dependence of temperature for pure water and for seawater.

Dynamic viscosity is given in $N\,s\,m^{-2}$ or Pa s. In many cases dynamic viscosity is expressed in Poise (P), that is, dyns cm^{-2} or in centi-Poise (cP) where $1.0\,dyn = 10^{-5}\,N$ and $1.0\,P = 10^{-1}$ Pa s. Dividing dynamic viscosity by the solution's density ($kg\,m^{-3}$) the kinematic viscosity is derived:

$$\nu = \eta/\rho \qquad (2.75)$$

expressed in $m^2\,s^{-1}$ or Stokes ($cm^2\,s^{-1}$, $1.0\,Stoke = 10^{-4}\,m^2\,s^{-1}$). In desalination systems, dynamic viscosity applies in general, which for routine operations may be calculated by the following empirical equations for pure water and seawater, respectively (Khan, 1986):

1. *For pure water*

$$\ln(\eta_{pw}) = -3.79418 + \frac{604.129}{139.18 + t_{68}} \qquad (2.76)$$

where, temperature t_{68} is given in °C, and dynamic viscosity in cP. The equation has an accuracy of $\pm 1\%$.

2. *For sea water solutions* by Fabuss and Korosi (1967) and Korosi and Fabuss (1968)

Sea water solutions (concentrates and dilutes) are temperature and salinity functions, thus viscosity of seawater η_{sw}, is:

$$\eta_{sw} = \eta_r \cdot \eta_{pw} \qquad (2.77)$$

where, η_r is relative viscosity. It is represented by the ratio of salt solution viscosity to the viscosity of pure water η_{pw} for the same temperature and is calculated from the following empirical equation (Isdale et al., 1972):

$$\eta_r = \eta_{sw}/\eta_{pw} = 1 + aS_p + b \cdot Sp^2 \qquad (2.78a)$$

where parameters a and b are functions of solution's temperature t_{68}°C. They are formulated by using pure water viscosity by Fabuss and Korosi (1967) and Korosi and Fabuss (1968):

$$a = 0.001474 + 1.5 \times 10^{-5} \cdot t_{68} - 0.003927 \times 10^{-5} \cdot t_{68}^2 \qquad (2.78b)$$

$$b = 1.0734 \times 10^{-5} - 8.5 \times 10^{-8} \cdot t_{68} + 0.00223 \times 10^{-7} \cdot t_{68}^2 \qquad (2.78c)$$

By using the normalized pure water viscosity data (IAPWS-2008) the parameters a and b of Eq. (2.78a) have the following new values (Sharqawy et al., 2010):

$$a = 1.541 + 1.998 \times 10^{-2} \cdot t - 9.52 \times 10^{-5} t^2 \qquad (2.78d)$$

$$b = 7.974 - 7.561 \times 10^{-2} \cdot t + 4.724 \times 10^{-4} \cdot t^2 \qquad (2.78e)$$

where the normalized pure water viscosity is given by:

$$\eta_{pw} = 4.2844 \times 10^{-5} + [0.157(t + 64.993)^2 - 91.296]^{-1} \qquad (2.79)$$

The equation is valid for temperatures $0 \leq t \leq 180$°C, Accuracy is $\pm 0.05\%$. The data in Table 2.16 are based on values of Eqs. (2.77) and (2.78d), (2.78e).

For kinematic viscosity Chen et al. (1973) present the following expression based on salinity:

$$v = v_{pw} + aS^{1/2} + bS + cS^2 \qquad (2.80)$$

where v_{pw} is the kinematic viscosity of pure water and salinity S is expressed in % mass of salt. The parameters a, b, and c have the following values for temperature range between 10 and 150°C:

Table 2.16 Dynamic Viscosity 10^3 kg m^1 s^{-1}, of Seawater Solutions in Various Salinities and Temperatures

°C	0[a]	10	30	50	70	90	110	120
				Salinity (g kg^{-1})				
0	1.791	1.820	1.887	2.925	2.055	2.156	2.268	2.328
10	1.306	1.330	1.382	1.443	1.511	1.548	1.669	1.714
20	1.002	1.021	1.065	1.114	1.168	1.227	1.259	1.326
30	0.797	0.814	0.851	0.891	0.936	0.984	1.037	1.064
40	0.653	0.667	0.699	0.734	0.771	0.812	0.855	0.878
50	0.547	0.560	0.587	0.617	0.649	0.684	0.721	0.740
60	0.466	0.478	0.502	0.528	0.556	0586	0.618	0.635
70	0.404	0.414	0.436	0.459	0.484	0.510	0.538	0.553
80	0.354	0.364	0.383	0.404	0.426	0.449	0.474	0.487
90	0315	0.323	0.340	0.359	0.379	0.400	0.422	0.434
100	0.282	0.289	0.305	0.322	0.340	0.369	0.380	0.390
110	0.255	0.262	0.276	0.291	0.308	0.325	0.344	0.354
120	0.232	0.238	0.251	0.261	0.288	0.297	0.314	0.323

[a]Pure water, Accuracy ±1.5%.
Based on Isdale, J.D., Spencer, C.M., Tudhope, J.S. 1972. Physical properties of seawater solutions. Desalination 10, 319–328; Jamieson, D.T. 1986. Experimental methods for determination of the properties of saline water. Desalination 59, 219–240.

$$a = -1.464, \quad b = 205.4, \quad c = 153.0$$

2.4.1.9 Ionic Strength (I)

Ionic strength is a measure of the intensity of the electrical field created in a solution by an ionic activity α. It represents the variation of activity coefficient with concentration in electrolyte solutions. In dilute solutions ions react independently from each other but as concentration increases start an interaction between them and activity α decreases. Ionic strength I, is defined as half of the sum of terms obtained by multiplying molality (m_i) or concentration (c_i) of a constituent i present in the solution by the square of its valence (z) and is calculated from the following equation for water (Tsobanoglou and Schroeder, 1985):

$$I = \frac{1}{2} \sum \left(c_i z_i^2 \right) = \frac{1}{2} \sum (m z^2) \tag{2.81}$$

where

I	Ionic strength	mol kg^{-1}
c_i	Concentration of ions of component i	mol kg^{-1}
M_i	Molality of component i	mol kg^{-1}
z_i	Load of ions of species i	—

Ionic strength of a component i, I_i, may be calculated from the electrical conductivity E_c, by the following empirical equation:

$$I_i = 1.6 \times 10^{-5} E_c \qquad (2.82)$$

Ionic strength is linked to activity coefficient γ by the empirical equation:

$$\log I_{(i)} = 0.5 \cdot \gamma I / (1 + 1.5 \cdot \gamma I) + cI \qquad (2.83)$$

For $I < 0.1$ mol L^{-1} the term cI is negligible and can be omitted.

2.4.1.10 Enthalpy (H)—Specific Enthalpy (h)

Enthalpy expresses the results of thermodynamic processes in a system occurring under special conditions. It is a measure of the results due to pressure, volume, and internal energy changes in the system. Its value is given as change in the energy state of the system, that is, as difference between two thermal conditions in the system:

$$\Delta H = \Delta U + P \Delta V \qquad (2.84)$$

where

ΔH	The change in enthalpy of the system	mN = J
ΔU	The change of internal energy in the system	J
$P(\Delta V)$	Volume change of the system due to pressure	J

From the known equation of internal energy (first law of thermodynamics):

$$\Delta U = Q + W \qquad (2.85)$$

is calculated enthalpy in steady pressure conditions:

$$Q_p = \Delta U + P(\Delta V) = \Delta H \qquad (2.86)$$

Table 2.17 Specific Enthalpy h, (kJ kg^{-1}) of Various Salt Solutions at 100°C

Concentration (‰)	NaCl	KCl	MgCl$_2$	Na$_2$SO$_4$	MgSO$_4$	Seawater
Pure (H$_2$O)	4.214	4.214	4.214	4.214	4.214	4.214
10	4.164	4.164	4.155	4.170	4.170	4.164
20	4.116	4.113	4.096	4.127	4.129	4.114
40	4.023	4.013	3.980	4.132	4.025	4.019
60	3.933	3.932	3.866	3.968	3.964	3.925
80	3.845	3.822	3.749	3.895	3.890	3.841
100	3.673	3.730	3.630	3.825	3.818	3.758

where Q and W are heat and work respectively, that is, the forms of energy that are exchanged between a system and its surroundings. In Eq. (2.86) enthalpy change equals the heat effect of a reversible process when the system performs only PV work.

In practice, specific enthalpy h, is in use and refers to $J\ kg^{-1}$ or $J\ mol^{-1}$.

Table 2.17 presents the specific enthalpy h of seawater and various salts. In Steam Tables, specific enthalpy is referred as h_w for water, h_v for vapor, and $h_{ev} = (h_v - h_w) = h_{fg}$, for enthalpy of vaporization.

According to Gibbs energy equation, specific enthalpy equation for seawater is presented as:

$$h(s, t, p) = g + T_S \tag{2.87}$$

and the specific enthalpy of seawater h is given by:

$$h_{sw} = h(S_A, t, p) = g + (T_o + t)\eta = g - (T_o + t)\frac{\partial g}{\partial T}\bigg|_{S_A p} \tag{2.88}$$

At constant temperature and pressure enthalpy of seawater h_{sw} is expressed by the following equation (Sign and Bromley, 1973):

$$h_{ws} = (1 - x_s)\overline{h}_{pw} + x_s\overline{h}_s \tag{2.89}$$

where $\overline{h}_s$ and $\overline{h}_{pw}$ are partial specific enthalpies of salt and pure water in the seawater solution and x_s is the mass fraction of salt in the seawater sample.

Chou and Rowe (1969), give for 25°C and one atmosphere pressure the following equation for specific enthalpy of dilution, that is, for a solution diluted to infinity by adding moles $(\infty - n_1)$ of pure solvent (in this case pure water):

$$h_{w25} = (1 - x)\overline{h}_i^0 + xq_d \tag{2.90}$$

where x is mol fraction of solute, h_i^0 the reference state of solution based on the convention of zero enthalpy for saturated water at 0°C. It has the value of $104.4\ J\ g^{-1}$ at 25°C. The factor q_d is heat of dilution given as:

$$q_d = n_1/n_2\left(\overline{h}_1^0 - \overline{h}_1\right) + \left(\overline{h}_2^0 - \overline{h}_2\right) \tag{2.91}$$

where

n_1, n_2	Number of moles of pure solvent and solution respectively	—
$\overline{h}_1, \overline{h}_2$	Partial enthalpies of solvent and solute respectively	$J\ kg^{-1}$
$\overline{h}_1^0, \overline{h}_2^0$	Partial enthalpies in infinite dilution (reference state) of solvent and solute respectively	$J\ kg^{-1}$

For homogeneous systems of invariable composition they give for seawater enthalpy, the following expression:

$$h = h_o + \int_{T_o}^{T} c_p dT + \int_{p_o}^{p} \left[v - T \left(\frac{dv}{dT} \right)_p \right] dp \tag{2.92}$$

where v is specific volume ($m^{-3} kg^{-1}$) of seawater, T thermodynamic temperature (K) of the seawater, $h_o = h_{w25}$ reference enthalpy, and p pressure (Pa).

The best fitting specific enthalpy data calculated by Gibbs energy function is presented by IAPWS-2008 (Sharqawy et al., 2010). They are valid for a temperature range of 10–120°C.

$$h_{sw} = h_{pw} - S(a_1 + a_2 S + a_3 S^2 + a_4 S^3 + a_5 t + a_6 t^2 + a_7 t^3 + a_8 St + a_9 S^2 t + a_{10} St^2) \tag{2.93}$$

where

$\alpha_1 = -2.348 \times 10^4$, $\quad \alpha_2 = 3.152 \times 10^5$, $\quad \alpha_3 = 2.803 \times 10^6$, $\quad \alpha_4 = -1.446 \times 10^7$
$\alpha_5 = 7.826 \times 10^3$, $\quad\quad \alpha_6 = -4.417 \times 10^1$, $\alpha_7 = 2.139 \times 10^{-1}$, $\alpha_8 = -1.991 \times 10^4$
$\alpha_9 = 2.778 \times 10^4$, $\quad\quad \alpha_{10} = 9.728 \times 10^1$

The validity of this equation is for h_{sw} and h_{pw}, $10 \leq t \leq 120°C$ and $0 \leq S \leq 0.12 \ kg \ kg^{-1}$. Accuracy is $\pm 0.5\%$.

2.4.1.11 Enthalpy of Evaporation (h_ev or h_fg)

Vaporization enthalpy or latten heat of vaporization is the amount of heat required to transform 1 kg of water into water vapor, under normal conditions, that is, of boiling temperature of 100°C and atmospheric pressure. Evaporation enthalpy is affected by temperature and decreases with decreasing temperature. For water in critical point enthalpy becomes zero. Enthalpy of evaporation is expressed in $kJ \ kg^{-1}$ or in some older papers in $kcal \ kg^{-1}$ (1 kcal = 4.184 kJ). To evaporate m kg of water are needed Q kJ of heat the amount of which is calculated from the equation:

$$Q = h_{ev} m \tag{2.94}$$

Enthalpy of vaporization for seawater is given in Steam Tables. In case those Steam Tables are not available the following empirical equation gives good approach for calculating the enthalpy of evaporation:

$$h_{ev} = [597.49 - 0.56624t \times 10^{-1} + 0.15082t^2 \times 10^{-4} - 0.32764t^3 \times 10^{-6}] \times 4.184 \times 103 \tag{2.95}$$

where h_{ev} is evaporation enthalpy ($kJ \ kg^{-1}$) and t temperature (°C).

The enthalpy of vaporization of $NaCl-H_2O$ system is tabulated in Table 2.18 (Table 2.19).

Table 2.18 Enthalpy of Vaporization of NaCl-H$_2$O System (kJ kg^{-1})

°C	0% NaCl	1% NaCl	3% NaCl	5% NaCl
0	10.76	—	—	—
100	9.71	9.23	9.22	9.22
140	9.16	9.03	9.04	9.03
180	8.66	8.69	8.68	8.70
220	7.99	8.15	8.15	8.17
260	7.15	7.33	7.41	7.42
300	6.05	6.21	6.30	6.50
320	5.34	5.47	5.65	5.69
340	4.44	4.52	4.67	4.83
360	3.10	3.21	3.52	3.81
373.15	0.00	—	—	—
389	—	1.28	210	2.61
386	—	0.00	—	—
390	—	—	1.30	2.07
398	—	—	0.00	—

Table 2.19 Enthalpy (latten heat) of Evaporation, kJ kg^{-1}

°C	0[a]	10	30	50	70	90	110	120
				Salinity (g kg^{-1})				
0	2500.9	2475.9	2425.9	2375.9	2325.8	2275.8	2225.8	2200.8
10	2477	2452.5	2402.9	2353.4	2303.8	2254.3	2204.7	2180.0
20	2453.6	2429.0	2379.9	2330.9	2281.8	2232.7	2183.7	2159.1
30	2429.8	2405.5	2356.9	2308.3	2259.7	2211.1	2162.5	2138.2
40	2406.0	2381.2	2356.9	2285.7	2237.6	2189.4	2141.3	2117.3
50	2382.0	2358.1	2310.5	2262.9	2215.2	2167.6	2120.0	2096.1
60	2357.7	2334.1	2287.0	2239.8	2192.7	2145.5	2098.3	2074.8
70	2333.1	2309.8	2263.1	2216.4	2168.8	2123.1	20765	2053.1
80	2308.1	2285.0	2238.8	2192.7	2146.5	2100.4	2054.2	2031.1
90	2282.6	2259.7	2214.1	2168.4	2122.8	2207.1	2031.5	2008.7
100	2256.5	2233.9	2188.8	2143.7	2098.5	2053.4	2008.3	1985.7
110	2229.7	2207.4	2162.8	2118.2	2073.6	2029.0	1984.4	1962.1
120	2202.1	2180.1	2136.1	2092.0	2048.0	2003.9	1959.9	1937.9

[a]Pure water, Accuracy ±0.01%.

When seawater is treated as ideal solution a simple relation may be used for the heat of evaporation, a function of latten heat of vaporization of pure water and salinity (Sharqawy et al., 2010):

$$h_{ev-sw} = h_{ev-pw}[(1 - S/1000) \qquad (2.96)$$

Chou and Rowe (1969) report an equation for enthalpy vaporization derived from the definition of water activity a_w:

$$\left(\frac{\partial \ln a_w}{dT}\right)_{p,x} = -\frac{\overline{h}_w - \overline{h}_w^o}{RT^2} \tag{2.97}$$

The enthalpy of vaporization is then given by adding, in the above Eq. (2.97) the molar enthalpy of water vapor, h_{ev}, at temperature T and the heat of vaporization of infinite dilution:

$$(\overline{h}_{ev} - \overline{h}_w) = (h_{ev} - h_w^o) + RT^2(\partial \ln a_w/\partial T)_{p,x} \tag{2.98}$$

The last term of the above equation is the difference between heat of vaporization of water and the latten heat of pure water at the same temperature and pressure.

By using osmotic coefficient instead of activity coefficient enthalpy of evaporation may be calculated by the equation (Sharqawy et al., (2010)):

$$h_{ev-sw} = (1 - S/1000)\left(h_{ev-pw} - \frac{RT^2 \partial \phi}{1000 \partial T}m\right) \tag{2.99}$$

where

$h_{ev\text{-}sw}$	Enthalpy of evaporation of seawater	$J\,kg^{-1}$
$h_{wv\text{-}pw}$	Enthalpy of evaporation of pure water	$J\,kg^{-1}$
$\overline{h}_{ev}, h_{ev}$	Mean molar enthalpy of water vapor, molar enthalpy of water vapor, respectively	$J\,kg^{-1}$
$\overline{h}_w$	Mean enthalpy of water	$J\,kg^{-1}$
h_w^o	Reference water enthalpy = 104.4 J g^{-1} at 25°C and 101,325 Pa	$J\,kg^{-1}$
m	Total molality of solutes	$mol\,kg^{-1}$
φ	Osmotic coefficient	—

2.4.1.12 Vapor Pressure p—Boiling Point Elevation (ΔT_{el} or Δt_{el})

Vapor pressure is created over the free surface of a solution (as seawater) where the solvent is vaporized at the environmental temperature and pressure. Water is a volatile substance and dissolved material lowers its vapor pressure. This means that as salinity of seawater increases vapor pressure is decreasing. Vapor pressure and boiling temperature can be obtained by inverting boiling temperature or vapor pressure function respectively. For a normal seawater sample vapor pressure decrease is about 98% of pure water's vapor pressure.

Vapor pressure of seawater is dependent on the ionic strength I, of the solution. Since ionic strength and salinity are proportional it is preferable to use salinity in the corresponding equations.

Vapor pressure may be calculated from the solutions ionic strength I, according to the equation (Grunberg, 1970):

$$Log(p_{ev-sw}/p_{pw}) = aI - bI^2 \qquad (2.100)$$

where

P_{v-sw}	Vapor pressure of seawater solution	Bar
P_{v-pw}	Vapor pressure of pure water	Bar
I	Total ionic strength	g mol kg^{-1}
a, b	Parameters having the following values with values:	—

$a = -1.1057 \times 10^{-2}$ and $b = -9.1673 \times 10^{-4}$

Eq. (2.100) thus may be written as:

$$Log(p_{v-sw}/p_{v-pw}) = 1.057 \times 10^{-2}I - 9.1673 \times 10^{-4}I^2 \qquad (2.101)$$

Feistel (2008) presents the vapor pressure relation, based on Gibbs function, as:

$$g_v(t,p) = \mu_w(S_A, t, p) = g(S_A, t, p) - S_A\left(\frac{\partial g(S_A), t, p)}{\partial S_A}\right)_{t,p} \qquad (2.102)$$

Another empirical equation gives vapor pressure as salinity S, function based on vapor pressure measurements by Emerson and Jamieson (1967):

$$Log(p_{v-sw}/p_{v-pw}) = -2.1609 \times 10^{-4}S_p - 3.5012 \times 10^{-7}S_p^2 \qquad (2.103)$$

Eq. (2.103) is valid for temperatures $100 < t_{48} < 180°C$ and salinities $35 < S_p < 170‰$. Its accuracy is 0.07%. For approximate calculations the following empirical equation gives reliable results:

$$\ln(p_{v-pw}/p_{v-sw}) = 1 + 0.57357[S/(1000 - S)] \qquad (2.104)$$

Pure water vapor pressure is calculated by (ASHRAE, 2005):

$$\ln(p_v/p_{pw}) = a_1/T + a_2 + a_3 T + a_4 T^2 + a_5 T^3 + a_6 \ln T \qquad (2.105)$$

where the terms a have the following values:

$$a_1 = -5800, \quad a_2 = 1.391, \quad a_3 = -4.846 \times 10^{-2}, \quad a_4 = 4.176 \times 10^{-5}$$
$$a_5 = -1.445 \times 10^{-8}, \quad a_6 = 6.545 \qquad (2.106)$$

The accuracy is $\pm 0.1\%$, temperature range $273.15 < T < 473.15$, and pressure p is expressed in Pa.

Table 2.20 Vapor Pressure Lowering (Bar) of Seawater Solutions at Various Chlorinities

°C	Cl, 5‰	Cl, 10‰	Cl, 15‰	Cl, 17‰	Cl, 19‰	Cl. 21‰
5	0.03	0.06	0.09	0.10	0.11	0.13
10	0.06	0.11	0.17	0.19	0.22	0.24
15	0.10	0.20	0.31	0.35	0.40	0.45

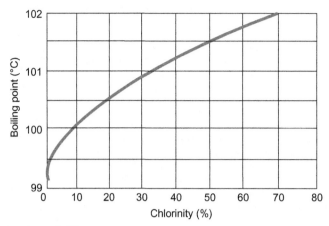

Figure 2.8 Boiling points of different seawater concentrates.

Dissolved salt in the aqueous solution will lower water vapor pressure at any given temperature. Table 2.20 presents vapor pressure lowering at various chlorinities of seawater solution and various temperatures.

Boiling point elevation is the progressive increase of boiling temperature following the increase of salt concentration, in concentrating seawater and brine. It is an important parameter for conventional seawater distillation methods because it can affect proper operation and economics of the system. During distillation of seawater, as water evaporates, brine concentration and boiling temperature are increased meanwhile pressure is decreased. Boiling point elevation is given by an empirical equation as Sharqawy et al., (2010):

$$T_{el} = aS^2 + bS°C \tag{2.107}$$

where

S	Seawater or brine salinity	kg kg^{-1}
T_{el}	Temperature elevation over 100°C of pure water	K
a,b	Temperature factors depending on the solution temperature	—

Factors a and b, can be calculated from the following empirical equations:

$$a = 17.95 + 0.2823t - 4.584 \times 10^{-4}t^2 \tag{2.108}$$

$$b = 6.56 + 5.267 \times 10^{-2} + 1.536 \times 10^{-4}t^2 \tag{2.109}$$

where t is temperature in degrees Celsius. Eq. (2.107) is valid for temperature range $0 < t < 200°C$ and a salinity range of $0 < S < 120$ g kg^{-1}. The higher the boiling point elevation of a solution the more the properties deviate from those of seawater and the laws of ideal solutions and thus from pure water properties. Vapor that forms at elevated temperature seawater has the temperature of the solution, that is, is superheated

Table 2.21 Vapor (Saturation) Pressure (kPa) of Seawater, its Concentrates, and Boiling Point Elevation (°C)

°C	0[a]	10	30	50	70	90	110	120
				Salinity (g kg^{-1})				
0	0.611	0.698	0.601	0.578	0.571	0.567	0.571	0.567
ΔT_{el}	0.000	0.067	0.213	0.373	0.547	0.736	0.939	1.046
10	1.228	1.221	1.207	1.192	1.177	1.162	1.147	1.139
ΔT_{el}	0.000	0.073	0.232	0.407	0.599	0.807	1.032	1.151
20	2.339	2.325	2.298	2.270	2.242	2.213	2.184	2.169
ΔT_{el}	0.000	0.163	0.251	0.442	0.652	0.880	0.128	1.258
30	4.247	4.222	4.172	4.070	4.070	4.018	3.965	3.938
ΔTel	0.000	0.085	0.272	0.479	0.707	0.929	1.225	1.358
40	7.384	7.341	7255	7.167	7.078	6.987	6.895	6.848
ΔT_{el}	0.000	0.092	0.293	0.517	0.764	1.033	1.325	1.480
50	12.351	12.279	12.135	11.988	1.839	11.687	11.532	11.44
ΔT_{el}	0.000	0.099	0.315	0.556	0.8.22	1.112	1.428	1595
60	19.946	19.829	19.596	19.359	19.118	18.749	18.624	1.47
ΔT_{el}	0.000	0.106	0.338	0.597	0.882	1.194	1.532	1.711
70	31.201	31.018	30.654	30.284	29.907	29.523	29.133	28.95
ΔT_{el}	0.000	0.114	0.362	0.639	0.944	1.277	1.639	1.831
80	47.415	47.139	46.585	46.022	45.449	44.866	44.273	43.972
ΔT_{el}	0.000	0.121	0.387	.682	1.007	1.363	1.748	1.952
90	79.182	69.776	68.957	68.124	67.701	66.846	65.975	65.09
ΔT_{el}	0.000	0.129	0.412	0.726	1.72	1.450	1.860	2.076
100	101.48	100.85	99.651	98.447	97.221	95.974	94.705	94.02
ΔT_{el}	0.000	0.138	0.438	0.772	1.139	1.540	1.973	2.203
110	143.36	142.58	140.84	139.12	137.40	135.67	133.82	132.8
ΔT_{el}	0.000	0.146	0.465	0.819	1.208	1.631	2.089	1.331
120	18.65	197.541	195.048	192.863	190.463	188.019	195.533	184.272
ΔT_{el}	0.000	0.155	0.493	0.868	1.278	1.725	2.207	2.462

[a]Pure water, Accuracy ± 0.07%, Saline water, Accuracy ± 0.018%.

in the same degree. In Fig. 2.8 experimental data of boiling point are plotted for various seawater salinities and its concentrates, as function of temperature and chlorinity. Table 2.21 presents vapor pressure at saturation calculated from Eq. (2.102). The boiling point elevation of seawater and its concentrates as function of salinity for temperatures from 0°C to 120°C (Clark et al., 1960; Jamieson, 1986; Sharqawy et al., 2010) are also presented in the same Table 2.21 depicted by brown color.

Brandani et al. (1985) give the following equation for predicting boiling point elevation:

$$\Delta T_{el} = \left(-h_{ev} + \sqrt{h_{ev}^2 - 4RT_o^2 \Delta c_p (\ln a_w)} \right) / (2\Delta c_p) \qquad (2.110)$$

where

α_w	Activity of water at temperature T_o before boiling	—
h_{ev}^o	Enthalpy of evaporation of pure water at temperature T_o before boiling	$J\,kg^{-1}$
Δc_p	Difference between specific heat of liquid water and water vapor	$J\,kg^{-1}\,K^{-1}$
R	The gas constant = 8.31441	$J\,mol^{-1}\,K^{-1}$
T_o	Temperature before boiling point elevation	K

2.4.1.13 Heat Capacity(C)—Specific Heat (c$_p$, c$_v$)

Heat capacity is the rate by which a system absorbs the heat at a given temperature:

$$C = dQ/DT \quad J\,K^{-1} \qquad (2.111)$$

Jamieson et al. (1969), Jamieson (1986), and Grunberg (1970) give the following expression for seawater heat capacity:

$$c_p = (a_1 + a_2 S_p + a_3 S_P^2) + (b_1 + b_2 S_p + b_3 S_P^2)T$$
$$+ (c_1 + c_2 S_p + c_3 S_P^2)T^2 + (d_1 + d_2 S_p + d_3 S_P^2)T^3 \qquad (2.112)$$

were S_p is practical salinity in $g\,kg^{-1}$ and T temperature in K. Constants a, b, c, and d have the following numerical values:

$\alpha_1 = 5.328$	$b_1 = -6.913 \times 10^{-3}$
$\alpha_2 = -9.76 \times 10^{-2}$	$b_2 = 7.351 \times 10^{-4}$
$\alpha_3 = 4.04 \times 10^{-4}$	$b_3 = 3.15 \times 10^{-6}$
$c_1 = 9.6 \times 10^{-6}$	$d_1 = 2.5 \times 10^{-9}$
$c_2 = -1.927 \times 10^{-6}$	$d_2 = 1.666 \times 10^{-9}$
$c_3 = 8.23 \times 10^{-9}$	$d_3 = -7.125 \times 10^{-12}$

Specific heat capacity (c) or simply specific heat is the heat capacity per unit mass of water. It is a function of temperature and change in pressure or volume, according to enthalpy. It is distinguished in c_p and c_v ($J\,kg^{-1}\,K^{-1}$) for steady pressure and steady volume or at saturation c_{sat} respectively. At low pressures saturation specific heat c_{sat} equals specific heat at steady pressure ($c_{sat} \approx c_p$). For desalination purposes in design work c_p of natural water is used as its specific heat is almost unchanged up to 100°C. During ice melting c_p changes from 2.072 to 4.228 $Jg^{-1}\,K^{-1}$.

Specific isobaric heat capacity c_p, of seawater is computed from Gibbs function as (Feistel, 2008):

$$c_p(S_A, t, p) = -(T_o + t)\left(\frac{\partial^2 g}{\partial t^2}\right)_{S_A, p} \qquad (2.113)$$

where T_o is Celsius zero point in Kelvin.

Specific heat of electrolyte solutions is one of the most important thermodynamic properties for heat balance calculations. For low and intermediate concentrations are

Table 2.22 Specific Heat Capacity c_p (J kg^{-1} K^{-1}), of Seawater and Its Diluents and Concentrates

				Salinity S (g kg^{-1})				
°C	0[a]	10	30	50	70	90	110	120
0	4206.8	4142.1	4929.1	3907.8	3805.2	3712.4	3629.3	3591.5
10	4196.7	4136.7	4022.8	3916.9	3819.2	3729.5	3647.9	3610.1
20	4189.1	4132.8	4025.3	3924.5	3839.4	3743.0	3662.3	3624.5
30	4183.9	4130.5	4027.8	3930.8	3839.4	3753.6	3673.3	3635.3
40	4181.0	4129.7	4030.7	3936.4	3846.7	3761.8	3681.6	3643.2
50	4180.6	4130.8	4034.1	3941.5	3852.9	3768.3	3687.8	3649.0
60	4182.7	4133.7	4038.3	3946.5	3858.3	3773.7	3692.6	3653.4
70	4197.1	4138.5	4943.6	3951.9	3863.6	3778.5	3696.7	3657.0
80	4194.0	4145.3	4050.1	3958.1	3869.2	3783.5	3700.8	3660.7
90	4203.4	4154.2	4058.3	3965.4	3875.7	3789.1	3705.6	3665.0
100	4215.2	4165.4	4068.2	3974.3	3883.6	3796.0	3711.7	3670.8
110	4229.4	4178.8	4080.2	3985.1	3893.3	3804.9	3719.9	3678.6
120	4246.1	4194.7	4094.6	3998.2	3905.4	3816.2	3730.7	3689.4

[a]Pure water, Accuracy ±0.28%.

less than that of pure water. In desalination specific heat c_p is usually used (J kg^{-1} K^{-1}) for constant pressure p, or $c_{p\text{-}mol}$ (J mol^{-1} K^{-1}). Under operation conditions of desalination plants, at almost saturation conditions, practically the values of c_p and c_v are identical. Eq. (2.84) can be written as function of specific heat:

$$\int_{T_1}^{T_2} dH = \int_{T_1}^{T_2} c_p dT = \Delta H = \Delta Q \tag{2.114}$$

Chou and Rowe (1969) give the pressure dependence of c_p, of Eq. (2.113) as:

$$c_p = c_p^o - T \int_{p_o}^{p} \left(\frac{\partial^2 v}{\partial T^2} \right)_p dp \tag{2.115}$$

where c_p^o is reference specific heat, the rest given previously. Specific heat is referred as the rate of heat capacity to the mass unit of water at 15°C. According to this definition the following empirical equation is valid (Khan, 1986):

$$c_p = a + bt + ct^2 + dt^3 \tag{2.116a}$$

where t is solution temperature (°C) and a, b, c, and d are numerical parameters depending on the salinity of the solution given below:

$$a = 4206.8 - 6.6197S + 0.012288S^2 \tag{2.116b}$$

$$b = -1.1262 + 5.4178 \times 10^{-2}S - 2.2719 \times 10^{-4}S^2 \tag{2.116c}$$

$$c = 1.2026 \times 10^{-2} - 5.3566 \times 10^{-4}S + 1.8906 \times 10^{-6}S^2 \tag{2.116d}$$

$$d = 6.8774 \times 10^{-7} + 1.517 \times 10^{-6}S - 4.4268 \times 10^{-9}S^2 \tag{2.116e}$$

Specific heat can be compiled from tables. The above empirical equations are for routine calculations when values in Tables are not available and/or are not referred for the required temperatures. Table 2.22 contains smoothed values of specific heat capacities of seawater for salinities $0 < S < 120 \text{ g kg}^{-1}$ and temperatures $0 < t < 120°C$ (Sharqawy et al. 2010) calculated using equation by Grunberg (1965); Jamieson et al. (1969) and Jamieson (1986) after adopting IAPWS parameters. The specific heat of pure water is calculated as (Sun et al., 2008):

$$c_{p-pw} = a_1 + a_2 t + a_3 t^4 + a_5 p + a_6 pt + a_7 pt^3 + a_8 p^2 + a_9 p^2 t + a_{10} p^2 t^2 \qquad (2.117)$$

Where the numerical values of the terms a_1 to a_{10} are given in tables.

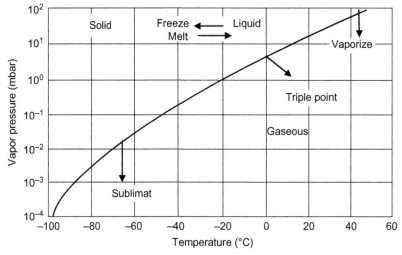

Figure 2.9 Pure water phase diagram and the area where the change from solid to vapor is possible (Oetjen, 1999).

Table 2.23 Freezing (and Melting) Point of Water and Seawater (°C) at Various Salinities (S)

S (‰)	T (°C)	S (‰)	T (°C)	S (‰)	T (°C)	S (‰)	T (°C)
1	−0.055	11	−0.587	21	−1.129	31	−1.683
2	−0.108	12	−0.640	22	−1.184	32	−1.740
3	−0.161	13	−0.694	23	−1.239	33	−1.797
4	−0.214	14	−0.748	24	−1.294	34	−1.853
5	−0.267	15	−0.802	25	−1.349	35	−1.910
6	−0.320	16	−0.856	26	−1.405	36	−1.967
7	−0.373	17	−0.910	27	−1.460	37	2.024
8	−0.427	18	−0.965	28	−1.516	38	−2.081
9	−0.480	19	−1.019	29	−1.572	39	−2.138
10	−0.534	20	−1.074	30	−1.627	40	−2.196

2.4.1.14 Freezing Point of Water and Seawater

Pure water freezes at $0°C$. The phase diagram of pure water is presented in Fig. 2.9. As dissolved salts are increasing, freezing point decreases and seawater, for example 35‰, at normal conditions, solidifies at $-1.91°C$. The same time the colligative properties such as vapor pressure, boiling point elevation, and osmotic pressure also change by the freezing point depression at maximum density. Table 2.23 gives the below zero freezing indicative temperatures as a function of salinity. At freezing point solid and liquid exist in equilibrium state and have the same vapor pressure. Fig. 2.10 presents the indicatively freezing point depression as function of vapor pressure of liquid and solid phase. The curves refer to vapor pressure of pure water (A), of an aqueous solution (B), and of solid state (C). As the concentration increases, for example curve (B), freezing point decreases and freezes at temperature T.

This phenomenon explains why icebergs consist of pure ice. Generally in the seas the dense brine solution sinks away, ice is growing as pure water ice and in the solution the number of ice molecules and of the solute are n_{ice} and n_{sol} respectively. The mole fraction of ice is then given as $x_{ice} - n_{ice}(n_{sol} + n_{ice})$ or as $n_{sol} \gg n_{ice}$, $x_{ice} = n_{ice}(n_{sol} + n_{ice}) \approx n_{ice}/n_{sol}$. The relation between solute concentration and freezing point depression is:

$$T_o - T_{sol} = (RT_{fr}^2/h_f)(n_{ice}/n_{sol}) = C_f \cdot m \tag{2.118}$$

where h_f is the enthalpy difference between sublimation and vapor vaporization $(h_f = h_s - h_{ev})$, C_f is a solvent characteristic constant, and m is the molality.

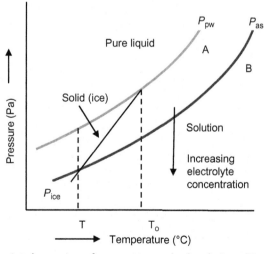

Figure 2.10 Freezing point depression of pure water and salt solutions (Horne, 1969).

Constant C_f for water has the value of 1.855 K. Eq. (2.118) is valid for dilute solutions.

where

C_f	Characteristic constant = 1.855 for water	$K\ kg\ mol^{-1}$
h_f	Enthalpy of ice fusion	$J\ mol^{-1}\ K^{-1}$
M	Molality of the solution	$mol\ kg^{-1}$
n_{ice}, n_{sol}	Number of ice and solution moles	—
T_o	Pure water freezing temperature	K
T_{sol}	Solution temperature	K
T_{fr}	Freezing temperature of the solution	K
x_{ice}	Mole fraction of ice	—

Pure water phase diagram and the area where the change from solid to vapor is possible are shown in Fig. 2.9.

Seawater freezing presents some peculiarities. By freezing seawater very slowly crystallization of pure water ice starts. By lowering temperature, electrolytes start to crystallize selectively from the concentrated solution. At $-8.2°C$ crystallizes Glauber salt $(Na_2SO_4 \cdot 10H_2O)$ and by $-23°C$ crystallizes sodium chloride (NaCl). The remaining brine freezes at $-55°C$. If seawater freezes by a very slow procedure, up to $-30°C$, ice and the remaining brine have the following composition:

Solid ice crystals

931.9 g   ice crystals   3.95 g   crystalline Na_2SO_4
20.23 g   crystalline NaCl   traces of $CaCO_3$

Table 2.24 Melting Enthalpy/Specific Enthalpy of Sea Ice, in kg J kg^{-1}

T (°C)	S (0‰)	S (2‰)	S (4‰)	S (6‰)	S (8‰)	S (10‰)	S (15‰)
−1	335 J	301 J	264 J	209 J	192 J	155 J	67 J
−2	339 J	322 J	301 J	285 J	251 J	247 J	201 J
			Specific	Enthalpy			
−2	2.0	10.75	19.37	28.03	36.65	45.31	66.98
−4	2,0	4.184	6.27	8.32	10.42	12.51	17.74
−6	2,0	3.050	4.02	5.02	5.98	6.94	9.37
−8	2.0	2.636	3.18	3.68	4.22	4.77	6.11
−10	2.0	2.38	2.68	2.97	3.26	3.55	4.26
−12	1.966	2.30	2.468	2.68	2.84	3.05	3.55
−14	1.966	2.23	2.385	2.55	2.677	2.845	3.22
−16	1.966	2.22	2.385	2.51	2.677	2.80	3.18
−18	1.966	2.22	2.34	2.42	2.55	2.677	2.97
−20	1.966	2.175	2.30	2.38	2.51	2.59	2.84
−22	1.934	2.175	2.26	2.34	2.426	2.51	2.72

Remaining brine

23.31 g	water	0.39 g	Ca^{2+}
1.42 g	Na^+	7.03 g	Cl^-
1.31 g	Mg^{2+}	0.08 g	Br^-
0.38 g	K^+	0.03 g	SO_4^{2-}

Nevertheless sea ice is not totally free of salts, the salinity of which depends on the salinity of seawater and/or from the rate of freezing. On the other hand by freezing seawater rapidly part of the electrolyte solution is trapped into small ice holes giving some salinity to solid ice formed. At the same time ions are trapped, selectively, on the surface of ice crystals. This results in the development of electrical potentials.

By the start of ice formation, colligate properties, as referred previously, boiling point, vapor pressure, and osmotic pressure also chance, but at the freezing point, where solid and liquid are in equilibrium have the same vapor pressure. Table 2.24 gives the melting enthalpy of ice at temperatures -1 and $-22°C$ as well as the specific enthalpy at various temperatures.

Phase diagram of seawater may be studied by the phase diagram of various concentrations of sodium chloride solution for temperatures from 10 to $-30°C$, as it is presented in Fig. 2.11. Solutions eutectic point is about $-21.2°C$.

Specific enthalpy of pure ice is temperature dependent but varies within narrow limits. Sea ice varies within salt or brine content and temperature, thus having more variable properties. For melting of sea ice the necessary amount of heat depends on the ice salinity.

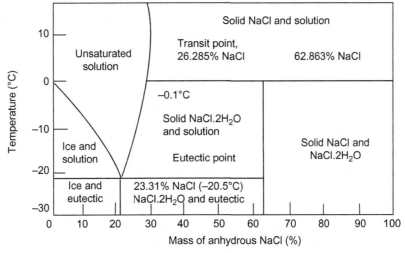

Figure 2.11 Phase diagram of the system sodium chloride-water.

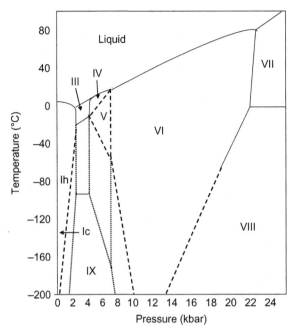

Figure 2.12 Phase diagram of water. Ih represents hexagonal and Ic cubic ice formation, including the various ice configurations (Oetjen, 1999).

The freezing temperature $t_{fr}(S_A,p)$ of seawater of absolute salinity S_A at pressure p obeys the thermodynamic equilibrium between seawater and hexagonal ice phase Ih, that the chemical potential of water must have in both phases the same potential μ (Feistel, 2008):

$$\mu_w(S_A, t_{fr}, p) = \mu_{Ih}(t_{fr}, p) \tag{2.119}$$

where chemical potential of ice equals specific Gibbs energy:

$$\mu_{Ih}(t_{fr}, p) \equiv g_{Ih}(t, p) \tag{2.120}$$

The Gibbs function of ice for the freezing point is (IARWS-2006):

$$g_{Ih}^g(t_{fr}, p) - g_w(t_{fr}, p) = g^S(S_A, t_{fr}, p) - S_A \left(\frac{\partial g_s}{\partial S_A} \right)_{t_{fr},p} \tag{2.121}$$

Fig. 2.12 presents the ice crystal phases, where hexagonal ice phase is noted by Ih and cubic phase by Ic.

Enthalpy of melting may be calculated by the following empirical equation:

$$Q = 335 \times (1 - S/S_t) + 2.09(T_o - T_{sw-fr}) \text{ kJkg}^{-1} \tag{2.122}$$

where:

S	Salinity of the ice	‰
S_t	Salinity of brine at temperature T	‰
T_o	Temperature of pure ice $\sim 0°C$	K
$T_{sw\text{-}fr}$	Freezing point of seawater	K

Additionally, 2.09 kJ kg^{-1} is approximately the amount of enthalpy to increase the temperature of the pure ice and brine from T to T_{sf}, and 335 is enthalpy of fusion kJ kg^{-1}.

The fusion heat of ice at 0°C is 333.69 kJ kg^{-1} and heat of sublimation 2838 kJ kg^{-1}. Molal heat capacity of ice down to $-140°C$ is given by the equation:

$$C_{ie-m} = 2.369 + 0.130 \ T \text{ kJ K}^{-1} \qquad (2.123)$$

And thermal conductivity, at temperature T can be calculated as:

$$\lambda_{ice} = (488.19/T) + 0.4685 \qquad (2.124)$$

For pure water the melting heat Q_{ml} to be withdrawn for freezing can be calculated by the following equation:

$$Q_{ml} = c_{p-pw}(T_1 - T_o) + Q_{ml-ice} + c_{p-ice}(T_o - T_2) \qquad (2.125)$$

where

c_{p-pw}	Specific heat capacity of pure water	J kg^{-1} K^{-1}
c_{p-ice}	Specific heat capacity of ice	J kg^{-1} K^{-1}
Q_{ml-ice}	Melting heat of ice	J kg^{-1}
T_o	Freezing temperature of ice	K
T_1	Initial temperature of ice	K
T_2	Final temperature of ice	K

To solve Eq. (2.125) the initial temperature T_1 and final temperature T_2 must be known. The freezing rate v_{fr} (°C/min) is of importance for the size of ice crystals. In order to produce large ice crystals:

- The rate of ice nucleation should be small, therefore the subcooling should be small and freezing should take place in quasi-equilibrium state
- Since v_{fr} is inversely proportional to the size of crystals, the time of crystallization has to be increased
- The crystals grow according to the function $e^{-1/T}$, thus temperature should be as high as possible.

In case where freezing takes place under high pressure only few crystals are produced. If freezing rate is very high a large degree of supercooling is then produced, without ice crystals.

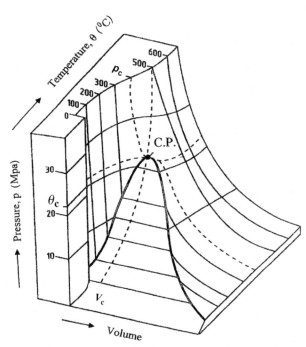

Figure 2.13 The *p-V-T* diagram for water (Ullman, 1998).

2.4.1.15 *The p-V-T Diagram of Water*

Water is the only inorganic substance, found in nature that occurred in three phases, at ambient conditions.

Fig. 2.13 presents a three-dimensional *p-V-T* diagram, for ordinary water (Ullman, 1998). The curve in bold presents the boundary where both liquid–vapor phases exists. The point C.P is the triple point at 273.16 K (or 0.01°C) where the three phases, liquid, vapor, and ice, exist simultaneously.

2.4.1.16 *Electrical Conductance (E$_c$)/Electrical Resistance (R$_c$)*

Electrical conductance (or electrical conductivity) is the ability of a solution to conduct an electrical current. By inserting two electrodes into distilled water a direct current does not flow through, but if the solution contains electrolytes the electrical current flows through the salt solution. This ability of an electrolytic solution is called "*solution conductance.*" The more concentrated the solution in ions the higher the conductance. Conductance, for a given electrolyte solution increases by increasing temperature. The conductance, in micro-Siemens, is given by the following equation:

$$E_c = (C_i \cdot f_{ai}) \, \mu S \, cm^{-1} \tag{2.126}$$

Table 2.25 Coefficient of Specific Electrical Conductivity (f_{ai}) of Ions in Seawater, $\mu S\ cm^{-1}$

Ion		meq L^{-1}	mg L^{-1}
Calcium	Ca^{2+}	52.0	2.60
Magnesium	Mg^{2+}	46.6	3.82
Potassium	K^+	72.0	1.84
Sodium	Na^+	48.9	2.13
Hydrogen carbonates	HCO_3^-	43.6	0.715
Carbonates	CO_3^-	84.6	2.82
Chlorides	Cl^-	75.9	2.14
Sulfates	SO_4^-	73.9	1.54
Nitrates	NO_3^-	71.0	1.15

where

E_c Specific electrical conductance $\mu S\ cm^{-1}$
C_i Concentration of ions of component i   ppm
f_{ai} Specific coefficient of ion conductance $\mu S\ cm^{-1}$

Table 2.25 gives the coefficient of specific ion conductance, f_{ai}, for the main elements in seawater. Table B.18, in Appendix B, gives the specific electrical conductance of seawater and its concentrates for chlorinities $1-22\ g\ kg^{-1}$ and temperatures $0-25°C$.

Electric conductivity increases by increasing salt content. It fluctuates from 21 for natural waters up to $56\ mS\ cm^{-1}$ for seawater. At $27°C$, has a mean value of $42\ mS\ cm^{-1}$ in Pacific Ocean regions and about $72\ mS\ cm^{-1}$ in the region of Persian Gulf where salinity is of about 45‰. Electrical conductivity of brackish waters, depends on salt content and fluctuates from 32 to $36\ mS\ cm^{-1}$.

Electrical resistance, on the contrary, is decreasing by increasing concentration of salts in water. Conductance is the reciprocal of resistance and can be expressed in micromhos (the reciprocal of micro-ohms). 1.0 micromho equals 10^{-6} mho and presents the flow of electric current of 1.0 Ampere (A) when the potential difference is 1.0 Volt. It is the measure of TDS in waters.

2.4.1.17 Dissolved Gases and the Carbonate System

A. *Dissolved gases.* Natural waters contain small quantities of dissolved gases. They are originated from the earth's atmosphere. Almost all gases exist in seawater most of them in very small quantities. For desalination purposes of importance are dissolved carbon dioxide CO_2 and oxygen O_2. Carbon dioxide is corrosive even by the absence of oxygen. Oxygen is corrosive by itself but in the presence of carbon dioxide its corrosiveness increases. Dry vapor containing free oxygen is also corrosive, but vapor condensate has no intensive corrosive action. Oxygen solubility in water is double than nitrogen's. Nitrogen is the most abundant gas in seawater but chemically a totally inert. Carbon

Table 2.26 Solubility of CO_2, O_2, N_2 in Natural Waters per m^{-3} of Pure Water at Pressure 1.0 bar (10^5 Pa)

°C	CO_2	O_2	N_2	°C	CO_2	O_2	N_2
0	1690	10.19	18.45	30	637	5.24	10/15
5	1400	8.91	16.30	40	491	4.48	8.68
10	1170	7.87	14.50	50	284	3.85	7.55
15	996	7.04	13.07	60	293	3.28	6.50
20	855	6.35	11.91	80	–	1.97	4.03
25	733	5.75	10.96	100	–	0.0	0.0

Table 2.27 Henry's Constant for Carbon Dioxide and Oxygen

Temperature (°C)	K_H for CO_2 (bar^{-1})	K_H for O_2 (bar^{-1})
0	0.001397	0.0000391
5	0.001137	0.0000330
10	0.000967	0.0000303
15	0.000823	0.0000271
20	0.000701	0.0000244
25	0.000611	0.0000222
40	0.000413	0.0000188
60	0.000286	0.0000159

dioxide is highly soluble in seawater because it reacts chemically with water to produce carbonic acid which dissociates to form hydrogen and bicarbonate ions.

$$CO_2 + H_2O \leftrightarrows H_2CO_3 \leftrightarrows H^+ + HCO_3 \qquad (2.127)$$

Or some of the bicarbonate dissociates to form hydrogen ions and carbonate:

$$H_2CO_3 \leftrightarrows 2H^+ + CO_3^{2-} \qquad (2.128)$$

The solubility of various gases in natural waters at 1 bar is shown in Table 2.26.

Oxygen is extremely corrosive for all metals except for noble ones. Waters having low pH and high temperature increase oxygen corrosive activity but in the same time higher temperatures decrease its solubility. The content of both oxygen and carbon dioxide in natural seawater is not stable as their solubility depends from various parameters.

Concentration of gases generally increases with pressure and decreases with temperature. Dissolved gases follow Henry's low (see sub-section 2.2.1) which refers to the equilibrium of a dissolved gas in the solution having the free gas onto the surface of the solution, that is, the vapor tension of the gas. It is given by the expression:

$$x_i = K_H p \qquad (2.129)$$

where x_i is the ratio of dissolved gas i, p the partial gas pressure (Pa), and K_H Henry's constant (1/Pa), that is, the equivalent constant at saturation point. Henry's constant is a function of gas type, temperature, and concentration of the various species in the solution. In Table 2.27 the Henry's constant for carbon dioxide and oxygen for temperatures from 0°C to 60°C is presented. Tables B.19—B.22 in Appendix B give solubility of oxygen, carbon dioxide, and nitrogen in seawater respectively (Tsobanoglou and Schroeder, 1985).

B. *The carbonate equilibrium.* The main inorganic components of natural water and seawater include the following species:

Strong electrolyte cations Ca^{2+}, Mg^{2+}, Na^+, K^+
Strong electrolyte anions Cl^-, SO_4^{2-}, NO_3^-
Carbonic acid species CO_2, H_2CO_3, HCO_3^-, CO_3^{2-}
Water dissociation products H_2, H^+, OH^-

The carbonate system in natural waters, that is, CO_2, H_2CO_3, HCO_3^-, and CO_3^{2-} is the most complex equilibrium system. Chemical species composing the carbonic system include carbonic acid (H_2CO_3), aqueous carbonic dioxide $(CO_2)_{liq}$, gaseous carbon dioxide $(CO_2)_g$, carbonate (CO_3^{2-}), and bicarbonates (CO_3^{2-}). They control the pH of the aqueous solution.

This carbonate system is of importance for all natural waters but especially applies for desalination purposes, at the equilibrium state where:

$$x_{CO_2} = K_H \cdot p_{CO_2} \qquad (2.130)$$

where

x_{CO_2} Mole fraction of carbon dioxide at equilibrium in liquid phase $(-)$
K_H Henry's constant, given in Table 2.27 (bar^{-1})
p_{CO_2} Partial carbon dioxide pressure in surrounding atmosphere (bar)

Concentration of CO_2, HCO_3, and CO_3^{2-} depends on the carbonic acid dissociation. The following hydration and dissociation phenomena take place when CO_2 dissolves in seawater:

$$CO_2 \leftrightarrows (CO_2)_{aq} \qquad (2.131)$$

$$(CO_2)_{aq} + H_2O \leftrightarrows H_2CO_3 \qquad (2.132)$$

It is difficult to distinguish $(CO_2)_{aq}$ in a solution and H_2CO_3 in waters is present at very low concentrations, thus it is better to use the effective carbonic acid, that is:

$$(H_2CO_3)_{ef} = (CO_2)_{aq} + H_2CO_3 \qquad (2.133)$$

Table 2.28 Thermodynamic Dissociation Constants of Carbonic Acid as a Function of Temperature

Temperature (°C)	Constant (K_1 mol L^{-1})	Constant (K_2 mol L^{-1})	Constant K_{sp} (mol^2 L^{-2})
0	2.64×10^{-7}	2.36×10^{-11}	
5.0	3.04×10^{-7}	2.77×10^{-11}	
10.0	3.44×10^{-7}	3.27×10^{-11}	8.13×10^{-9}
15.0	3.81×10^{-7}	3.71×10^{-11}	7.08×10^{-9}
20.0	4.16×10^{-7}	4.20×10^{-11}	6.03×10^{-9}
25.0	4.44×10^{-7}	4.69×10^{-11}	$4,57 \times 10^{-9}$
30.0	4.71×10^{-7}	5.13×10^{-11}	
40.0	5.07×10^{-7}	6.03×10^{-11}	3.09×10^{-9}
60.0	5.07×10^{-7}	7.24×10^{-11}	1.82×10^{-9}

Carbonic acid dissociates in two steps forming in first step bicarbonate and then carbonate according to the equations:

$$H_2CO_3 \leftrightarrows H^+ + HCO_3^- \tag{2.134}$$

$$HCO_3^- \leftrightarrows H^+ + CO_3^{2-} \tag{2.135}$$

The thermodynamic equilibrium constants of the above dissociation phenomena are characterized by the ionic product of water:

$$K_1 = \frac{[H^+] \times [HCO_3^-]}{[(H_2CO_3)_{ef}]} \quad \text{and} \tag{2.136}$$

$$K_2 = \frac{[H^+] \times [CO_3^{-2}]}{[HCO_3^-]} \tag{2.137}$$

Dissociation constants K_1 and K_2 are functions of temperature and ionic strength I, $K_1(T,I)$, $K_2(T,I)$ and are given in Table 2.28. Both constants increase by increasing temperature.

The sum of carbonic acid species is presented by total inorganic carbon (TIC):

$$TIC = (CO_2) + (HCO_3^-) + (CO_3^{2-}) \tag{2.138}$$

The relative amount m_r (which lies in the range $0 < m_r < 2$) of carbonic acid species is expressed by the equation:

$$m_r = (HCO_3^-) \times 2(CO_3^{2-})/TIC \tag{2.139}$$

Eqs. (1.130) and (1.132) may be combined to:

$$(CO_2)_{aq} + H_2O \leftrightarrows H_2CO_3^{2-}, \quad K_m = [H_2CO_3]/[CO_2] \tag{2.140}$$

K_m at 25°C has for water a value of 1.56×10^{-3}.

If a carbonate system is in equilibrium with a solid phase, for example calcium carbonate ($CaCO_3$), then the equilibrium constant, involves simultaneously precipitation which is called "solubility product" for example:

$$CaCO_3 = Ca^{2+} + CO_3^{2-}, \quad K_{sp} = [Ca^{2+}][CO_3^{2-}] \tag{2.141}$$

2.4.1.18 Osmosis—Osmotic Coefficient

The term osmosis means the free transport of a solvent across a semipermeable membrane, from a solution of lower salt concentration to a higher concentrated solution until a kinetic equilibrium is achieved.

Concern a solution having different salt concentrations. With the assumption that there exists no stirring, a mutual movement of ions will take place by diffusion until the two parts of the solution come in dynamic equilibrium. Consider now two solutions with different concentrations separated by a physical barrier, for example a membrane which is permeable to the solvent but not to the solutes. Diffusion stops and a transport takes place of solvent, here water, through the membrane's pores, from the dilute to the concentrated solution. This movement is called "osmosis." It comes from the Greek word ὠσμος (osmos) which means push. Such a membrane is called "semipermeable."

Fig. 2.14 presents a schematic of osmosis. In Fig. 2.14B, on the left side compartment pure water is separated from an aqueous solution (right side compartment) by a semipermeable membrane. Solvent molecules will pass the membrane entering the solution compartment. Pure solvent solution will decrease meanwhile the amount of the solution will increase. At a certain point the hydrostatic pressure will be in

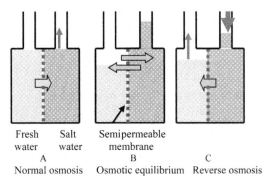

Fresh water	Salt water	Semipermeable membrane
A	B	C
Normal osmosis	Osmotic equilibrium	Reverse osmosis

Figure 2.14 The concept of osmosis and reverse osmosis.

equilibrium, that is, solvent molecules will cross both directions of the membrane, at equal rates, and pressure will be steady. This pressure is called "osmotic pressure (Π)" of the solution. The same results exactly will be achieved if a counter pressure is applied to the solution side (Fig. 2.14C), the pressure by which osmosis stops is the osmotic pressure of the solution. By applying to the solution a counterpressure higher than the osmotic, the procedure is reversed and solvent is transferred from the solution to the water or from the higher concentration, through the membrane, to the lower concentration solution. Thus in side of higher pressure the solution becomes more concentrated. This is the principle of reverse osmosis (RO), method for desalination of brackish and sea waters. In order to achieve a fixed solvent flow pressure must be applied stepwise and this procedure can be stopped at any desired solution concentration.

Osmosis derives really from the decrease of free energy (potential energy), of water in the solution, relatively to pure water. As a consequence there is a decrease in vapor pressure, a depression of freezing point, and an increase of boiling point.

The term "semipermeable membrane"was first established by van't Hoff in 1886 who also gave the corresponding equation for ideal liquids. Membrane selectivity toward various species is not absolute. A semipermeable membrane selective to for example only water, permits trace amounts of other components to diffuse through the membrane pores, according the type of membrane and size of its pores. Van't Hoff's equation of osmotic pressure for ideal liquids is given as:

$$\Pi = n R T m_w M_w / \overline{v}_w \qquad (2.142a)$$

For nonideal liquids a correction factor ϕ, is added, the so-called osmotic coefficient which gives a measure of deviation from ideal conditions:

$$\Pi = n \phi R T m_w M_w / \overline{v}_w \qquad (2.142b)$$

where

Π	Osmotic pressure	Pa
N	Number of ions of the electrolyte solution	—
ϕ	Osmotic coefficient	—
R	The universal constant of gases	$m^3 \, Pa \, mol^{-1} \, K^{-1}$
T	Thermodynamic temperature	K
M_w	Molecular mass of water (or solvent)	$kg \, mol^{-1}$
m_w	Molality	$mol \, kg^{-1}$
v_w	Mean molar volume of water	$m^3 \, mol^{-1}$

During dynamic equilibrium state, chemical potential is equal in both sides. From Eq. (2.142a) we derive the following relation:

Table 2.29 Osmotic Pressure Π, of Seawater and Its Concentrates, in Atmospheric Pressure ($= 1.0315 \times 10^5$ Pa)

% Salts	25°C	40°C	60°C	80°C	100°C
1.00	7.11	7.42	7.80	8.13	8.42
2.00	14.29	14.93	15.70	16.37	16.94
3.45	25.15	26.34	27.74	28.94	29.92
5.00	37.49	39.35	41.51	43.32	44.79
7.50	59.30	62.41	65.98	68.69	71.17
10.0	83.97	88.53	94.69	97.82	100.97
15.0	144.93	152.81	161.53	168.27	173.16
20.0	228.59	239.86	252.06	261.17	267.47
25.0	348	361	376	386	393

Table 2.30 Osmotic Coefficient ϕ

°C	Salinity (g kg^{-1})							
	10	30	40	50	70	90	110	120
0	0.894	0.897	0.901	0.906	0.921	0.941	0.967	0.982
10	0.898	0.901	0.905	0.911	0.926	0.928	0.975	0.991
20	0.900	0.904	0.908	0.914	0.930	0.953	0.982	0.999
30	0.902	0.906	0.910	0.916	0.933	0.957	0.988	1.006
40	0.903	0.907	0.911	0.917	0.935	0.960	0.992	1.010
50	0.903	0.906	0.911	0.917	0.935	0.961	0.994	1.013
60	0.902	0.905	0.910	0.916	0.935	0.961	0.994	1.014
70	0.900	0.903	0.908	0.914	0.933	0.959	0.993	1.013
80	0.898	0.901	0.905	0.911	0.930	0.956	0.991	1.011
90	0.894	0.897	0.901	0.908	0.926	0.952	0.986	1.006
100	0.890	0.883	0.897	0.903	0.921	0.947	0.981	1.001
110	0.886	0.888	0.891	0.897	0.915	0.940	0.974	0.993
120	0.880	0.882	0.885	0.891	0.908	0.933	0.965	0.985

Accuracy $\pm 1.4\%$.

$$\Pi = p_{os} = -(RT \ln a_w / \overline{v}) \tag{2.143}$$

By substituting in van't Hoff's Eq. (2.140), it gives:

$$\Pi = \phi c_{ms} R T \tag{2.144}$$

The osmotic pressure Π, of seawater and its concentrates, in atmospheric pressure is shown in Table 2.29.

And the osmotic coefficient is given as a function of the water activity:

$$\phi = \ln a_w / \overline{v} c_{ms} \tag{2.145}$$

where, c_{ms} is the molar salt concentration.

Table 2.30 gives osmotic pressure of sea water and its concentrates at atmospheric pressure. In Fig. C.6 in Appendix C the curves of osmotic pressure as function of concentration of seawater are presented.

Osmotic coefficient, based on data by Bromley et al. (1974) is presented by Sharqawy et al. (2010) formulated by the IAPWS-08 Gibbs energy function as:

$$\phi = a_1 + a_2 t + a_3 t^2 + a_4 t^4 + a_5 S + a_6 S t + a_7 S t^3 + a_8 S^2 + a_9 S^2 t + a_{10} S^2 t^2 \qquad (2.146)$$

The values of parameters a, are:

$a_1 = 8.9453 \times 10^{-1}$, $\quad a_2 = 4.1561 \times 10^{-4}$, $\quad a_3 = -4.6262 \times 10^{-6}$, $\quad a_4 = 2.2211 \times 10^{-11}$,
$a_5 = -1.1445 \times 10^{-1}$, $\quad a_6 = -1.4783 \times 10^{-3}$, $\quad a_7 = -1.3526 \times 10^{-8}$, $\quad a_8 = 7.0132$,
$a_9 = 5.696 \times 10^{-2}$, $\quad a_{10} = -2.8624 \times 10^{-4}$

The range of temperature is $0 \le t \le 200°C$ and of salinity $10 \le S \le 120$ g kg^{-1}.

The osmotic coefficient describes the change of the chemical potential of water per mole of added salt, expressed as multiples of the thermal energy $R(T_o - t)$ (Feistel, 2008):

$$\mu^W(0, t, p) = \mu^W(S_A, t, p) + m_{sw} R(T_o + t)\phi \qquad (2.147)$$

Osmotic coefficient can be calculated from the saline Gibbs function g^S, by means of the Feistel (2008) equation:

$$-m_{sw} R T \phi = g^S - S_A \left(\frac{\partial g^S}{\partial S_A} \right)_{t,p} \qquad (2.148)$$

or as:

$$\phi = -\frac{1 - S_A}{N_A S_A kt} \left[g^S - S_A \left(\frac{\partial g^S}{\partial S_A} \right)_{t,p} \right] \qquad (2.149)$$

Table 2.31 Classification of Suspended Material in Seawater

Type of Material	Remarks	Particle Size
Dissolved in solution	Simple and complex inorganic ions, inorganic and organic molecules, not dissociated solutes and polyelectrolytes	$<10^{-9}$ m
Colloidal dispersions	Macromolecules, mineral substances, hydrolysis, and precipitation products	10^{-9} to 10^{-7} m
Fine particles	Mineral substances, bacteria, plankton, Microorganisms, coagulated particles	10^{-7} to 10^{-5} m
Coarse dispersions (visible)	Mineral substances, macromolecules, all types of organisms	$>10^{-5}$ m
Turbidity	Fine inorganic and organic materials, coming from decompositions, silica. (They are carried by river waters and winds)	$\gg 10^{-3}$ m

where

M_{sw}	Molality $= S_A/[(1-S_A)M_s]$, moles of salt per mass of water	mol kg^{-1}
K	The Bolzman constant $= 1.3806488 \times 10^{-23}$	J K^{-1}
M_S	The mean atomic mass of sea salt	mol^{-1}
N_A	The Avogado constant $= 6.0221412 \times 10^{23}$	mol^{-1}
R	The molar gas constant $= kN_A = 8.314472$	J mol^{-1} K^{-1}

There exist, as it is obvious, a big variety of equations derived mostly from experimental work. The selection of a relationship depends on the parameters and the accuracy needed for each procedure.

2.5 SUSPENDED PARTICULATE MATERIAL IN SEAWATER

In addition to dissolved substances seawater contains various suspended materials that are visible or invisible to naked eye. They consist of coarse and fine particle minerals and of colloidal dispersions, as macromolecules, precipitation products, etc., from plankton and other microorganisms. Suspended substances are of importance especially to desalination methods. A classification of the suspended material in seawater is shown in Table 2.31.

Organic suspended material is in general in excess to inorganic particulate material. Shallow seawaters near coast contain more suspended substances that can be as much as about 11% which are decreasing with increasing depth. In deep waters suspended material can be as low as 0.002% or less.

Another important element in seawater is silicon. It is the most variable element in the sea and can be found as dissolved silicate ion ($H_3SiO_4^-$) and as suspended silica. Its exact form is not very clearly known but is clear that it is not in colloidal form.

Suspended silica has particles in the rage of 10^{-3} cm of diameter. It flows to the sea by incoming river waters.

2.5.1 Suspended Matter Evaluation

The presence of suspended matter in seawater may affect the proper and smooth operation of desalination systems. It can vary considerably in amount and type of material depending from the place of a region. Of importance for desalination systems the items that have to be determined are:
- Dissolved organic carbon (DOC)
- Total dissolved organic carbon (TOC)
- The estimation of very fine suspended or colloidal matter in a seawater sample. It is measured by the silting index, $S = (t_3 - 2t_2)/t_1$ where t is the time needed for three seawater samples of volume V_1, V_2, and V_3 to pass through a membrane filter of pore size $<0.8\ \mu m$ at constant pressure of 3.5 bar

Table 2.32 WHO's Specifications for Drinking Water

Constituents		Higher Accepted (mg L^{-1})	Higher Permissible (mg L^{-1})
TDS	TDS	500	1500
Chlorides	Cl$^-$	200	600
Sulfates	SO$_4^{2+}$	200	400
Calcium	Ca^{2+}	75	100
Magnesium	Mg^{2+}	30	150
Nitrates	NO$_3^-$	<50	100
Fluorates	F$^-$	0.7	1.7
Copper	Cu^{2+}	0.05	1.5
Iron	Fe^{3+}	0.10	1.0
Sodium chloride	NaCl	250	—
Hydrogen as pH	pH	7.0–8.5	6.5–9.2

- The colloidal or fouling index K_{col}, in % per min. It is a measure of very fine suspended solids and of colloidal substances. It is determined by passing a seawater sample of 100 or 500 ml through a membrane filter of defined pore size at constant pressure of 2.0 bar (Heitman, 1970):

$$K_{col} = \frac{(1 - t_1/t_2) \times 100}{t} \qquad (2.150)$$

where t_1, t_2 is filtration time of water volumes V_1 and V_2 and t is testing time.

2.6 QUALITY OF DRINKING AND UTILIZATION WATER

Natural waters may possess few characteristics that made them suitable for drinking, domestic, or other uses. Exactly the same characteristics may be possessed by desalinated water, which after desalination is posttreated to get the properties, Thus quality of fresh water, natural or desalinated, is of great importance (WHO, 2004). For drinking water specifications are very stringent according to the WHO. The highest accepted TDS are 500 mg L^{-1}.

Except the WHO many nations have their own local specifications. Table 2.32 presents the recommended and the higher permitted amounts, without health problems, of main constituents in drinking water. Chloride content higher than 600 mg L^{-1} deteriorates water test and is corrosive to the pipes of the distribution network.

Table 2.33 Water Characteristics that Affect Its Quality

Characteristic	Cause		Remarks
Hardness	Calcium and magnesium ions in water	Affects suitability for drinking and in some industries, as textile, paper, steam boilers, etc.	Classification: Soft: 0−60 moderate hard: 61−120 Hard: 121−180 very hard $\ll$ 180
pH	Molecule dissociation, acids and bases in water	Affects the use and treatment of waters	pH = 7.0 neutral pH < 7.0 acidic pH > 7.9 basic
Specific electrical conductance	Comes from substances that form ions when dissolving	The larger the conductance the more ions in water (the more mineralized)	It is measured by electrical conductivity, in micromhos, in 1 m^3 water at 25°C
TDS	All mineral substances dissolved in water	<500 ppm is suitable for drinking and domestic use and industrial processing	$\ll$ 1000 ppm, fresh $1-3 \times 10^3$ slight saline $3-10 \times 10^3$ moderate $10-35 \times 10^3$ very saline $\ll 35 \times 10^3$ Brine

Table 2.34 Natural and Added Pollutants in Waters

Polluting Element	Concentration c (mg L^{-1})	Polluting Element	Concentration c_{max}
Nitrate as N	10.0	Turbidity	1−5 NTU
Fluorides	11.4−2.4	Bacteria	1/100 mL
Barium	1.0	Radium 226 & 228	5 pCi L^{-1}
Silver	0.05	Strontium 90	8 pCi L^{-1}
Arsenic	0.05	Radiation á	15 pCi L^{-1}
Lead	0.05	Radiation â	50 pCi L^{-1}
Chromium	0.05	Carbohydrate Chlorides, mg L^{-1}	0.1−0.00002
Cadmium	0.01		
Selenium	0.01		
Mercury	0.002		

For salt content in natural waters the expression ppm (parts per million) is used, which equals $mg\,L^{-1}$. The expression can be applied, according to US Geological Survey (1984), up to 7000 ppm. For concentrations higher than this value water constituent has to be divided by density. In Appendix B, Table B.24 the conversion factors are given to convert ppm into eq pm and vice versa.

Some major characteristics exist that affect the drinking water quality but also the utilization suitability for many industries. They are presented in Table 2.33. The same characteristics are valid for desalinated waters. Table 2.34 gives the main chemical elements that pollute drinking water and their highest permissible amount. These elements may create biological anomalies to kidneys, liver, skin, and eyes and may be responsible for cancer as well.

2.7 CORROSION AND SCALE FORMATION

Other important factors are the corrosion, scale formation, and fouling mechanisms that are caused by various water types and seawater used in desalination systems. Operating thermal desalination units, especially at elevated temperatures, the potential of severe scaling on the metallic surfaces in contact with hot seawater and/or brine exists. Formation of scale reduces the thermal conductivity and lowers considerably the productivity. Scaling, fouling, and corrosion are major problems for desalination equipment having metallic surfaces, such as evaporators, heat exchangers, pipes, pumps, etc. Prevention of scale by appropriate treatment is thus of importance for these systems.

2.7.1 Corrosion

Of prime importance for desalination system is the selection of suitable metals. Inadequate metallic material could lead to increased maintenance and replacement cost that affects the overall economy of the system. Seawater is slightly alkaline having a pH close to 8.0, a pH region where electrolyte solutions cause galvanic and crevice corrosion. The curves of Fig. 2.15 present typical corrosion rates of iron and other metals as function of brine pH (Drake, 1986). The rate of corrosion is also dependent on the velocity of the seawater flow, the degree of flow rate turbulence, the seawater temperature, and the presence of oxygen. Severe localized attacks may be seen in hot sites where turbulence is high, as, for example, in bends.

Pure water, as condensate, is also aggressive, especially in the presence of dissolved oxygen, carbon dioxide, or other noncondensable gases. Condensate

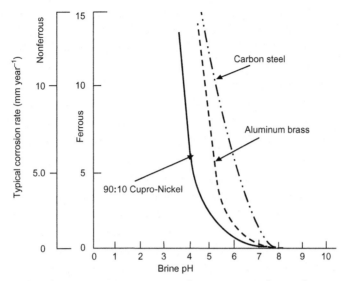

Figure 2.15 Typical indicative corrosion rates of various metals as function of brine pH (Drake, 1986).

and pure water are especially aggressive to copper and brasses but in lesser extent to copper–nickel alloys. Alkaline aqueous solutions are as well aggressive to the above metals and alloys.

The corrosion parameters that affect all metallic surfaces of seawater distillation units, in general, are:

- Chemical composition of seawater and its concentration
- Amount and type of suspended particles, organic and inorganic, as sand, algae, bacteria, etc.
- pH of feed seawater or treated seawater feed
- Seawater operation temperature
- Type and amount of gas contained, especially oxygen and carbon dioxide
- Flow rate of seawater
- Local turbulence caused in flow rate directions
- Various chemicals and additives of the feed water pretreatment
- Type of scale, fouling, and bacteria presence in raw seawater.

Brine corrosion is usually less severe than dilute solutions due to decrease of oxygen content as brine is concentrating.

2.7.2 Scale Formation and Fouling

Scale, is the insoluble material formed on the metallic surfaces of distillation units, as evaporators, heat exchangers, etc., due to salt precipitation from brine. Scale may include suspended organic and inorganic material in seawater and metallic particles from corrosion activities. Scale refers especially to hard, adherent, and normally crystalline deposits which are difficult to be removed.

Sludge, is soft, amorphous scale deposit a part of which remains suspended in the brine. If deposited, in general, is easily removed from the metallic surfaces.

Fouling, is accumulating deposits due to organic biological material, such as bacteria, plankton, organic colloid substances, etc.

Scale deposits on heat transfer surfaces decrease heat transfer and consequently productivity of the system. Formation of scale deposits and fouling on heat transfer surfaces is one of the most serious problems of desalination equipment that operate with sea or brackish water. The mechanism of this formation is very complicated and depends on various parameters, as temperature of seawater intake, which is related to organic material or fine suspended substances, temperature and type of feed water pretreatment, operation temperature, local surface area superheating, etc. Inorganic salt scaling deposits may be alkaline and/or acid (sulphate scale).

Precipitation of calcium carbonate is the main reason of hard scale. It is estimated from the solubility product K_{sp}, for $CaCO_3$ which is presented in Table 2.28, last column. Solubility product is of great importance for species forming scale on desalination equipment surfaces. There exists interdependence between carbonate balance and solubility of calcium carbonate. The state of saturation, which is the important point for scale formation, is characterized by the "Saturation Index-(SI)":

$$SI = \log[Ca^{2+}][CO_3^{2+}]/K_{sp} \qquad (2.151)$$

Saturation index is a function of temperature and alkalinity. Positive values of saturation index indicate precipitation of calcium carbonate from the solution and negative values that precipitated $CaCO_3$ can be dissolved. Saturation index points out the tendency of seawater or its concentrates to form scale, that is, is the measure of declination from the solubility limit of calcium carbonate.

Fig. 2.16 presents the Langelier saturation index diagram for calcium carbonate (Langelier et al., 1950; Meller, 1984), with an example on how the index is calculated. Saturation Index is the difference between real pH at 25°C and pH at the saturation point. It predicts $CaCO_3$ scale formation.

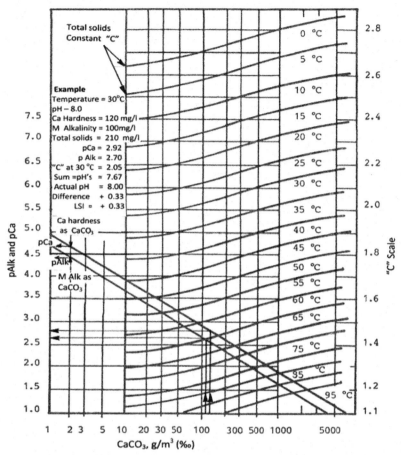

Figure 2.16 Saturation Index (SI) diagram of Langlier, for calcium carbonate ($CaCO_3$) and alkalinity, expressed as calcium carbonate (Meller, 1984).

2.7.2.1 Alkaline Scale Formation

Alkaline scale normally consists of calcium carbonate, magnesium hydroxide, or admixture of both compounds, which crystallize on the heat transfer surfaces. Alkaline scale is formed in waters containing bicarbonates, but the mechanism of scale formation is very complex depending on many parameters, such as salinity of feed water, temperature, and mode of operation, etc., which determine the amount and composition of the scale (Doodly and Glater, 1972).

During the evaporation progress as salt concentration increases, the critical solubility point may be reached, at which solubility limit of some salts is exceeded and scale formation starts. The factors that favor scale formation are:

- Nucleation, which once formed induces further scale formation
- Sufficient contact time of formed nucleus with the solution
- Local supersaturation of the brine.

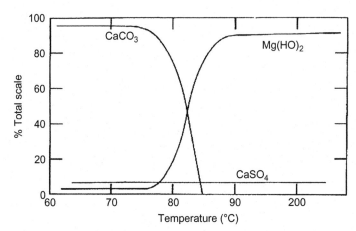

Figure 2.17 Alkaline scale composition as function of temperature (Delyannis and Delyannis, 1974).

Alkaline scale results from the decomposition of hydrogen carbonate by heating seawater up to about 82°C, where hydrogen carbonate decomposes forming calcium carbonate. Both alkaline scale constituents, $CaCO_3$ and $Mg(OH)$, have negative solubility. The curves of Fig. 2.17 show the effect of temperature on scale composition. About 77°C calcium carbonate precipitations predominate, in the range of 77–85°C both constituents are deposit and over 85°C deposition of magnesium hydroxide predominates.

2.7.2.2 Sulfate Scale, or Acid Scale

This type of scale results from the deposits of three forms of calcium sulphate; the anhydrite ($CaSO_4$), the hemihydrate ($CaSO_4.0.5H_2O$), and dehydrate calcium sulfate or gypsum ($CaSO_4.2.H_2O$). Precipitation is affected by concentration, pH, temperature, and solubility of calcium sulfate which in turn is affected by the presence of other ions. Calcium sulfate solubility increases as chloride concentration in the solution approaches 4–5% and then decreases to values similar to those in chloride-free solutions, in the range where chloride concentration approaches 15–19%.

Fig. 2.18 presents the solubility of the three forms of calcium sulfate in pure water as a function of temperature and in Fig. 2.19 is given the solubility of the same forms in seawater as a function of temperature and concentration factor of normal seawater. Table 2.35 gives the solubility of calcium sulfate anhydrite in various chloride concentrations on brine and temperatures from 50°C up to 160°C (Furby et al., 1968).

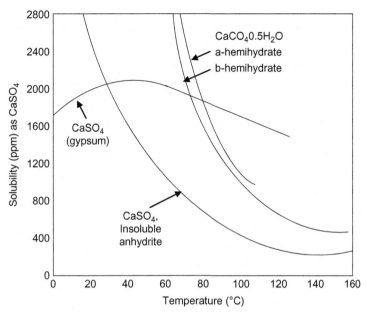

Figure 2.18 Solubility of the three forms of calcium sulfate in pure water (Hasson and Zahavi, 1970).

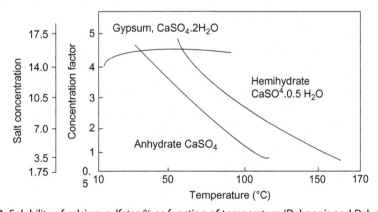

Figure 2.19 Solubility of calcium sulfates % as function of temperature (Delyannis and Delyannis, 1974).

For the prevention of scale formation numerous studies have been carried out and many antiscaling agents are commercially available today for feed water pretreatment. Some of the most-in-use procedures are very shortly described below.

2.7.3 Scale Prevention

In addition to alkaline and sulfate scale in the sludge that is formed during operation of thermal units, organic material and chemicals are included derived from various

Table 2.35 Molar Solubility Product of $CaSO_4$ Anhydrite in Sea Salt Solutions, for $10^4 K_s$[a]

‰Cl	$10^2 X_{SO4}$	50°	75°	100°	110°	120°	130°	140°	150°	160°	180°
10	0.923	7.5	4.44	2.43	1.85	1.41	1.08	0.82	0.63	0.487	0.296
20	1.882	13.1	8.04	4.62	3.60	2.80	2.18	1.71	1.34	1.06	0.676
30	2.877	17.6	11.2	6.68	5.27	4.16	3.10	2.62	2.98	1.68	1.11
40	3.911	21.2	13.8	8.50	6.78	5.41	4.33	3.49	2.81	2.30	1.56
50	4.987	23.7	15.9	10.0	8.05	6.48	5.25	4.27	3.48	2.88	2.01
60	6.106	25.4	17.3	11.2	9.06	7.36	6.01	4.95	4.07	3.41	2.44
70	7.272	26.1	18.2	12.0	9.80	8.03	6.62	5.54	4.60	3.90	2.87
80	8.488	26.2	18.5	12.5	10.29	8.50	7.09	3.99	5.10	4.33	3.27
90	9.756	25.9	18.7	12.8	10.65	8.85	7.46	6.38	5.45	4.75	3.68
100	11.08	35.5	18.7	13.0	10.87	9.09	7.72	6.68	5.78	5.12	4.07
110	12.47	24.7	18.5	13.2	11.1	9.37	8.07	7.04	6.16	5.50	4.44
120	13.92	28.7	18.2	13.3	11.3	9.72	8.44	7.44	6.57	5.88	4.80
130	15.43	22.5	17.8	13.4	11.5	10.05	8.60	7.83	5.94	6.29	5.26
140	17.03	21.2	17.3	13.5	11.8	10.46	9.29	8.30	7.40	6.74	5.74
150	18.70	20.0	16.7	13.6	11.9	10.74	9.61	8.66	7.83	7.24	6.36

[a]$K_s = \{[Ca^{++}][SO_4^{--}]\}$.

additives during pretreatment of feed water. Precipitated sodium chloride, silica, and metallic substances, as copper and iron from corrosion activities are included in the sludge which are not scaling substances but in general they are considered as trapped material. As scaling is not totally prevented by the pretreatment methods, periodic shut-downs of the plant would be required for cleaning. Thus efficient pretreatment of seawater feed is the best prevention for scale formation (Standiford and Sinek, 1961). Scale prevention by pretreatment is achieved by various procedures, depending on seawater analysis, operation temperature, and other parameters. The most-in-use methods will be described shortly.

2.7.3.1 Acid Injection

Any acid is suitable for injection but usually, due to its low cost, sulfuric acid is used, although calcium sulfate concentration is increasing. The acids are added in proportion corresponding stoichiometrically to the hydrogen carbonate ion concentration. Hydrogen carbonate breaks down to calcium carbonate and then is transformed to calcium sulfate, by the sulfuric acid:

$$CaCO_3 + H_2SO_4 \rightarrow CaSO_4 + H_2O + CO_2 \tag{2.152}$$

After treatment with an acid, seawater is decarbonated by counter current air striping. Decarbonation is followed by vacuum deaeration and steam stripping to remove dissolved oxygen and noncondensable gases.

2.7.3.2 Injection of Carbon Dioxide

The carbon dioxide is injected in the cold stream of seawater. This procedure prevents alkaline scale formation and the kinetics has been described by Langelier et al. (1950). Before introducing feed seawater into the evaporation system carbon dioxide is removed by steam stripping in a packed column where alkaline scale precipitates and periodically removed.

2.7.3.3 Polyphosphates as Scale Inhibitors

Polyphosphates act as sequestrates for calcium and magnesium ions. Other additives, as lignin sulfonic acid, tannin, starch, and others are added as dispersants which prevent crystal growth and the adherence of scale on the surfaces is inhibited. Polyphosphate-based additives are restricted to low temperature treatment, below 88°C. Treatment with polyphosphates is a rather economic procedure as the required facilities are simple and low cost.

2.7.3.4 The Seeding Technique

This technique refers to the addition in seawater of relatively low concentration finely divided seed crystals. When seawater is heated above its saturation temperature it will cause scale forming compounds, which precipitate with the seeding particles than on the heat transfer surfaces, which will remain clean.

Many other scaling inhibition techniques exist. In many cases the type of pretreatment system and method is included with the evaporation system by the company that provides the installation.

2.8 CONCLUSION

Large desalination systems use as raw material seawater and/or brackish waters discharging simultaneously back to the sea or to water streams large to huge amounts of concentrated brine which has an impact on the region's ecosystem. Seawater salt concentration is affected by various parameters, such as from the region, the geographic location, the season of the year, and from local condition, as local currents.

For each intake site, salinity and other fluctuating parameters that affect appropriate operation of the desalination system must be examined very carefully as fluctuations of salinity; density and temperature are of importance for the smooth and efficient operation of any desalination system.

REFERENCES

Abouzaid, H., Cotruvo, J.A., 2004. WHO/EMRO Guidance on Desalination for a Safe Water Supply: Health and Environmental Aspects Applicable to Desalination. All Health Factors Related to the Composition of Desalinated Water. WHO, Geneva, 2007.

ASHRAE (American Society of Heating, Refrigerating and Air-Conditioning Engineers) Handbook, Fundamentals (2005).

Bernal, J.D., 1965. The structure of water, Proc. 1rst Inter. Symposium on Water Desalination, Washington D. C., 3–9 October, 1965, vol. 1, 371–381.

Brandani, V., Del-Re, G., Di Giacomo, G., 1985. A new model for predicting thermodynamic properties of sea salt solution. Desalination. 56, 299–313.

Bromley, L.A., Sign, D., Ray, P., Sridharand, S., Read, S.M., 1974. Thermodynamic properties of sea salt solutions. AICHE J. 20, 326–335.

Castelli, V., Stanley, E., Fisher, E., 1974. The thermal conductivity of seawater as a function of pressure and temperature, Deep-Sea-Research, Part I, Oceanographic Research Papers, 21 (4), 311–319.

Chen, S.F., Chan, R.C., Read, S.M., Bromley, L.A., 1973. Viscosity of seawater solutions. Desalination. 13, 37–51.

Chou, J.C.S., Rowe Jr, A.M., 1969. Enthalpies of aqueous sodium chloride solutions. Desalination. 6 (1), 105–115.

Clark, R.L., Nabavian, K.J., Bromley, L.A., 1960. Heat of concentration and boiling point elevation of seawater. Advances in Chemical Series, Saline Water Conversion. 27, 21–26.

Cotruvo, J.A., Lattemann, S., 2008. WHO guidance on desalination. Int. Desalination & Water Reuse. 1, 47–52.

Delyannis, E., Belessiotis, V., 1995. Methods and Systems of Desalination (in Greek), 460 pp.

Delyannis, A., Delyannis, E., 1974. Gmelin Handbuch der Anorganischen Chemie, "Sauerstoff," Anhang "Water Desalting,". Springer Verlag, p. 340.

Doodly, R., Glater, J., 1972. Alkaline scale formation in boiling seawater brines. Desalination. 11, 1–16.

Drake, F.A, 1986. Measurement and control in flash evaporator plants. Desalination. 86, 241–262.

Drost-Hansen, W., Anomalies in the properties of water, Proc. 1rst Inter. Symposium on Water Desalination, Washington D. C., 3–9 October, 1965, vol. 1, 382–412.

Emerson, W.E., Jamieson, D.T., 1967. Some physical properties of sea water in various concentrations. Desalination. 3 (2), 213–224.

Fabuss, B.M., Korosi, A., 1967. Thermodynamic properties of seawater and its concentrates. Desalination. 2, 271–278.

Feistel, R., 1993. Equilibrium thermodynamics of seawater revised. Prog. Oceanogr. 31, 101–179.

Feistel, R., 2003. A new extended Gibbs thermodynamic potential of seawater. Prog. Oceanogr. 58, 43–114.

Feistel, R., 2008. A Gibbs function for seawater thermodynamics for −6 to 80°C and salinity up to 120 g kg^{-1}. Deep Sea Res. 1 (55), 1639–1671.

Feldman, S.E., 2007. WHO Report on "Desalination for Safe Water Supply. Guidance for Health and Environmental Aspects Applicable to Desalination," WHO, p. 4.

Forch, C., Knudsen, M., Sörensen, S.P.L., 1902. Berichte über die Konstantenbestimmungenn zur Aufstellung der Hydrographischen Tabellen (Report on constants determinati9on for the preparation of hydrographic Tables), D. Kgl Danske Vidensk, Selsk, Skrifter, 6, Raekke, naturvidensk. Or mathem., Afd XII. 1151 p.

Frenkel, V., 2008. Brackish vs seawater desalination. Which one is for you? Int. Desalination & Water Reuse. 17 (4), 47–50.

Furby, E., Glueckauf, E., McDonald, L.A., 1968. The solubility of calcium sulfate in sodium chloride and sea salt solutions. Desalination. 4, 264–276.

Grunberg, L., Some properties of water and salt solutions, Proc. 1rst Inter. Symposium on Water Desalination, Washington D. C., October 3–9, 1965, vol. 1, 157–179.

Grunberg, L., 1970. Properties of seawater concentrates, Proc. 3rd Int. Symposium on "Fresh Water from the Sea," vol. 1, 31–39.

Hampel, C.A., 1950. Densities and boiling points of seawater concentrates. Ind. Eng. Chem. 42 (2), 383–386.

Hasson, D., Zahavi, J., 1970. Mechanism of calcium sulfate scale deposition on heat transfer surfaces. I.EC Fundamentals. 9 (1), 1–10.

Heitman, H.G., 1970. Saline Water Processing. VCH Verlasgesselschaft, mbH, Weinheim.

Horne, R.A., 1969. Marine Chemistry. Wiley Interscience, New York, p. 568.

IAPWS, 1995. Release on the IAPWS Formulation 1995 for the Thermodynamic Properties of Ordinary Water Substance for General Scientific Use, available in www.iapws.org/relguide/iapws95.pdf.

IAPWS, 2006. International Association for Properties of Water and Steam, Supplementary, Release on an Equation of State for H_2O-ice Ih., IAPWS, Witney, UK, September 2006, available in www.iapws.or/relquide/ice.pdf.

IAPWS, 2008. International Association for Properties of Water and Steam, Release on the IAPWS Formulation in 2008 to the Thermodynamic Properties of Seawater, Berlin, Germany, 2008, 19 pp., available in http://www.iapws.org.

IAPWS, 2009. International Association for Properties of Water and Steam, Supplementary Release on a Computational Efficient Thermodynamic Formulation for Liquid Water for Oceanographic Use, Doorwerth, The Netherland, 2009, available in http://www.iapws.org.

IOC, SCOR & IAPSO, 2010. The international thermodynamic equation of seawater—2010: Calculation and use of thermodynamic properties. Intergovernmental Oceanographic Commission, Manual Guides No 56, UNESCO (English), 196 p. Available from: http://www.teos-10.org/pubs/TEOS-10_Manual.pdf.

Isdale, J.D., Morris, R., 1972. Physical properties of sweater solutions: density. Desalination. 10 (4), 329–339.

Isdale, J.D., Spencer, C.M., Tudhope, J.S., 1972. Physical properties of seawater solutions. Desalination. 10, 319–328.

Jamieson, D.T., 1986. Experimental methods for determination of the properties of saline water. Desalination. 59, 219–240.

Jamieson, D.T., Tudhope, J.S., 1970. Physical properties of seawater: Thermal conductivity. Desalination. 8, 393–401.

Jamieson, D.T., Tudhope, J.S., Morris, R., Cartwright, C., 1969. Physical properties of seawater solutions: Heat capacity. Desalination. 7 (1), 23–30.

Khan, A.H., 1986. Desalination Processes and Multistage Flash Distillation Practice. Elsevier, Oxford.

Knudsen, M., 1901. Hydrographisen Tabellen (Hydrographic Tables), G.E.C. Gad, Copenhagen, 63 p.

Korosi, A., Fabuss, B.M., 1968. Viscosity of liquid from 25°C to 150°C. J. Anal. Chem. 40, 157–162.

Kretzschmar, H., Feiste, R., Wagner, W., Miyagawa, K., Harvey, A.H., Cooper, J.R., et al., 2015. The IAPWS industrial formulation for the thermodynamic properties of seawater. Desalin. Water Treat. 55 (5), 1177–1199.

Langelier, W.F., Caidwell, W.B., Lawrence, W.B., 1950. University of California, Berkeley, Inst. Eng. Res. Report Ser. 4 Issue No 2.

Lewis, E.L., Perkin., R.G., 1978. Salinity: Its definition and calculation. J. Geophys. Res. 83, 466–478.

McDougall, T.J., Feistel, R., Millero, F.J., Jackett, D.R., Wright, D.G., King, B.A., et al. Calculation of thermodynamic properties of seawater, Global ship-based Report Hydrographic Manuel, IOCCP Report No 14, CPO Publication Series No 134 (2009).

McKetta, J.J., 1994. Executive Editor, Encyclopedia of Chemical Processing and Design, vol. 44 & 66. Marcel Dekker Inc, New York.

Meller, F.H., 1984. Electrodialysis (ED) and electrodialysis reversal (EDR) technology, Ionics Incorporated, 66 p.

Millero, F.J., Huang, E., 2009. The density of seawater as a function of salinity 5-70 g kg^{-1} and temperature (273.15 to 363.15°C). Ocean Sci. 5, 91–100.

Millero, F.J., Poison, A., 1981. International one atmosphere equation state of seawater. Deep-Sea-Res. 28A (6), 625–629.

Millero, F.J., Feistel, R., Wright, D., McDougall, T., 2008. The composition of standard seawater and the definition of the reference-composition salinity scale. Deep-Sea-Res. I. 55, 50–72.

OCN-623-Chemical Oceanography, Chemical composition of seawater: salinity and the major constituents, available in Salinity 2012 Web.

Oetjen, G.W., 1999. Freeze Drying. Wiley-VCH, Weinheim, Germany.

Robinson, R.A., Stokes, R.H., 1959. Electrolyte Solutions. Butterworths, London, p. 559.

Sharqawy, M.H., Lienhard, V.J., Zubair, S.M., 2010. Thermophysical properties of seawater, a review of existing correlations and data. Desalination and Water Treatment. 16, 354–380.

Sign, D., Bromley, L.A., 1973. Relative enthalpies of sea salt solutions at 0 to 75°C. J. Chem. Eng. Data. 18 (2), 174–181.

Spiegler, K.S., El-Sayed, 1994. A Desalination Primer, Balaban Desalination Publications, 1994, 215 p.

Standiford, F.C., Sinek, J.R., 1961. Stop scale in seawater evaporators. Chem. Eng. Prog. 57 (1), 5–63.

Sun, H., Feistel, R., Koch, M., Markoe, A., 2008. New equations for density, heat capacity and potential temperature of a saline thermal fluid. Deep-Sea-Res. 55, 1303–1310.

Tsobanoglou, G., Schroeder, E.D., 1985. Water Quality. Addison-Wesley Publ., Comp., Menlo Park, CAL.

THEOS-10, The International Thermodynamic Equation and the use of seawater-2010, http://www.teos-10.org/ in IOC, SCOR & IAPSO, 2010: The international thermodynamic equation of seawater-2010: Calculation and use of thermodynamic properties. Intergovernmental Oceanographic Commission, Manual Guides No 56, UNESCO (English), 196, http://www.teos-10.org/pubs/TEOS-10Manual.pdf.

Ullman, 1998. Fifth Edition. Encyclopedia of Industrial Chemistry, Vols., A.28, B8. VCH Verlag, Weinheim, Germany (English version).

US Geological Survey, Water Survey, Paper 2220, (1984).

Walton, E.G., 1974. Handbook of Marine Science, Vol. 1. CRC press, Cleveland, Ohio.

Wagner, W., Kretzschmar H.-J. International Steam Tables, Properties of water and steam based on industrial formulation, IAPWS I F, 97, 2nd Edition, Springer Verlag, 2008, 392 p.

WHO, 2004. Environmental health Criteria, Document 170. Assessing human health risks of chemicals: derivation of Guidance values for health-based exposure limits. World Health Organization, Geneva.

Wieser, M.E., Coplen, T.B., 2011. Atomic weights of elements 2009 (IUPAC technical Report). Pure Appl. Chem. 83 (2), 359–396.

CHAPTER THREE

Solar Distillation—Solar Stills

3.1 INTRODUCTION

Solar distillation refers to the evaporation of aquatic solutions by means of solar energy and to the simultaneous condensation of the vapors, created by the activity of solar thermal energy. It is a simple procedure of humidification—dehumidification that takes place in one and the same device, called *"solar still."* It is known from the early antiquity as a physical procedure.

There exist many brief historical reviews on solar distillation starting from antiquity, where many trials were performed to produce fresh water from salty waters. Many of the descriptions about desalination are perfect ideas but the primitive knowledge of technology and construction designs is very poor at the time they were expressed for achieving any practical application.

The most extensive and informative works of antiquity, however, are those of Aristotle (384—322 BC) (1962), the well-known philosopher and scientist. Von Lippman (1910, 1911) and Briegel (1918) discuss the desalination references by Aristotle who described in a surprising correct way the origin and the properties of natural, brackish, sea waters, and some ways to desalinate salty waters.

The practice of distillation was developed in Alexandria, Egypt, during the Hellenistic period and Bittel (1959) gives a detail description of various distillers (alembics) developed that time. There were developed various types of alembics. The two pots alembic ($\Delta\iota\beta\iota\kappa\varsigma$) was used for vapor condensation. The head of the still in Greek was called "ambix." This name was applied often to the still as a whole. The Arabs named it "Al-Ambiq" from which the worldwide known "alembic" was developed.

During the mediaeval times, solar energy was used to fire alembics in order to concentrate dilute aquatic solutions, thus solar distillation was mainly used to concentrate alcoholic solutions, herbal extracts for medical applications, to produce wine, various perfume oils, etc.

Nebbia and Nebia-Menozzi (1966, 1967), in their historical desalination review, say that the most important scientist of Renaissance was undoubtedly Giovanni Batista Della Porta (1535—1615) who mentions three desalination systems and a solar distillation apparatus for brackish waters. From the time of Della Porta, no special reference can be found on solar distillation, but in 1870 the first American patent

Thermal Solar Desalination

on solar distillation was granted to N.W. Wheeler and W.W. Evans (1879). The patent, based on experimental work, describes almost all basic operations and features of solar still including corrosion problems. Two years later, in 1872, the first large-scale solar distillation plant was erected in Chacabuco (Las Salinas), Chile. It was the first worldwide industrial size solar distillation plant. Harding (1883) gives a detail description of the plant and Telkes (1955) reports that the plant was in operation continuously for about 40 years.

Since then and until the beginning of the decade of 1960, many references exist about small-scale solar stills but not of large solar distillation plants. The 50's American Office of Saline Water (OSW) erected in Daytona Beach, FL, USA, a Solar Station to study operation parameters of various types of solar stills such as thin water layer still, deep basin still, glass-covered still, and plastic-covered still. The Station was operated by Battelle Memorial Institute, Cleveland, OH. The experience obtained and the solar distillation plants constructed by this experience are outlined in Reports and papers by Bloemer et al. (1961, 1965a,b, 1966, 1970). A detail description of these stills, up to 1970, and many other worldwide detailed studies can be found in the Battelle Memorial Institute report by Talbet et al. (1970). In fact since that time few improvements have been made in solar stills.

Several brief or more extended historical reviews on Solar Distillation, Solar Desalination and in Desalination in general are available in the literature: Delyannis and Piperoglou (1967a) give titles and short description from antiquity up to 1940. Nebbia and Nebia-Menozzi (1966, 1967) give a more detail description on history of desalination and El-Nashar and Delyannis (eolss) describe the highlights of renewable energies for desalination applications. Belessiotis and Delyannis (2000b) and Delyannis (2003) make reference on the achievements on conversion of solar energy to thermal energy and finally Birket (1984, eolss) makes an extend overview on desalination history from antiquity.

3.1.1 Definitions

The following definitions have been compiled from books on solar energy of Duffie and Beckmann (2006, 2013) and the Kumar's paper on solar distillation (Kumar, eolss). They present the terms that are used in solar energy systems. Solar energy symbols and units are given according to International Solar Energy Society (ISES), www.ises.org/ises.nsf/primarypages/SEJEditorial and SI (2008).

3.1.1.1 Absorptivity–Transmittance–Reflectance of Cover Material

Cover Absorptivity (a_c) is the ability of a material, opaque or transparent to absorb the solar radiation. It is given as the ratio of absorbed solar radiation to the incident radiation: $a_c = G_{abs}/G_{inc}$.

Cover Emittance or Coefficient of Emissivity (ε_c) refers to the ratio of intensity emitted by the cover ε_c at fixed temperature, to the intensity of a black body emittance ε_{bl}, of exactly the same shape and temperature. Emittance is given as $\varepsilon = \varepsilon_c / \varepsilon_{bl} = (1/\sigma T^4) \int_0^\infty \varepsilon_\lambda \varepsilon_{bl} d_l$, where ε_λ is the wavelength emittance, σ is the Stefan–Boltzmann constant, and d_l is a shape parameter.

Cover Reflectivity (ρ_{cnc}) refers to the ratio of reflected solar radiation to the incident: $\rho_c = G_{ref} / G_{inc}$.

Cover Transmissivity (τ_c) is the property of a transparent material to permit the sun's radiation to be transmitted through its path length with simultaneous refraction of the rays. It is presented by the ratio of solar radiation passing the cover to the incident solar radiation: $\tau_c = G_{ra} / G_{inc}$.

3.1.1.2 Solar Radiation

Black body refers to a body having ideal solar radiation absorptivity.

Incident solar radiation (G_g) is the radiant solar energy that hits the earth's surface and is referred as "global radiation" on a surface (W m^{-2}). Total or global solar radiation consists of:

- **Beam (or direct) Radiation (G_b)** is the part of total radiation that reaches earth's surface directly without being scattered by the atmosphere (W m^{-2})
- **Diffuse Radiation (G_d)** is the part of total radiation that reaches earth's surface after a change of its directions due to scattering by the atmosphere (W m^{-2}).

Incident solar radiation to normal refers to solar radiation falling perpendicular on a surface, ie, having an angle of 90° to the surface.

Global irradiance (G) is the total solar flux density (W m^{-2}).

Intensity of solar radiation is the transfer rate of the beams energy across the unit area of a body (W m^{-2}).

Irradiance (E, H) is called the rate by which the radiant solar energy hits the unit area of a surface. Irradiance may be beam (G_b), diffuse (G_d), or spectral (G_s). In G_o, subscript o refers to radiation above the earth's atmosphere, ie, extraterrestrial irradiance. By integrating irradiance over a fixed time, usually an hour or a day derive the incident solar energy per unit area of surface which is called "*irradiation* or specifically *insolation.*"

Insolation is expressed as instantaneous (G, W m^{-2}), hourly (I, J m^{-2} h^{-1}), or daily (H, MJ m^{-2} d^{-1}). It is referred as beam or diffuse insolation. Subscripts *inc*, *n* or *(0)* and *til* are used for incident, normal to the direction of propagation and for tilted surfaces, respectively. In many books, the incident hourly radiation on a horizontal surface is represented by the symbol *I*.

Radian energy (Q) is the energy transmitted by electromagnetic waves from a hot body to a colder one (J).

Sky radiation refers to the heat exchange between a surface and the sky. The sky is considered as a black body of temperature T_{sky}. The net thermal loss from the surface, at temperature T and of emittance ε, to the sky is given as $Q = \varepsilon A \sigma (T^4 - T_{sky}^4)$, where σ is the Stefan—Boltzmann constant. It is applied to solar stills and flat plate collectors especially during night operation.

3.1.1.3 Solar Stills

They are small devices containing saline water which simultaneously collect solar energy, evaporate the water, and condensate the generated vapor. To avoid confusion of terminology on solar stills, due to the variety of terminology used in bibliography, we use the most common and widely accepted terms which are described below.

Conventional solar still is a simple device as presented in Fig. 3.1. It is termed "greenhouse type" as its operation is based on the greenhouse effect. Other expressions are *"roof type," "tent type,"* etc. It can be symmetrical (double slope, Fig. 3.1A) or asymmetrical (single slope) as shown in Fig. 3.1B and C.

Conventional, multiple effect solar stills are types of stills having more than one effects into the still's chamber, mostly small basin into the main one. Evaporation surface is increased thus increasing the productivity.

Solar stills with wick. In these stills the black liner is replaced by a black porous hydrophilic fabric, presenting large surface area to water flow. The wicks are materials with high porosity that sucks water by capillary action.

Solar stills of various geometric forms are solar stills having various geometric and technical shapes and/or additional external or internal parts and in many cases incorporate external or internal reflectors, focusing collectors, etc. There is a big variety of solar still configurations having all possible geometric forms.

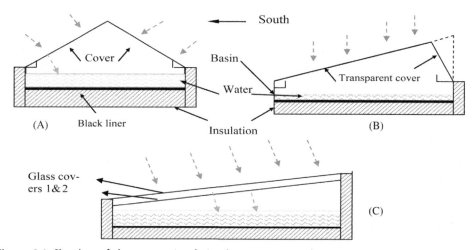

Figure 3.1 Sketches of the conventional simple symmetric and asymmetric stills: (A and B) with one cover glass and (C) with two cover glasses.

According to operation mode, they are divided into:

- **Passive solar stills** which are simple stills that accomplish their operation without circulation of feed water. The basin is refilled early, every morning, about sunrise time
- **Active solar stills** are conventional or not conventional solar stills where the water in the basin circulates, in very low flow rate, by the use of a pump, running by height difference from a storage tank or usually by connection to a solar collector (Belessiotis, eolss) which increases as well the salt water feed temperature.

3.1.1.4 Surface Conditions Inside the Still

Selective surface (black) refers to a black surface coated with a selective material of highest possible absorptivity of solar radiation, mainly ≥ 0.95. It is called black liner. There exist many configurations of the black absorbing surfaces such as wicks, sponges, and addition of black colors into the salty water.

Stagnant conditions are characterized by the absence of fluid circulation in solar collectors and absence of water in the basin of the solar stills. In such conditions, black absorbing surface area receives considerable solar radiation increasing temperature inside the corresponding chamber which can be destructive. Stagnant conditions may also occur during the shutdown of operation.

3.2 SOLAR STILLS

A conventional solar still is a simple device which acts as an absorber for solar radiation and simultaneously as a converter of solar energy into thermal energy which evaporates the water in the still's basin. Created vapor condenses on the cooler inside glass cover surface and is collected into gutters inside the still. They are oriented towards South in the Northern hemisphere and towards the North in the Southern. They can be constructed easily, by experience, by using inexpensive materials. Life time depends on the material and proper maintenance. In fact the desalination systems are used by poor people in Africa, the Caribbean islands, and other remote poor regions on earth. A conventional solar still consists of:

- **A rectangular basin** and seldom from a round basin or other geometry
- **A black liner** at the basin's bottom to increase solar radiation absorption and transform it into thermal energy. Black surface may be a black paint, a sheet of a black plastic material, or any material that increases radiation absorption
- **A top case** to receive the transparent cover
- **A transparent cover material** such as glass or a durable plastic material
- **Insulation material** which covers the basin bottom and sides of the still
- **Feeding and blow down pipes** comprising joints fittings, etc.

- **Pumps** for feed water circulation and brine blow down
- **Adhesive material** to fix the transparent cover to the case and eventually the liner to the basin or other components of the still together.

The heart of still's operation is the air tide space between basin and cover in which salt water evaporation and vapor condensation are performed simultaneously; the vapor is condensing at the cooler inside cover surface. Solar radiation penetrates the airtight space through the tilted transparent cover and is partially absorbed by the salt water in the basin; below this space the rest being absorbed by the black liner. The water in the basin is heated to a temperature higher than the temperature of the cover but lower than the boiling point of salty water. As radiation strikes the black liner surface changes wavelength and becomes long-wave heat.

In general for conventional solar stills, the following may be taken in consideration:

- They are of symmetric or double sloped (of roof type) and of asymmetric configuration (Fig. 3.1A and B). Both are oriented South—North. In asymmetric stills, the northern wall has either a small inclination or is vertical to water surface (Fig. 3.1B)
- Solar stills may be single effect or multiple effect. Single effect may be of single basin or double basin, ie, double transparent cover (Fig. 3.1C). Multiple effect stills are more complicated configurations than single ones
- They operate either in batch (passive) or in continuous (active) mode.

1. **Batch operation (passive solar stills)**

   In batch operation mode, the salt water feeding occurs every morning about sunrise time. Due to inertia during the morning hours, productivity is low until the proper temperature is achieved. During day or night operation, the salt water in the basin is evaporated decreasing the amount of mass. The brine in the basin every second or third day, according to the depth of the water, is totally evacuated to avoid scale or algae formation and refilled with fresh salt water.

   The water into the basin may be of a thin or deep layer. Thin layer water has higher daily output during sunshine hours but distillation stops after sunset. On the contrary deep layer basins act as sensible heat storage units continuing to produce distillate during night time, although in decreasing rate, up to sunrise.

2. **Continuous operation (active solar stills)**

   In continuous or active mode operation, feed water flows continuously in very low rates in a way that keeps height of the water in the basin almost stable. Their characteristics are described below.

   a. They are sized as still units for individual and family house use
   b. They are sized in arrays as solar distillation plants for community, agricultural, or other purposes utilization of the distilled water

    c. They are suitable for regions without fresh water resources but abundant solar intensity radiation or for places having brackish water wells or they are surrounded by seawater

    d. Conventional solar stills may be connected to solar collectors to increase their efficiency (increase feed temperature)

    e. Feed water may be waste water from any industrial waste

    f. Due to operation temperatures in the range of 50–80°C, distilled water is disinfected. As stills are operating in lower temperatures, eg, $\sim >50°C$, a disinfection posttreatment may be necessary

    g. Efficiency of conventional solar stills is low, 30–50%, depending on the intensity of solar radiation, the still construction, its geometric details, and the time of the year (ie, winter, summer, etc.)

    h. From references of many investigators has been found that the mean value of distillate output is about $2–2.5 \text{ L m}^{-2}\text{ d}^{-1}$ in winter time and $5–6 \text{ L m}^{-2}\text{ d}^{-1}$ for summer, according to the intensity of solar radiation and to the clearness of the day.

Solar radiation intensity falling on unit surface of evaporation is assessed by nature. The daily productivity M, $\text{L m}^{-2}\text{ d}^{-1}$, of a conventional solar still is:

$$M = \sum Q_{wg}/h_{ev} = \eta \sum Q_{inc}/h_{ev} \qquad (3.1)$$

where $\sum Q_{wg}$ (J m^{-2}) is the sum of daily total thermal energy amount given from the water to the inside of the transparent cover surface and h_{ev} is vaporization enthalpy of water (J kg^{-1}). $\sum G_{inc}$ is the sum of total daily incident solar radiation on the water surface (MJ m^{-2}) and η is its efficiency.

It should be noted that:

- To have a scientifically realistic overview of the still's efficiency, operation measurements have to be performed round the year in order to get mean values
- Efficiency during winter time may be increased by constructing external canals for rain catchment.

Most of the other still configurations are based on these two types: single effect or multiple effect conventional solar stills. The main parameters that influence their effective operation and consequently the effective radiation collection are pointed out by Cooper (1969a):

    1. Day of the year

    2. Latitude of site

    3. Orientation

    4. Cover slope

    5. Percentage diffuse radiation

6. Profile of day clearness
7. Glass thickness
8. Cover's extinction coefficient
9. Reflectance of still liner
10. Geometry of the still.

3.2.1 Manufacture—Maintenance

Many solar stills are empirical or semiempirical constructions, especially for domestic use, with all consequences deriving from these constructions, ie, lower efficiencies, shorter life, etc. Nevertheless, a solar still must be constructed very carefully and its components should be perfectly fitted in order to avoid heat loss and leakage from the edges. The following parameters are noted with respect to manufacture and maintenance of solar stills.

- A well-constructed and well-maintained solar still has a life of 20—25 years for the glass-covered systems and some years less for the plastic material of the cover (Tleimat and Howe, 1969)
- Glass or plastic cover is better to be constructed with an inclination of 10—20° for better efficiency and higher output collection. Experience showed that inclination less than 10 degrees helps formation of big droplets falling back to the salty water of the basin. Inclination higher than 20—25 degrees adds construction material, as glass and frame without important productivity increase, but the most important is that space inside the still increases, increasing also the distance between water surface and cover glass (Morse and Read, 1968). As a consequence, convection heat losses are increasing considerable decreasing efficiency. The still itself, for horizontal systems, must have an inclination of 1.5—3° to allow easy brine drainage flow
- After construction and starting operation, the still should keep in continuous operation, especially during day time. Long time stops of operation, ie, stagnant conditions may destroy the black liner, the black paint, the adhesive material and may be from algae or hard scale on the liner, if some brine has been left. If the still has to stop operation, for any reason, it should be emptied, cleaned, and be opened from one side, to keep inside temperature low (Fig. 3.2)
- The inside surface of glass cover, before installation, should be cleaned carefully. Dirty glass surface do not allow free flow of the condensate to the gutters
- Broken glass must be replaced immediately in order to avoid vapor loss, ie, heat and product loss
- The outside glass surface of the cover has to be cleaned occasionally as accumulates dust and other materials, such as dry leaves, soil, and sand, due to weather conditions, which decrease radiation penetration and thus decreasing of distillate output (Fig. 3.3).

Figure 3.2 Movable back glass wall for maintenance and cleaning the stills of solar distillation plant in Porto Santo, Madeira, Portugal. *Photograph by E. Delyiannis.*

Figure 3.3 Accumulation of dried leaves and other polluting materials onto the external surface of an array of stills in a solar plant. *Photograph by E. Delyannis, The Kimolos Island, Solar Stills plant, Greece.*

If these rules are kept, the system needs very low maintenance work and operation is smooth and efficient.

3.2.2 Materials of Construction

For the construction of family size solar stills or still arrays, a variety of material from very cheap ones up to more expensive for longer life of the system have been used. As solar distillation installations are addressed to small poor communities, or to individuals, the scope is to construct the system with cheap but durable materials, locally available, thus avoiding expensive transportation. Independently from the material to be used, the construction must be very careful. The general properties of construction material are (Lawand, 1968, 1975; Lawand and Ayoub, (eolss); Delyannis and Belessiotis, 2004):
- Should be easily assembled at the site
- Should be inexpensive and easily maintained or repaired. The use of very cheap materials is better to be avoided as they shorten the life of the still

- Should have long life
- Should resist weather conditions such as strong winds, hail, and sand winds
- Must be corrosion resistant to saline water, to environmental corrosion attacks, as salt spray from seawater and especially to product water as distilled water is very corrosive
- If in the region rain periods, even low ones, exist, it is necessary to provide, outside of the still, rain catchment canals in order to collect this water and thus increase product output. Additional expenses for these parts are very low.

3.2.2.1 Basin Materials

Basin is the basis of the still. A variety of materials have been used by scientist investigating solar stills and also by experienced construction personnel such as:

- Suitable concrete, seawater resistant. It is formed on site by purring the concrete in prefabricated molds
- Wood impregnated in epoxy resins to withstand deterioration
- Aluminum—magnesium alloy, stainless steel
- Hard plastic material, durable to UV, as plexiglass. Plastic material, in general, has less time period of life than concrete or metal
- In some cases, the absorbing black material acts as basin itself, by placing it directly on the soil, without a support or on sand which acts as insulation. In this case, the material should not deteriorate in contact with the soil
- A canal outside the basin constructed from the same material may collect rain water increasing monthly output during winter time.

3.2.2.2 The Absorbing Black Material

For the solar radiation absorbing material, various media are used such as black paint, wicks painted black, and black sponges. For the conventional basin type solar stills, a black liner is usually used, which may have selective coating properties:

- Black plastic sheets durable to higher temperature and to sun's rays such as butyl rubber and black polyethylene
- For cheaper construction, the bottom of the basin is painted black with ordinary or selective black paint or black color is spread into the water. The black paint should not have any volatile material that may distill with water vapor or contaminate the condensate
- Tar impregnated jute, asphalt, is also a cheap absorbing material used often.

  Any suitable black material may be used as a liner. This material should have the following properties:

- Should be waterproof
- Must withstand the operational temperatures up to about 90°C without any deterioration

- absorptivity (a) should be ≥ 0.95
- Should be free of toxic or other harmful substances. Its surface is better to be smooth. Experience showed that a smooth surface is cleaned easily and in addition scale deposits are avoided. Rough surfaces act as crystallization nucleus forming hard scale difficult to be cleaned.

3.2.2.3 The Frames

The frame is better to be constructed from the same material as the basin. Its form should fit perfectly to the glazing. For safety reasons and for smooth operation and maintenance, the following should be taken into consideration:

- The construction material has to be heavy enough in order to withstand extreme weather conditions such as strong winds and sand storms. In regions which suffer from strong storms, the frame is better to be anchored to the ground, keeping the whole system safe
- Some parts have to be easily removed to permit access inside the still for cleaning and maintenance. Fig. 3.2 gives an example of movable still component. Back, glass-covered wall, is opened to clean or to maintain the inside of the still.

3.2.2.4 Glazing Material

Most solar stills are using glass as cover material. This has a longer life than the other components of the system itself and the only disadvantage is its breakability. Normally low iron (0.01% Fe_2O_3) thin glass sheet is the most suitable. Its refraction index is 1.53 and has extinction coefficient $K = 4$ (Kumar, eolss).

Plastic cover material is inferior to glass concerning the efficiency and life time period, although is more flexible permitting wider geometrical choice of still configuration and has expansion coefficient higher than glass. Phadatare and Verma (2009) who studied the behavior of glass and plastic cover in solar stills found that glass cover gives 30–35% more productivity than solar stills with plexiglass. Some transparent plastic material should be treated in the inside surface to become wettable in order to permit condensate to flow free to the gutters. Treated Tedlar (polyvinylchloride) and Mylar (tetraphalate-polyethylene) are some of the commercial plastic sheets used as transparent cover. Their life does not exceed 15 years.

Absorptivity, transmittance, and reflectivity depend on the angle of solar rays striking glazing and water in the basin. Falling onto the glazing's surface at 90°, about 90% of radiation is transmitted inside the still. Table 3.1 presents the percentage reflectance, absorptance and transmittance of the cover, the water in the basin and the basin liner (Cooper, 1973).

Table 3.1 Effect of Angle of Incidence of Solar Radiation on the Absorptivity, The Transmissivity and the Reflectivity of the Stills Components

Angle of incidence, θ, deg.	0	30	45	60
Glass cover				
Absorptivity (α) %	5	5	5	5
Transmissivity (τ) %	90	90	89	85
Reflectivity (ρ) %	5	5	6	10
Water in the basin				
Absorptivity %	30	30	30	30
Transmissivity %	68	68	67	64
Reflectivity %	2	2	3	6
Basin bottom (liner)				
Absorptivity %	95	95	95	95
Transmissivity %	0	0	0	0
Reflectivity %	5	5	5	5

3.2.2.5 Insulation Material

Insulation material is used beneath the basin's bottom in order to reduce as much as possible the heat loss to the ground or to surroundings for inclined stills. The side walls of the basin may also be insulated. Any suitable material can be used having the following characteristic properties:
- Should be waterproof
- Should be easily applicable and easily replaced, if necessary
- To be lightweight but of high strength to support the weight of the still
- To have high thermal resistance R (very low thermal conductivity, λ)
- To withstand high temperature up to $\sim 90°C$ without any deformation
- To fit easily to the bottom, the side walls of the basin and to the edges of the still.

3.2.2.6 The Sealing Material

It is used to seal either one to another glazing or glazing to other components, as to the frame, keeping the still's inside air tide. It takes up expansion between cover and frame or other sealed parts. Sealants should have the following characteristics:
- Easy application and easy replacement
- Very low solar radiation absorptivity
- Should not impact odor or any bad test to distilled water
- Should not be toxic.

Some of the sealants in use are window or black patty for high temperatures, tar plastic, silicone, synthetic rubbers, latex, etc.

3.2.2.7 Auxiliary Components

Auxiliary components comprise piping, fittings, pumps, rain collecting canals, and storage tanks. Pipes and fittings preferably should be from the same material. In use are plastic pipes such as PVC (polyvinyl chloride), black polyethylene, or any other plastic material that does not degrade easily by the sun's rays and the high operating temperatures. Metallic pipes may be used as well but only for the feeding of salt water, eg, galvanized steel, copper, and for product water, stainless steel pipes. Stainless steel has longer life than plastic but in general is more expensive. The auxiliary components should have the following characteristics:

- Auxiliary parts coming in contact with distilled or rain water should have a protective coating
- Internal canals, eg, gutters and pipes, should be constructed in single pieces to avoid joints.

To store the product water any kind of water reservoir is suitable, such as concrete, plastic, and metallic or an existing storage tank in the site may be used.

3.2.3 Rain Catchment Canals

Solar stills are excellent rain catchment areas. The rain falling onto the still's cover, glass, or plastic can be collected in canals attached around the still's base. Only a small additional investment is required. As rainfall coincides with the period of low distillation output production, this may increase by the rain water collected. It is obvious that according to the amount of rain in each region, sufficient storage capacity has to be provided to store the excess of water. A filtration system to retain debris, leaves, and various materials that accumulate on the transparent cover is necessary. There exist many references on rain catchments referring to the percentage of rain collection, eg, Howe (1968a,b) gives a collection of about 80%, Delyannis and Piperoglou (1967b) about 70%, and Lawand (1968, 1975) a range from 90% up to 100% collection. Brace Research Institute (1993) in a Report for individual, family sized desalination systems reports few rules about correct rain collection. The rain water collection during winter time is shown in Fig. 3.4 (Delyannis and Piperoglou, 1967b).

Rainfall collection was in practice in Mediterranean area, especially in dry islands and remote regions, for centuries. They collected rain water into terraces having small inclination. Canals and pipes drained the rain water to the cisterns built underneath the houses in the basement.

3.2.4 Conditions for Proper Installation/Operation of Solar Stills

To install successfully a solar still or an array of stills and to achieve the best operational conditions and proper productivity, some studies are necessary to be made before the erection of the system. These studies are referring mainly to solar distillation plants,

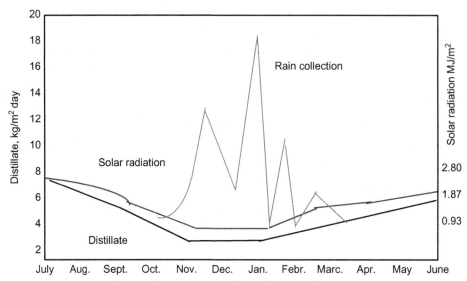

Figure 3.4 Rain water collection during winter time (Delyannis and Piperoglou, 1967b).

ie, for a series of still arrays which are exposed to various environmental conditions, as heavy rains, hail, and strong winds. For such solar distillation plants, the study must comprise:

1. The environmental conditions of installation side. Yearly data of solar radiation intensity, sunshine hours, wind speed, and eventual existence of strong winds or sand winds have to be collected. Hail is destructive not only for plastic but also for glass covers

2. Installation site must be as possible near to the feed water source to avoid unnecessary length of piping. The central storing cistern or storage tank must also be near the site of installation. The site must have a flat smooth surface area without trees, buildings, or any obstacle that can cast shadow on the surface of the stills. A future extension for new still arrays needs to be considered during the design stage

3. The design and/or selection of the type of solar stills. Local material availability is of importance for the selection of the still's type as well as the availability of skilled personnel

4. Monthly productivity must be calculated according to monthly mean meteorological data taking into consideration that, according to the region, at summer time consumption may be increased considerably due to the increase of inhabitants, especially for regions near the sea

5. If in the region rains are available, a study to add external rain collection canals may be helpful in order to increase productivity

6. Maintenance is of importance. Solar stills do not need important maintenance but they have to be watched for smooth operation, ie, to be sure that there is no any glass cover broken and no scale or algae is formed inside the water basin. Broken glasses will decrease efficiency as there will be vapor losses. Broken glass may be happen due to hale or any other activity (eg, playing boys). The basin of the still must be always covered by water to avoid stagnant conditions. If there is salt or algae accumulation inside the basin, it is wise to shut down the operation and clean the black liner as layers of salts decrease considerably the absorptivity of solar radiation. The outside surface of the still must also be cleaned from accumulated dry leaves and other material (Fig. 3.3)

7. Before installing a solar distillation system some cost calculations are necessary. Capital cost, installation, maintenance cost, and cost of installation land, if it not donated free by the community, are the main items to be calculated. In general, operating cost is very low for solar stills or solar plants. From these data, the cost of produced distilled water may be calculated

8. Various other parameters must be taken in account such as social parameters, security of the installation, cost of distilled water transportation, and distribution.

3.3 OPERATION PRINCIPLES OF SOLAR DISTILLATION—SOLAR STILLS

The principle is the warming up of a mass of salty water in the air tide chamber of the still. As the water warms up, currents are creating inside the water mass due to density differences, forming steam in the surface which escapes to the air above. Steam generation takes place in the temperature range of 30°C to about 80°C, the mean range of operating temperatures of a still. Steam generation is a more smooth operation than vaporization near boiling point temperature. Boiling point vaporization may create bubbles that carry some salts or other contaminants to the distillate.

Evaporation of water is an energy consuming process. It takes $\sim 4218\,kJ\,kg^{-1}\,K^{-1}$ to change phase from water to vapor at boiling point (100°C) and $\sim 4194\,kJ\,kg^{-1}\,K^{-1}$ for mean water temperature in the basin of 75°C.

There exist a big variety of solar stills, simple or more complicated, having internal or external parts which help to increase productivity in some extent due to increase of evaporation area. A lot of mathematical models were also presented by various researches. Most of the models are formulated for each special still design, taking in consideration specific design details of the still. In general, solar stills have to be constructed and operated easily at very low price. A very few mathematical models that are applicable to all geometries of solar stills exist. In Africa, the Caribbean Islands,

India, and in some other places, thousands of small, individual solar stills that just produce fresh water from salty well water or seawater exist. The people over there do not take care about efficiency, etc. They construct the stills with very cheap materials and proper construction is done by experience only, without using mathematical models. Their main target is only to have fresh water supply.

3.3.1 The Phase Movement Inside the Still

Fig. 3.5 presents schematically the vapor/air mixture circulation inside the air tide space of the still. By increasing the water temperature in the basin the air-vapor mixture above the basin's water surface has higher temperature and lower density than the air—vapor mixture immediately beneath the glass cover. Thus convection currents are formed between the cover and water surface. The air—vapor mixture moves upwards by the action of buoyancy due to density difference, coming in contact with the cooler inside cover surface where it becomes saturated and partially condenses. The condensate thus formed runs along the covers' inside surface and is collected into the gutters. The process takes place in narrow multiple layers between the two surfaces. The bulk of air mass does not participate due to low diffusion and heat conductance. For this reason, it is advantageous to keep the distance between cover and water surface and the angle between cover and the horizontal as small as possible, but high enough to allow free flow of condensate by gravity, along the cover surface. The larger the temperature difference between water surface and cover the more intense becomes air—vapor mixture circulation. This is an unsteady state operation procedure (Delyannis and Delyannis, 1974).

This type of vapor circulation inside the still applies for passive and for active solar stills, but in the case of continuous feed water circulation, even in small flow rates the whole procedure takes place in low turbulent flow.

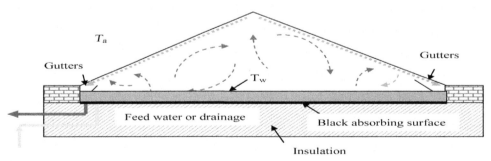

Figure 3.5 Sketch of evaporation/condensation function into the air tide chamber of a single effect solar still.

3.3.2 The Cover Inclination

The transparent cover in a solar still must have a certain inclination to permit the free flow of the condensate from the inside surface of the cover to the gutters. The exact required inclination has not been totally fixed yet. For solar collectors, this inclination has to be about ±1.5 the latitude of the region but for the inclination of solar stills some other parameters count as well. The production of distilled water for a fixed inclination fluctuates according to the latitude and also from the season of the year, as the incident angle of the solar rays has different values each season. Many studies on the correct cover inclination exist but the conclusions on which must be the right inclination are very different and in many cases opposing and conflicting.

The flow distribution and heat transfer inside an asymmetric, greenhouse type, solar still by numerical solutions of the turbulent flow and energy equations for natural convection over a wide range of Rayleigh numbers ($Ra = 10^{-7} - 10^{10}$) has been studied by Papanicolaou et al. (2002). This covers a corresponding range of temperature differences between basin and transparent cover during a 24-h cycle of operation.

The computed streamlines and isotherms, at steady state and for Ra number $= 10^9$, based on the width of the still are presented in Fig. 3.6. The results are for 15° and 25° cover inclination angle and for 24th hour cycle operation. In general, the conclusions are that cover inclination does not affect seriously the performance of the still.

The first who studied the impact of the cover inclination to the productivity of the still is Cooper (1969a). The study refers to inclination angles of 15°, 30°, 45°, and 60°. Tiwari and Tiwari (2005a) studied about the same inclination angles of 15°, 30°, and 45°. They conclude that the best angle for efficient heat conduction

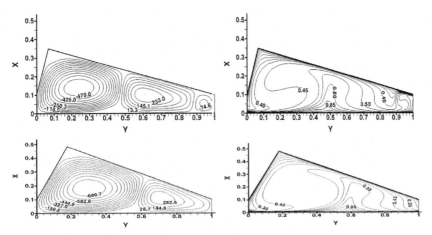

Figure 3.6 Streamlines (left) and isotherms (right) at steady state for $Ra_L = 10^9$ and for cover inclination of 25° (below) and 15° (on the top) (Papanikolaou et al., 2002).

and evaporation is 30°. They point out that inclination angle has an important impact on the coefficient of heat transfer by conduction and on the productivity. The same investigators, Tiwari and Tiwari (2005b), studied the same inclination angles, of the stills cover as above, for Indian climatic conditions. For these environmental conditions, the maximum productivity was achieved for cover inclination of 15°.

Khalifa and Hamood (2009) give a function that resulted from the experimental study of many investigators. The function correlates the productivity of the still to its inclination and to the inclination of incident solar rays:

$$\dot{m}_d = -0.025(\beta)^2 + 0.156(\beta) + 0.843, \quad R^2 = 0.734 \tag{3.2}$$

$$\dot{m}_d = 0.0036(H)^2 + 0.0701(H) + 0.2475, \quad R^2 = 0.762 \tag{3.3}$$

where β is the inclination angle of the transparent cover (°) and H the daily solar radiation in MJ m^{-2}.

Setoodeh et al. (2011) present results of a study for the proper design, the analysis of parameters, and the construction of solar stills. They studied heat transfer coefficients in a conventional solar still by a three-dimensional two panel model for one-stage basin solar still using a computational fluid dynamics technique (CFD). The study concerns steady temperature inside the still and an inclination of the cover of 45°. The whole procedure is a useful tool for solar distillation plant installation.

The following equations are based on the continuity of momentum, energy, and mass transfer conservation principles at steady state conditions (Fig. 3.7).

1. **Continuity flow equations (Coherence equations)**
   1.1 Gas phase:

$$\nabla(r_v \rho_v V_v) + \dot{m}_{lv} = 0 \tag{3.4}$$

   1.2 Liquid phase:

$$\nabla(r_l \rho_l V_l) - \dot{m}_{lv} = 0 \tag{3.5}$$

2. **Momentum equations**
   2.1 Gas phase:

$$\nabla[r_v(\rho_v V_v V_v) = -r_v \nabla p_v + \nabla[r_v \eta_v(\nabla V_v + (\nabla V_v)^T)] + r_v \rho_v g - M_{vl} \tag{3.6}$$

   2.2 Liquid phase:

$$\nabla[r_l(\rho_l V_l V_l)] = -r_l \nabla p_l + \nabla[r_l \eta_l(\nabla V_l + (\nabla V_l)^T)] + r_l \rho_{lv} g - M_{vl} \tag{3.7}$$

where r is the volume ratio of the two phases, V is the velocity vector (m s^{-1}), $\dot{m}_{lv}$ (kg m^{-3} s^{-1}) is the mass transfer rate from liquid to gas phase and vice versa, on air–vapor interface. The term M_{vl} refers to interface momentum transfer (kgm^{-2} s^{-2}) and η_{g-lam}, η_{l-lam} (Pas) is dynamic viscosity of gas and liquid phases at laminar flow.

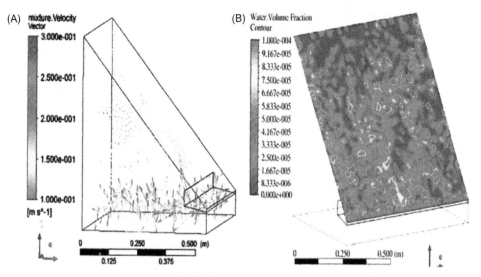

Figure 3.7 (A) Air and vapor movement on a plane inside the solar still. (B) Water volume fraction contour on glass of the cover (Setoodeh et al., 2011).

Subscripts g and l refer to gas (vapor) and liquid phases, respectively. Mass transfer equations must satisfy the local balance, $\dot{m}_{lg} \equiv \dot{m}_{gl}$.

3. **Energy equations**

    3.1 Gas phase:

$$\nabla(r_v \rho_v V_v h_v) = \nabla q + (\dot{Q}_{lv} + \dot{m}_{lv} h_{lv}) \tag{3.8}$$

    3.2 Liquid phase:

$$\nabla(r_l \rho_l V_l h_l) = \nabla q - (\dot{Q}_{lv} + \dot{m}_{lv} h_{lv}) \tag{3.9}$$

4. **Mass transfer equations**

    Mass transfer equations are based on the fraction of the light element A in liquid phase X_A. In gas phase, this fraction is given as Y_A.

    4.1 Gas phase:

$$\nabla[r_v(\rho_v V_v Y_A) - \rho_v D_{Av}(\nabla Y_A)] - \dot{m}_{lv} = 0 \tag{3.10}$$

    4.2 Liquid phase:

$$\nabla[r_l(\rho_l V_l X_A) - \rho_l D_{Al}(\nabla X_A)] + \dot{m}_{lv} = 0 \tag{3.11}$$

where h_v and h_l are enthalpies of vapor and liquid phases, respectively, and h_{vl} is enthalpy at the interface. $\dot{Q}_{lv}$ refers to heat transfer between phases (W m^{-3}) and D is diffusion coefficient (m s^{-2}). Energy equations must fulfill the phase balance, ie, $\dot{Q}_{lv} = -\dot{Q}_{vl}$.

3.3.3 The Optical Behavior of the Transparent Cover

Reflectance–transmittance and absorptance. These parameters concern the optical behavior of the transparent cover that has an important impact on heat balance inside and outside the still and consequently on mass balance inside the still. Reflectance, transmittance, and absorptance are functions of the incident radiation, refraction index, extinction coefficient of the nature of transparent cover and its thickness. When a solar beam strikes a transparent surface a small portion is reflected (ρ_g) at the same incident angle θ_{inc}. A small portion transverse the thickness path of glass cover is absorbed (a_g) by its mass. The rest is transmitted (τ_g) and simultaneously refracted through the glass at an angle θ_{ref} (Fig. 3.8). The sum of these three components is:

$$\rho_g + \alpha_g + \tau_g = 1.0 \tag{3.12}$$

A solar ray striking a smooth transparent surface on angle θ_{inc}, with a refraction index of the surrounding air n_a ($n_a \approx 1.0$), is partially reflected to the environment and its energy is lost. The rest is transmitted through the path length of glass thickness, of refraction index n_g, where partially is absorbed by the mass of the material. The rest of the solar radiation flows through the air–vapor mixture and strikes the water surface at the basin where the same procedure is repeated with the corresponding refraction indices and angles. Solar radiation penetrated into the still's air-vapor space is partially absorbed by the water mass and the rest is transmitted to the liner surface where is absorbed warming further the water meanwhile a small portion is reflected back to the water mass (Duffie and Beckmann, 2006, 2013; Kalogirou 2009, 2014).

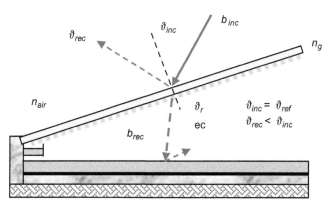

Figure 3.8 Reflection, absorptance and transmittance of a solar ray striking onto an inclined glass surface.

Reflectance (ρ). Each interface (glass−air, air−vapor mixture, and liner−water) is governed by the Fresnel's relation for parallel and perpendicular reflection components of unpolarized radiation rays passing from medium A to medium B, eg, in our case from air to glass cover and from humid air to water. An unpolarized ray, r, striking a surface with an incident angle θ_{inc} and refraction angle θ_{ref} is presented by the parallel and perpendicular components, $r_\parallel$ and $r_\perp$ of reflexion. The corresponding equations are:

$$r_\perp = \frac{\sin^2(\theta_{inc} - \theta_{ref})}{\sin^2(\theta_{ref} + \theta_{inc})} \tag{3.13}$$

for the parallel component, and for the perpendicular is:

$$r_\parallel = \frac{\tan^2(\theta_{inc} - \theta_{ref})}{\tan^2(\theta_{ref} + \theta_{inc})} \tag{3.14}$$

The average reflection of unpolarized radiation is half the sum of the perpendicular and parallel components:

$$\bar{r} = G_{ref}/G_{inc} = 1/2(r_\parallel + r_\perp) \tag{3.15}$$

$$\bar{r} = G_{ref}/G_{inc} = \frac{1}{2}\left[\frac{\sin^2(\theta_{inc} - \theta_{ref})}{\sin^2(\theta_{ref} + \theta_{inc})} + \frac{\tan^2(\theta_{inc} - \theta_{ref})}{\tan^2(\theta_{ref} + \theta_{inc})}\right] \tag{3.16}$$

The relationship between incident and refraction indices is given by Shell's law as:

$$\frac{n_a}{n_g} = \frac{\sin \theta_{ref}}{\sin \theta_{inc}} \tag{3.17}$$

where

$r_\parallel, r_\perp$	Parallel/perpendicular reflectivity component at each interface	−
$\bar{r}$	Combined components	−
$\overline{G}_{inc}, \overline{G}_{ref}$	Intensity of incident and refracted solar radiation, respectively	W m^{-2}
n_a, n_g	Refraction index of surrounding air and glass, respectively	−
$\theta_{inc}, \theta_{ref}$	Incident and refraction angles, respectively	°

In the case where incidence solar radiation is at normal angles, θ_{inc} and θ_{ref} are zero. Combining Eqs. (3.15) and (3.17) results in:

$$r_{(0)} = \frac{G_{ref}}{G_{inc(0)}} = \left(\frac{n_g - n_a}{n_g + n_a}\right)^2 \tag{3.18}$$

where subscript (0) means incident radiation at normal. For the case of air—cover interface ($n_{air} = \sim1$), and incidence at normal applies:

$$r_{(0)} = \left(\frac{n_g - 1}{n_g + 1}\right) \tag{3.19}$$

For example, for glass of refraction index $n_{(0)} = 1.53$ at normal incidence of unpolarized radiation, the reflection component is:

$$r_{(0)} = \left(\frac{1.53 - 1.0}{1.53 + 1.0}\right)^2 = \left(\frac{0.53}{2.53}\right)^2 = 0.04383$$

Table 3.2 presents the average refraction index of some transparent materials used in solar stills and/or solar collectors (Duffie and Beckmann, 2006, 2013).

The absorptivity (α) of solar radiation at the glass cover is controlled by the extinction coefficient K expressed by the initial (G_{inc}) and the final (G_{fin}) radiation intensity and is described by the Bouguer's law, based on the assumption that absorbed fraction of incident radiation passing through the transparent cover is proportional to the local intensity G in the medium and the traveled distance x:

$$dG = -GKdx \tag{3.20a}$$

$$G_{tras} = G_{inc}e^{Kx} \tag{3.20b}$$

where K the extinction coefficient, a proportional constant, is assumed to be constant in the solar spectrum.

The absorbed radiation by the cover material is the lost portion of the incident on the transparent cover radiation. Taking into account only absorption losses, the transmission through the glass cover of path length l (m) may be given as:

$$\tau_\alpha = \exp(-Kl) \tag{3.21}$$

As path length l extends from zero to $l/\cos\theta_{ref}$, Eq. (3.21) becomes:

$$\tau_a = \frac{G_{tras}}{G_{inc}}\exp\left(-\frac{Kl}{\cos\theta_{ref}}\right) \tag{3.22}$$

Table 3.2 Average Refractive Index n of Some Transparent Cover Materials

Transparent Material for Covers	$\bar{n}$
Glass	1.526
Polymethyl methacrylate	1.49
Polyvinylchloride (Tedlar)	1.45
Polyfluorinated-ethylene propylene	1.34
Polyethane fluoroethylene	1.37

or in terms of absorptivity α, which is more common in solar distillation practice, is:

$$a = 1 - e^{-Kl} \tag{3.23}$$

The same is applied for the water mass in the basin, where absorption depends as well on the path length of solar beam transmitting the water mass.

Transmittance—Transmissivity (τ) of the glass cover is a function of the amount of reflected and absorbed incident global solar radiation G_{inc}. In the same way as for reflectance, transmittance is calculated from the parallel and perpendicular components of reflectance:

$$\tau_{\parallel} = (1-r_{\parallel})^2 \sum_{n=0}^{\infty} r_{\parallel}^{2n} = \frac{(1-r_{\parallel})^2}{1-r_{\parallel}^2} = \frac{1-r_{\parallel}}{1+r_{\parallel}} \tag{3.24}$$

$$\tau_{\perp} = (1-r_{\perp})^2 \sum_{n=0}^{\infty} r_{\perp}^{2n} = \frac{(1-r_{\perp})^2}{1-r_{\perp}^2} = \frac{1-r_{\perp}}{1+r_{\perp}} \tag{3.25}$$

and the combined components are given by the equation:

$$\tau_r = \frac{1}{2}\left(\frac{1-r_{\parallel}}{1+r_{\parallel}} + \frac{1-r_{\perp}}{1+r_{\perp}}\right) \tag{3.26}$$

These transmittance equations derived from multiple absorption/transmittance through the thickness of the glass cover and are valid for no absorption (or with minimum absorption) glass covers. In the case of more glass covers (n parallel curves) and of the same glass material, Eq. (3.26) transmittance can be expressed as:

$$\tau_{r-n} = \frac{1}{2}\left(\frac{1-r_{\parallel}}{1+(2n-1)r_{\parallel}} + \frac{1-r_{\perp}}{1+(2n-1)r_{\perp}}\right) \tag{3.27}$$

Thus for a solar still of two glass covers, which is a usual case in solar distillation:

$$\tau_{r-2} = \frac{1}{2}\left(\frac{1-r_{\parallel}}{1+3r_{\parallel}} + \frac{1-r_{\perp}}{1+3r_{\perp}}\right) \tag{3.28}$$

All the above equations are formulated for flat plate collectors and are accepted for the solar stills transparent covers as well. Nevertheless, some small differences between them exist. In solar collectors, the air tide space between the inside cover surface and the absorbing black surface is filled by stagnant dry air. In a solar still, this space consists of humid air in natural movement and in addition in the inside glass surface a thin water layer of flowing condensate is formed reducing reflectance and simultaneously increasing the transmittance.

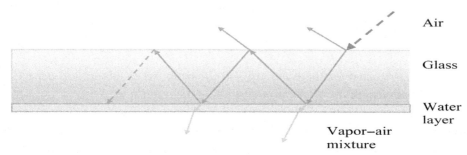

Figure 3.9 Multiple reflexion/absorption pattern for glass cover of a solar still according to Cooper (1969a).

No studies have been traced for calculating transmittance in multiple reflections/absorptions except a study by Cooper (1969a) who gives an equation for glass cover and condensate thin membrane layer (Fig. 3.9), but the complex form of the function and the large number of variables made impossible the prediction of the mean effective absorptance.

Cooper (1969a) presents also an equation to predict the film thickness at a point of the surface x meters from the still's apex:

$$s = \sqrt[3]{\frac{\eta \dot{m}_c x}{g \rho^2 \sin \beta}} \qquad (3.29)$$

where

β	Inclination of still's cover	$^{\circ}$
g	Gravitational constant	$\mathrm{m\,s^{-2}}$
η	Dynamic viscosity of condensate	Pas
$\dot{m}_c$	Vapor condensate flow rate	$\mathrm{kg\,m^{-2}\,s^{-1}}$
ρ	Density of flowing condensate	$\mathrm{kg\,m^{-3}}$
s	Thickness of condensate film at distance x from apex	m
x	Distance of the point from apex	m

It is assumed that no radiation is absorbed. Cooper (1969a) also gives an example to show that condensate film absorptivity is negligible. For a cover inclination of 15°, a peak distillation rate (or condensation rate) of $0.14\,\mathrm{L\,m^{-2}\,s^{-1}}$ at operation temperature of $\sim 38°\mathrm{C}$ and at a point $x \approx 0.61$ m from apex the film thickness is 0.02 mm and hence its absorptivity is $\sim 1\%$.

Solar rays transmitted through the glass cover become partially polarized. In the case of polarization conditions, for the parallel and perpendicular components, absorptance, reflectance, and transmittance can be determined by ray tracing technique and are given by the following equations:

$$\tau_\perp = \frac{\tau_{ab}(1-r_\perp)^2}{1-(r_\perp\tau_{ab})^2} = \tau_{ab}\frac{1-r_\perp}{1+r_\perp}\left[\frac{1-r_\perp^2}{1-(r_\perp\tau_{ab})^2}\right] \tag{3.30}$$

$$\tau_\| = \frac{\tau_{ab}(1-r_\|)^2}{1-(r_\|\tau_{ab})^2} = \tau_{ab}\frac{1-r_\|}{1+r_\|}\left[\frac{1-r_\|^2}{1-(r_\|\tau_{ab})^2}\right] \tag{3.31}$$

$$\rho_\perp = r_\perp + r_\perp\frac{\tau_{ab}^2(1-r_\perp)^2}{1-(r_\perp\tau_{ab})^2} = r_\perp(1+\tau_{ab}\tau_\perp) \tag{3.32}$$

$$a_\perp = (1-\tau_{ab})\left(\frac{1-r_\perp}{1-r_\perp\tau_{ab}}\right) \tag{3.33}$$

$$\tau \cong (\tau_{ab}\tau_{ref}) \tag{3.34}$$

Similar equations are valid for the parallel components. Eq. (3.17) can be simplified giving the simple product, ie, the product of transmittance due only to reflection losses τ_{ref} and of transmittance due only to absorptance losses τ_{ab}. The components r are of the 0.1 order and τ_{ab} in the order of ~0.9 or more. Thus Eqs. (3.32) and (3.33) can be simplified taking in consideration that $\rho = 1-\alpha-\tau$:

$$\rho \cong \tau_{ab}(1-\tau_{ref}) = \tau_{ab}-\tau \tag{3.35}$$

$$\alpha = (1-\tau_{ab}) \tag{3.36}$$

The above equations are formulated mainly for glass covers. Edlin and Willauer (1961) present reflectance and transmittance of tedlar plastic film, wettable and dry, which were used for transparent cover in solar stills. Grange (1966) presents as well some reflectance values for wettable tedlar plastic covers.

3.3.4 Thermal Behavior of Solar Stills

The productivity of a solar still is affected by various parameters such as the solar radiation intensity, environmental temperature, and wind velocity outside the transparent cover. These parameters are uncontrollable and cannot be altered. Other parameters, such as temperature differences between water surface and transparent inside cover surface ΔT_{w-g}, temperature differences between water surface and black liner surface ΔT_{w-bl}, cover inclination angle β, feed water inlet temperature T_{fw} and water depth in the basin, which are important parameters, may be controlled or changed accordingly.

Simultaneously all these parameters affect the heat flow rate and losses which in turn, partially, depend on the geometry and the construction of the still.

It has been referred previously that solar stills are partly semiempirical constructions. In order to improve efficiency and achieve the best economic construction and operating conditions, numerous studies have been conducted, mostly theoretical. Mathematical models have been formulated, mainly for the special geometry of each type of solar still. Very few general models applied to all solar stills exist. It has to be noted that solar distillation equipment must be simple in construction, installation, and operation. Tiwari et al. (2003a) concluded that *"only the simple, double slope FRP (fiber-reinforced plastic) conventional solar still is the most economical solar still to provide cheap drinking water."* There exist many complicated still geometries for productivity improvement but capital cost may be more expensive than simple solar stills and increased productivity may not balance the economics of the system.

Here we shall try to describe to some extent the developed geometries and models of various solar stills and their basic mode of operation.

3.3.4.1 Heat Balance—Heat Transfer—Heat Transfer Coefficients

As incident solar radiation G_g (W m^{-2}), ie, direct and diffuse, falls onto the transparent cover (Fig. 3.10) and transverse the covers thickness path, a portion of it ($G_{\alpha g}=q_{abs}$) is absorbed by the cover and the rest ($G_{\tau g}=q_{pen}$) penetrates into the air tide space of the still. The water system absorbs energy equal to $G_{\alpha w}\tau_g=q_{wab}$ and warms up. At temperature T_u, small portions of the water heat are reflected from the water surface back to the cover by radiation q_{rwg}, by convection q_{cwg}, and by evaporation/condensation currents q_{ewg}. At the same time, heat fluxes are created from the cover surface to the ambient by radiation, q_{rga}, and by convection, q_{cga}, which are in fact an energy loss. Despite the insulation, there exist some heat loss from the bottom, q_b, and the sides of the still q_{sd}. Vapor leakage to the ambient q_{leak} may occur in non air tide stills. Some energy is also lost, as sensible heat, with the distillate and brine blow down (q_{shb}).

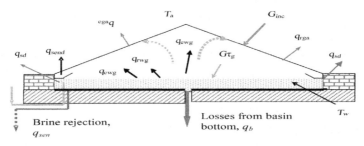

Figure 3.10 Schematic of heat fluxes in a conventional symmetric greenhouse-type solar still.

For air tide conventional stills constructed correctly, the typical heat distribution based on mean daily solar radiation of 22.61 MJ, ie, for summer conditions, and as result of mean value of many investigators is reported by Delyannis and Delyannis (1974).

Evaporation	40%	Internal convection	5%
Absorbed radiation by cover	5%	Edge losses	5%
Reflected radiation by cover	10%	Bottom losses	5%
Reflected radiation by liner	5%	Vapor leakage	5%
Reflected radiation by brine	15%	Miscellaneous	5%

The above values are indicative and may fluctuate according to still design, construction details, proper operation, and type of operation (eg, shallow or deep basin, passive or active) of solar radiation intensity and the time of the year.

Heat Balance

For heat balance calculations, the still may be divided into two sections: cover-distillate and brine/liner. In an idealized system having no leakage and operating without temperature gradients in glass cover and water, the energy balance is given by the following equations, presented initially by Cooper (1969b).

For the transparent cover the energy balance is:

$$(Q_{rga} + Q_{cga})A_g + C_g A_g \frac{dT_g}{dt} = (Q_{rwg} + Q_{cwg} + Q_{ewg})A_w + Ia_g A_g \qquad (3.37)$$

and for the water system in the basin the energy balance is:

$$(Q_{rwg} + Q_{cwg} + Q_{ewg})A_w + Q_b A_w + 2Q_{sd}A_{sd} + C_w A_w \frac{dT_w}{dt} = Ia_w A_w \qquad (3.38)$$

where

A_g, A_w, A_{sd}	Glass cover, water, and side surfaces	m^2
a_g, a_w	Absorptivity of glass cover and water, respectively	—
C_g, C_w	Glass and water heat capacity per unit surface area, respectively	$\mathrm{J\,m^{-2}\,K^{-1}}$
I	Incident solar radiation per hour	$\mathrm{J\,m^{-2}\,h^{-1}}$
Q_{rwg}, Q_{cwg}, Q_{ewg}	Heat flow between water in the basin and the cover by radiation, convection, and evaporation, respectively	$\mathrm{J\,m^{-2}\,h^{-1}}$
Q_{wgb}, Q_{rga}	Heat flow between cover and the environment by convection and radiation, respectively (heat losses from top)	$\mathrm{J\,m^{-2}\,h^{-1}}$
Q_{sd}, Q_b	Heat flow (heat loss) from the side walls and the bottom, respectively	$\mathrm{J\,m^{-2}\,h^{-1}}$
T_g, T_w	Cover and water temperature, respectively	K
t	Time interval	h

Many other investigators present small modifications of the above primary equations (3.37) and (3.38) based mainly on other heat transfer parameters. We shall not analyze all existing models but according to our opinion we shall describe the most simple and/or practical ones. Sodha et al. (1980) and Kumar (2000) present simplified expressions for the fraction of incident solar radiation absorbed by the cover, water mass, and basin black absorber. The water level in the basin is assumed to be stable. They assume that the global amount of energy available in the cover is the sum of energy absorbed by the cover and the reflected energy from the water in the basin by radiation and convection. They present the following equations for the cover, the water in the basin, and the black liner, respectively:

$$C_g \frac{dT_g}{dt} = \tau_1 G + (q_{rwg} + q_{cwg} + q_{ewg}) \frac{A_w}{A_g} - (q_{rga} + q_{cga}) \qquad (3.39)$$

$$C_w \frac{dT_w}{dt} = \tau_2 G + q_w - (q_{rwg} + q_{cwg} + q_{ewg}) - \frac{A_{sd}}{A_w} q_{sd} \qquad (3.40)$$

$$q_w + q_b = \tau_3 G \qquad (3.41)$$

If in the above equations, the term q is replaced by the corresponding term $h (T_2 - T_1)$, the new equations are expressed as:

$$C_g \frac{dT_g}{dt} = \tau_1 G + U_{wg}(T_w - T_g) \frac{A_w}{A_g} - U_{ga}(T_g - T_{sky}) \qquad (3.42)$$

$$C_w \frac{dT_w}{dt} = \tau_2 G + U_{bl}(T_{bl} - T_w) - U_{wg}(T_w - T_g) - \frac{A_{sd}}{A_w} h_{sd}(T_{bl} - T_a) \qquad (3.43)$$

$$\tau_3 G_{bl} = h_{wbl}(T_{bl} - T_w) + U_{bl}(T_{bl} - T_a) \qquad (3.44)$$

where

A_{bl}, A_c, A_w	Absorptivity of black liner, cover, and water, respectively	—
G	Incident solar radiation on top of transparent cover	$W\,m^{-2}$
h_{blw}	Heat transfer coefficient from black liner to water	$W\,m^{-2}\,K^{-1}$
h_{wbl}	Heat transfer coefficient from water to black liner	$W\,m^{-2}\,K^{-1}$
h_{sd}	Heat transfer coefficient from the sides of the still	$W\,m^{-2}\,K^{-1}$
$q_{rwg}, q_{cwg}, q_{ewg}$	Heat fluxes between basin water and glass cover by radiation, convection, and evaporation, respectively	$W\,m^{-2}$
q_{cga}, q_{rga}	Heat fluxes (heat loss) between glass cover and the ambient by convection and radiation, respectively	$W\,m^{-2}$
q_{sd}, q_b	Heat fluxes (heat loss) from the side walls and bottom, respectively	$W\,m^{-2}$
ρ_g	Reflectivity of glass cover	—
T_g, T_w, T_{bl}	Temperature of glass, water mass surface, and black liner, respectively	K
t	Time interval	s

τ_1, τ_2, τ_3	Refer to absorption and reflection terms, $\tau_1 = (1 - \rho_y)a_c$,	—
	$\tau_2 = (1 - \rho_g)(1 - \alpha_c)\alpha_w$ and $\tau_3 = (1 - \rho_g)(1 - \alpha_c)(1 - \alpha_w)\,\alpha_{bl}$	
U_{ug}	Overall heat transfer coefficient into the air tide still's chamber	$\mathrm{W\,m^{-2}\,K^{-1}}$
	from water to glass $= h_{rwg} + h_{cwg} + h_{ewg}$	
U_{ga}	Overall heat transfer coefficient from the inner side of glass	$\mathrm{W\,m^{-2}\,K^{-1}}$
	cover to the ambient $= h_{cga} + h_{rga}$	
U_{bl}	Overall heat transfer coefficient from the basin bottom to the	$\mathrm{W\,m^{-2}\,K^{-1}}$
	environment (looses)	

Of importance are the terms $C_g(dT_g/dt)$ and $C_w(dT_w/dt)$ only when heat capacities of transparent material or water are large. For the cover these terms can be omitted and for basin water only for deep basin solar stills is of importance.

Tiwari and Singh (eolls) and Tiwari (2002) present a slightly different mode, similar to the previous equations, based on fraction of solar absorption and refraction, of the heat balance equations for the various components of the still. They are based on the absorption equations of solar radiation of the glass cover, ie, $\rho_g G(t)$ and $a_g(G - \rho_g)G(t)$.

For the glass cover:

$$a_g' G(t) + (q_{rwg} + q_{cwg} + q_{ewg}) = q_{rgw} + q_{cga} \tag{3.45}$$

For the water in the basin:

$$a_w' G(t) + q_{bl} = (m_w c_{pw})\frac{dT_w}{dt} + (q_{rwg} + q_{cwg} + q_{ewg}) \tag{3.46}$$

and for the liner in the basin:

$$a_{bl}' G(t) = q_{bl} + \left[q_{cfa} + q_{sd}\left(\frac{A_{sd}}{A_{bl}}\right) \right] \tag{3.47}$$

where

A_{ss}	Surface area of the solar still	$\mathrm{m^2}$
A_{sd}	Surface area of the still's sides (of water column)	$\mathrm{m^2}$
a_g	Absorptivity of glass cover	—
a_{bl}	Absorptivity of the black liner in the basin	—
a_w	Absorptivity of the basin water	—
$a_g' =$	$a_g(1 - \rho_g)$	—
$a_w' =$	$a_w(1 - a_g)(1 - \rho_g)(1 - \rho_w)$	—
$a_{bl}' =$	$a_{bl}(1 - a_w)(1 - \rho_w)(1 - a_g)(1 - \rho_g)$	—
c_{pw}	Specific enthalpy of water	$\mathrm{J\,kg^{-1}\,K^{-1}}$
m_w	Water mass per unit surface	$\mathrm{kg\,m^{-2}}$
q_{bu}, q_{ba}, q_{sa}	Heat fluxes between basin and water, radiative and convective heat	$\mathrm{W\,m^{-2}}$
	fluxes between bottom and ambient, and radiative and convective	
	heat fluxes from the sides, respectively	

A detailed analysis of the above heat balance is given by the same authors (Tiwari and Singh, eolss).

Heat Transfer Rate Inside the Still

Heat transfer rate inside the still by convection and radiation has been expressed by the use of various parameters:

A. Convective heat transfer rate between water surface and inside glass surface is given by the heat transfer coefficient and temperature difference between the two surfaces as:

$$q_{cwg} = h_{cwg}(T_w - T_{gi}) \tag{3.48}$$

Dunkle (1961) first presented internal convective evaporative and radiative heat transfer rate equations as functions of vapor pressure. The equations, modified by Cooper and Read (1974) for SI, are:

$$q_{cwg} = 0.884\left[(T_w - T_{gi}) + \frac{p_w - p_{gi}}{268 \times 10^3 - p_w}T_w\right]^{1/3}(T_w - T_{gi}) = h_{cwg}(T_w - T_{gi}) \tag{3.49}$$

From Eq. (3.49), the convective heat transfer coefficient is calculated as:

$$h_{cwg} = 0.884\left[(T_w - T_{gi}) + \frac{p_w - p_{gi}}{2016 - p_w}\right]^{1/3} \tag{3.50}$$

Convection heat transfer coefficient h_{cwg}, between water surface and glass, for solar stills, is calculated from the dimensionless Nu number by the equation:

$$h_{cwg} = Nu\lambda_{ha}/l_{sp} = C(GrPr)_n \tag{3.51}$$

where the dimensionless Grassof's number for solar stills may be calculated as:

$$Gr = \Delta T(g\beta\rho^2 l^3/\eta^2) \tag{3.52}$$

B. Radiative heat transfer rate from water surface to inside cover surface may be expressed by the equation of radiation:

$$q_{rwg} = \sigma\varepsilon\left(T_w^4 - T_{gi}^4\right) \tag{3.53}$$

which can be also expressed by the equivalent radiative heat transfer coefficient, h_{rwg}:

$$q_{rwg} = h_{rwg}(T_w - T_{gi}) \tag{3.54a}$$

The radiative heat transfer coefficient in solar collectors from the outside transparent cover to the ambient is expressed by the relationship between a flat surface, of temperature T, and the sky, of temperature T_{sky}. It can be calculated from Eq. (3.54a) by replacing the corresponding temperatures:

$$q_{rwg} = \sigma\varepsilon\left(T_w^4 - T_{sky}^4\right) \tag{3.54b}$$

Combining Eqs. (3.53) and (3.54a), the following radiative heat transfers coefficient results:

$$h_{rgw} = \sigma\varepsilon(T_w^2 - T_{gi}^2)(T_w - T_{gi}) \tag{3.55a}$$

This equation for solar collectors was modified for solar stills (Sharma and Mullic, 1993) by introducing the sky temperature instead of ambient, as more realistic, especially for night operation conditions. Radiative heat transfer coefficient is then presented by:

$$h_{rgw} = \sigma\varepsilon\left(T_w^2 - T_{sky}^4\right)(T_w - T_{sky}) \tag{3.55b}$$

C. **Evaporative heat transfer rate** is the parameter that governs distillate output. It is expressed by the modified Dunkle (1961) equation for SI (Cooper and Read, 1974) as:

$$q_{ewg} = 16.276 \times 10^{-3}\frac{(p_w - p_{gi})}{(T_w - T_{gi})}q_{cwg} \tag{3.56}$$

$$q_{rwg} = 16.276 \times 10^{-3}(p_w - p_{gi})h_{cwg} \tag{3.57}$$

Evaporative heat transfer coefficient from water surface in the basin to the inside surface of the cover is given as (Sharma and Mullic, 1991, 1993):

$$h_{ewg} = \frac{9.15 \times 10^{-7}h_{cwg}(p_w - p_{gi})h_{ev}}{(T_w - T_{gi})} \tag{3.58}$$

Evaporative mass transfer is given as:

$$\dot{m}_{ev} = q_{ewg}/h_{ev} \tag{3.59}$$

The symbols in the above equations are

β	Expansion coefficient of wet air inside the still	K^{-1}
C	Constant factor of Eq. (3.51) calculated from experimental data	—
h_{cwg}	Convective heat transfer coefficient between water surface and inside cover surface	$W\,m^{-2}\,K^{-1}$
h_{ewg}	Evaporative heat transfer coefficient between water surface and inside cover surface	$W\,m^{-2}\,K^{-1}$
h_{rwg}	Radiative heat transfer coefficient between water surface and inside cover surface	$W\,m^{-2}\,K^{-1}$
h_{ev}	Enthalpy of evaporation	$J\,kg^{-1}$
Gr	Grashof number	—
L_{sp}	Average height between water surface and glass cover	m
l	Characteristic dimension (length, diameter, etc.)	m

$\dot{m}_e$	Evaporation mass flow rate	kg m^{-2} s^{-1}
η	Dynamic viscosity	Pas
Nu	Nusselt number	–
p_{gi}	Vapor pressure in the inner surface of glass cover	Pa
p_w	Vapor pressure at water surface	Pa
Pr	Prandl number	–
T_{gi}	Inner glass surface temperature	K
T_w	Water surface temperature	K
ε_w	Water emissivity	–
λ_{ha}	Thermal conductivity of humid air	W m^{-1} K^{-1}
σ	Stefan–Boltzmann constant = 5.67032×10^{-8}	W m^{-2} K^{-4}

D. Heat transfer outside the cover—The wind movement. The coefficient of convective heat transfer outside the still is an important factor for convective heat losses from the glass cover to the environment, as is affected by the wind movement outside the cover. This term has been pointed out first by Jurges (1924), who gave the following expression to calculate the coefficient h_{wd}:

$$h_{wd} = 5.7 + 3.8w \tag{3.60}$$

where w is wind velocity (m s^{-1}). The above equation is valid for wind velocities in the range $0 < w < 5$ m s^{-1} and includes convective and radiative heat transfers. Rowly et al. (1931), later on, excluded radiative heat transfer from Eq. (3.60) and presented an equation for only convective heat transfer:

$$h_{wd} = 2.8 + 3.0w \tag{3.61}$$

which is valid for wind velocities in the range of $0 < w < 7.0$ m s^{-1}. Watmuff et al. (1977) presents the plot of heat transfer coefficient versus wind speed from various sources (Fig. 3.11).

Using sky temperature, instead of ambient, coefficient calculations give more realistic approach for the radiative and convective heat losses, especially during night operation of the still. The convective heat transfer coefficient from outside surface of glass cover to the ambient is:

$$h_{cga} = h_{wd} \frac{(T_{go} - T_a)}{(T_{go} - T_{sky})} \tag{3.62}$$

Swinbank (1963) presented for solar collectors the sky temperature as:

$$T_{sky} = 0.05527^{1.5} \tag{3.63}$$

where T_{go} refers to temperature of the covers outside surface and h_{wd} refers to convective heat transfer coefficient due to wind movement over the outside surface of the transparent cover.

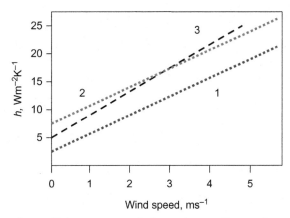

Figure 3.11 Heat transfer coefficients versus wind speed according to the equations: (1) (Jurges, Eq. (3.64)), (2) Duffie—Beckman (Eq. (3.65)), and (3) Rowly and Eckley (1932) (including radiation) (Watmurff et al., 1977).

Wind speed has an important impact on convective heat loss. Wind currents outside the still have a velocity typically ranging from 1.5 to 9.0 m s^{-1}. Morse and Read (1968) give an example of heat transfer coefficient increase due to wind velocity:

Wind velocity w (m s^{-1})	2.235	4.470	8.94
Corresponding convective heat transfer coefficient h_{cga} (W m^{-2} K^{-1})	14.75	23.25	40.84

Thus losses are increasing by increasing wind velocity, decreasing simultaneously the output of the still.

For radiative heat transfer loss outside the transparent cover, the following equations are valid:

$$h_{rga} = \varepsilon_g \sigma \left(T_g^2 + T_{sky}^2 \right) (T_g - T_{sky}) \tag{3.64}$$

$$q_{rga} = \varepsilon \sigma \left(T_g^4 - T_{sky}^4 \right) \tag{3.65}$$

The effect of wind speed w on the daily productivity of various active and passive solar stills was investigated, by computer simulation, by El-Sebaii (2004). He gives some interesting conclusions:

i. For the active solar stills, the basin type, the wick and the vertical solar stills the productivity increase by increasing wind velocity up to a typical value w_c, probably because for this type of solar stills night productivity is zero. Windy regions are suitable for this type of still

ii. For multieffect, horizontal, passive solar stills daily productivity m_d increase with increasing wind velocity up to the value of w_c, because the upper basin protects the lower part from heat loss

Table 3.3 Proposed Correlations for the Investigated Solar Stills

System	Season	α	b	c	Correlation	RC
(A)	Summer	8.87	− 0.037	− 0.045	$m_d = 8.87 m_w^{-0.037} w^{-0.045}$	0.977
	Winter	4.73	− 0.078	− 0.049	$m_d = 4.73 m_w^{-0.078} w^{-0.049}$	0.966
(B)	Summer	8.97	− 0.036	− 0.046	$m_d = 8.97 m_w^{-0.036} w^{-0.046}$	0.966
	Winter	3.67	− 0.085	− 0.041	$m_d = 3.67 m_w^{-0.086} w^{-0.046}$	0.968

iii. For passive single effect stills, a critical depth of water mass exists in the still's basin beyond which productivity decreases as wind velocity increases up to w_c. The critical depth of the single effect passive solar still was found to be 4.5 cm

iv. The typical value of w_c is independent of the geometric shape of the still, the mode of operation (passive or active), and the heat capacity of the brine. Its value on a typical summer or winter day found to be 8 and $10\ \mathrm{m\,s^{-1}}$, respectively.

El-Sebaii (2004) correlates the productivity m_m of the investigated solar stills for various capacities of basin water and refers that:

1. For passive stills the productivity is:

$$m_d = a m_w^b w^c \tag{3.66}$$

2. For active solar stills, the productivity is given either for constant water flow rate $\dot{m}_d$ or for constant height d_w of the water in the basin:

$$m_d = a d_w^b w^c \quad \dot{m}_w = \text{constant} \tag{3.67}$$

$$m_d = a m_w^b w^c \quad d_w = \text{constant} \tag{3.68}$$

where factor α and the exponents b and c are determined from numerical results using the least square method. As an example, the corresponding correlations and the regression coefficients (RC), as are proposed by El-Sebaii for a single basin asymmetric (single slope) solar still (A) and a double, asymmetric solar still (B), are shown in Table 3.3.

Overall Heat Transfer Coefficients

Sharma and Mullic (1993) present a mathematical model for instantaneous water thermal balance based on overall heat transfer coefficients considering the optical efficiency of the cover material η_o as well:

$$\frac{C_w dT_w}{A_w dt} = \eta_o G - U_{wa}(T_w - T_{sky}) - U_{bs}(T_w - T_a) \tag{3.69a}$$

The above equation was modified by Voropoulos (2003), who introduced the incident angle coefficient $k_p(t)$:

$$\frac{C_w dT_w}{A_w dt} = k_p(t)\eta_o G(t) - U_{wa}\left[T_w(t) - T_{sky}(t)\right] - U_{bs}[T_w(t) - T_a(t)] \qquad (3.69b)$$

where

A_w	Water surface area in the basin	m^{-2}
C_w	Heat capacity of water in the basin	$J\,K^{-1}$
G	Instantaneous solar radiation intensity	$W\,m^{-2}$
n_o	Optical efficiency of the cover material	–
$k_p(t)$	Incidence angle coefficient of solar radiation	–
T_a	Ambient temperature	K
T_{sky}	Equivalent black body sky temperature (given by Swinbank (1963) as $T_{sky} = 0.05527\,T_a^{1.5}$)	K
T_w	Temperature of water in the basin	K
t	Time interval	s
U_{wa}	Overall upward heat transfer coefficient factor between water and ambient	$W\,m^{-2}\,K^{-1}$
U_{bs}	Overall heat transfer coefficient through bottom and side walls of the solar still	$W\,m^{-2}\,K^{-1}$

The overall heat transfer coefficient U_{wa} between water in the basin and ambient is given as:

$$U_{wa}\left[\frac{1}{U_{wg}} - \frac{A_w}{A_g U_{ga}}\right]^{-1} = \left[\frac{1}{U_{wg}} + \frac{1}{A_r U_{ga}}\right]^{-1} \qquad (3.70)$$

where A_r is the ratio of glass to water surface area, $A_r = A_g/A_w$.

In Eq. (3.70), the overall heat transfer coefficients between water and glass cover, U_{wg}, and between glass cover and the ambient, U_{ga}, can be calculated by the following equations:

$$U_{wg} = h_{ewg} + h_{rwg} + h_{cwg} \qquad (3.71)$$

$$U_{ga} = h_{cga} + h_{rga} \qquad (3.72)$$

where radiative, evaporative, and convective heat transfer coefficients are presented in Eqs. (3.51), (3.55a), and (3.58).

3.3.4.2 Efficiency, Output, and Performance of Solar Stills
The efficiency of a single effect solar still η is given by the energy utilized for the water evaporation, at a time interval Δt, divided by the amount of incident

solar radiation collected at the same time interval. For example, mean daily efficiency is:

$$\overline{\eta}_d = \frac{\overline{Q}_{ev}}{\sum\limits_{day} G} = \frac{\overline{Q}_{ev}}{H} \tag{3.73}$$

Tiwari (2002) presents the following expression for the instantaneous efficiency η_{ins}:

$$\eta_{ins} = \frac{h_{ev}\dot{m}_{con}}{A_w G(t)} = \frac{\dot{q}_{ev}}{G(t)} = (a\tau)_w - U_{bs}\frac{(T_w - T_a)}{G(t)} \tag{3.74}$$

where in the above equation the right term $(a\tau)_w - U_{bs}[(T_w - T)/G(t)]$ represents the heat losses and U_{bs} is the overall heat transfer coefficient from water to the ambient.

Instantaneous efficiency n_{ins} plotted against the temperature term $(T_w - T_a)/G$ gives a straight line having as the slope is equal to $-U_{wa}$ and an intercept equal to $(\alpha\tau)_w$.

Eq. (3.74) can be transformed to (Tiwari, 2002; Tiwari et al., 2003a):

$$\eta_{ins} = F(a\tau)_w - U_L[\exp(-\beta\Delta t)]\frac{(T_w - T_a)}{G(t)} \tag{3.75}$$

The overall thermal efficiency of a still unit for passive and active modes of operation is expressed mathematically, respectively, as (Tiwari and Singh, eolss):

$$\eta_{pas} = \frac{\sum \dot{m}_w h_{ev}}{A_w \int G(t)dt} \times 100 \tag{3.76}$$

$$\eta_{act} = \frac{\sum \dot{m}_w h_{ev}}{A_w \int G(t)dt + nA_c \int G'(t)dt} \tag{3.77a}$$

where

$(\alpha\tau)_w$	Absorptivity—transmissivity product of water	—
β	$= U_L/(mc_p)_w$	—
Δt	Time interval	s
h_{ev}	Enthalpy of water evaporation at mean daily temperature	J kg
F	Solar still efficiency factor, similar to F_R for collectors	—
H_d	Diurnal solar radiation for still operation	MJ m^{-2}
G, G'	Solar radiation intensity on glass cover of the still and on the glass cover of the collector	W m^{-2}
$\overline{m}$	Mean water mass per unit area	
m_w	Mass of water per unit	kg m^{-2}
n	Number of collectors in the collector field	—
$\dot{q}_{ev}$	Rate of evaporation heat	W m^{-2}
Q_{ev}	Total diurnal energy used for the amount of evaporated water	J m^{-2}
U_L	Overall heat loss coefficient	W m^{-2} K^{-1}

For an operating temperature of 73°C, which can be considered as the mean temperature for solar distillation, and mean evaporation enthalpy of 2326 kJ kg^{-1}, the efficiency may be calculated from the following equation (Cooper, 1973):

$$\eta = \frac{2326m}{3.6 \times 10^3 \int_{t_1}^{t_2} G dt} \qquad (3.77b)$$

Total daily distillate, from sunrise to sunset, for an ideal solar still, may be calculated for any intermediate time, from t_1 up to t_2 (Cooper, 1973) by:

$$m_d = 3.6 \times 10^3 \int_1^2 m_{ev} dt \qquad (3.78)$$

The efficiency over the day is then:

$$\eta_{sr-ss} = 0.727 - 2.88 \times 10^2 \frac{t_{sr-ss}}{H_d} \qquad (3.79)$$

where

m_d	Mass of daily distilled water	kg m^{-2}
$\dot{m}_{ev}$	Evaporation rate of water	kg m^{-2} s^{-1}
η_{sr-ss}	Mean daily efficiency from sunrise to sunset	—
H_d	Mean daily solar radiation on the still from sunrise to sunset	MJ m^{-2}
t_{sr-ss}	Time period from sunrise to sunset	s

Morse and Read (1968) give a chart to predict the daily distillate production rate. The chart is based on known initial water temperature, known incident solar radiation, ambient conditions, and still characteristics, taking into account ΔT_w, the hourly water temperature change. Kalogirou (2005) presents the modifications suggested by various researches in order to achieve better performance of convectional solar stills.

3.3.5 Diurnal and Nocturnal Operation—the Impact of Water Depth

3.3.5.1 Diurnal Operation of a Simple Solar Still

Many investigators have demonstrated that the shallower the brine layer in the basin, the higher the daily output, all other parameters being equal. Loef et al. (1961) was the first who demonstrated that during peak sunshine hours, evaporation rate increases exponentially with increase in brine temperature. El-Haggar and Awn (1993) claim that the optimum performance is achieved at basin water depth of 0.03 m. The thin layer of water has lower heat capacity and achieves higher operation temperatures, increasing simultaneously heat losses. For shallow water mass during night time, distillation almost stops at about sunset time.

Deep basin stills, due to higher heat capacity, have lower productivity during day time but distillation continuous during night time, as the water mass acts as a sensible heat storage material. Distillation continuous as long the temperature difference between glass cover and water surface in the basin is enough to cause vapor condensation on the inside glass cover. In nocturnal operation, salt water depth and initial water temperature T_{wo} determine the output.

For the prediction of daily or night output, many studies have been performed and corresponding mathematical models were proposed. In fact, output is a function of solar incident radiation of heat fluxes inside the still, heat losses, and the design characteristics of the system. Belessiotis et al. (1996) and Voropoulos et al. (1996) give an equation predicting the daily output of single effect stills, based on the "*input—output*" method, ie, a method determining daily output, M_d, in relation to daily energy input H_d, a method that is used successfully in many cases of solar applications. The method takes into consideration all the design parameters. Thus the equation can be applicable to different designs of conventional stills as well. It splits the daily operation into three phases:

1. The starting phase, or inertia phase, early in the morning when operation starts
2. The pseudo-steady state phase
3. The end of the day, the saturation phase.

During the starting phase, the productivity is almost zero. The incident solar radiation is used to heat up the whole system. During this phase, inertia is the driving force. The phase starts 6 hours before noon and ends at the time when water temperature T_w starts increasing.

During the second phase, the pseudo-steady state, the production of distilled water is increasing with increasing incident radiation intensity; it reaches a maximum and then starts to decrease as solar radiation intensity decreases.

The last phase, the phase of saturation, has a decreasing productivity until it stops, if a thin water layer still exists in the basin, it continues to distill with decreasing rate, if the water has enough depth.

For the mean daily output, the still is assumed to be a black box and the mass of distilled water produced during day time is then predicted by the equation:

$$m_{out-d} = f_{1d}H_d + f_{2d}(T_{wd-in} - \overline{T}_{ad}) + f_{3d} \tag{3.80}$$

where $\overline{T}_{ad}$ and T_{wd-in} are mean daily ambient and water temperatures over Δt intervals of the daily pseudo-steady state phase and H_d is daily average solar energy input calculated as:

$$H_d = \int_0^{\Delta t} (G)dt \quad \text{MJm}^{-2} \tag{3.81}$$

The above equation results from Eq. (3.69b) of instantaneous thermal balance of the still and the mass flow rate equation:

$$\dot{m} = \frac{h_{ewg} A_g}{h_{ev}} \frac{U_{wa}}{U_{wg}} \left[\overline{T}_{wd} - \overline{T}_{ad} \right] \Delta t_d = F(\overline{T}_{wd} - \overline{T}_{ad}) \Delta t_d \tag{3.82}$$

The coefficients, f_{1d}, f_{2d}, and f_{3d}, are characteristic parameters of the still and express the efficiency, losses, and inertia (during early morning) of the still, respectively. These parameters indicate technical and operational data. Their analytical values are:

$$f_{1d} = \frac{F_{eff} + (F_{los} A_w n_o \overline{k}/2FC_w)}{1 + F_{los}(U_{wa} + U_{bw})A_w/2FC_w} \tag{3.83}$$

$$f_{2d} = \frac{1}{1 + F_{los}(U_{wa} + U_{bs})A_w/2FC_w} \tag{3.84}$$

$$f_{3d} = f_{1d} G_{ph-1} \tag{3.85}$$

where the parameters F_{eff}, F_{los}, and F are calculated as:

$$F_{eff} = \overline{k}_d \eta_o \frac{A_g A_w}{C_w} \frac{h_{ewg}}{h_{ev}} \frac{U_{wa}}{U_{wg}} (\Delta t)_{iii} \tag{3.86}$$

$$F_{los} = (U_{wa} + U_{bs}) \frac{A_g A_w}{C_w} \frac{h_{ewg}}{h_{wv}} \frac{U_{wa}}{U_{wg}} (\Delta t)_{ii}^2 \tag{3.87}$$

$$F = A_g \frac{h_{ewg}}{h_{ev}} \frac{U_{wa}}{U_{wg}} \tag{3.88}$$

where

A_g, A_w	Transparent cover and water surface, respectively	m^2
C_w	Heat capacity of water in the basin	J K^{-1}
$(\Delta t)_i$, $(\Delta t)_{ii}$, $(\Delta t)_{iii}$	Time intervals of phases I, II, and III	s
F	System parameter = $A_g(h_{ewg} U_{wa})/(h_{ev} U_{wg})$	J s^{-1} K^{-1}
F_{los}	System parameter for heat losses	kg K^{-1}
F_{eff}	System parameter for the efficiency	kg m^{-1} MJ^{-1}
h_{ewg}	Evaporating heat transfer coefficient—water—glass	W m^{-2} K
$h_{ev} = h_g - h_w = h_{fg}$	Evaporation enthalpy (latent heat of vaporization)	kJ kg^{-1}
H	Daily collected solar radiation	MJ m^{-2}
G_{ph-1}	Incident solar radiation during phase 1	W m^{-2}
n_o	Optical efficiency of the solar still	—
$\overline{k}_d$	Incidence angle coefficient of solar radiation	—
U_{wa}, U_{wg}, U_{bs}	Overall heat transfer flow factors, upward, between water and ambient, between water and glass cover and through base and sides, respectively	W m^{-2} K

The overall heat transfer flow coefficients are calculated from Eqs. (3.70), (3.71), and (3.72).

For long-term average data, the relationship between radiation intensity and output is almost linear. Short-term data may give small declination from linearity due to ambient temperature and wind fluctuations.

3.3.5.2 Nocturnal Operation

During night operation, Eq. (3.80) is still valid without the term of solar incident radiation (Belessiotis et al., 1996).

$$M_{out-n} = f_{2n}(T_{wn-int} - \overline{T}_{an}) + f_{3n} \tag{3.89}$$

At night, productivity is a function only of basin water temperature T_{wn} and of ambient temperature T_a (or T_{sky}).

$$f_{2n} = \frac{F_{los}}{1 + F_{los}(U_{wa} + U_{bs})A_w/2F_w C_w} \tag{3.90}$$

$$f_{2n} = (U_{wa} + U_{bw})\frac{A_g A_w h_{ewg}}{C_w h_{ev}}\frac{U_{wa}}{U_{wg}}(\Delta t_n)^2 \tag{3.91}$$

The corresponding mean temperatures of water in the basin and during day and night operation are:

$$\overline{T}_{wd} = \frac{T_1 + T_2}{2} = \frac{A_w n_o k_d F H_d - (U_{wg} + U_{bs})m_{out-d}A_w}{2FC_w} + T_1 \tag{3.92}$$

$$\overline{T}_{wn} = \frac{T_1 + T_2}{2} = \frac{-(U_{wg} + U_{bs})m_{out-n}A_g}{2FC_w} + T_1 \tag{3.93}$$

where T_1 and T_2 are temperatures of the water in the basin at the beginning and the end of the phase, respectively (K).

Voropoulos et al. (2000) studied the transport phenomena and modeled dynamically greenhouse type solar stills. They investigated various equations and showed that Dunkle's model can be used to correlate mass transfer due to evaporation/condensation with heat transfer. The equations investigated could form the basis of a simple analytical tool, which could be part of a test and evaluation methodology of each specific solar still. The characteristic coefficients of these equations can be deduced from measurements taken over short time tests period and in turn these equations can be used to predict the dynamic behavior of the still for any meteorological conditions. Further on, Mathioulakis et al. (1999) extended the previous study for night production and investigated the long-term performance of solar stills. Based on Eq. (3.69b), input–output equations give the average water temperature separately for day and night operation as:

$$\overline{T}_{wd} = k_1 H_d + k_2 \overline{T}_{ad} + k_3 T_{an} + k_4 H_{d-1} + k_5 T_{ad-1} \qquad (3.94)$$

$$\overline{T}_{wd} = l_1 \overline{T}_{an} + l_2 H_d + l_3 T_{ad} + l_4 T_{an-1} \qquad (3.95)$$

The water production for day time period and for night hours respectively is:

$$m_{wd} = a_1 H_d + a_2 H_d + a_3 \overline{T}_{ad} + a_4 \overline{T}_{an-1} - a_5 \overline{T}_{an-1} + a_6 \qquad (3.96)$$

$$m_{wn} = b_1 \overline{T}_{an} + b_2 H_d + b_3 \overline{T}_{ad} + b_4 \overline{T}_{an-1} + b_5 \qquad (3.97)$$

where

α_i, b_i	Coefficients of linear model for day and night water	–
$i = 1, 2, 3, \ldots, 6$	production, respectively (constants)	
k_i, l_i	Coefficients of linear model for daily and night mean	–
$i = 1, 2, 3, \ldots, 4$ (5)	water temperature (constants)	
H_d	Daily incident solar radiation	MJ m^{-2}
m_{wd}, m_{wn}	Produced distillate at specific time period of day or night, respectively	kg
$\overline{T}_{ad}$, $\overline{T}_{an}$	Day and night mean ambient temperature	K
$\overline{T}_{wd}$, $\overline{T}_{wn}$	Day and night mean water temperature	K

Malik and Van Vi Tran (1973) refer that night operation is function only of initial temperature T_{wo} of water mass in the basin at initial time $t=0$, if productivity is expressed by the ratio of total distillate m_{bo} at time $t=0$, to the water mass in the basin m also at time $t = 0$. The energy balance in transparent cover and the water mass is:

$$\frac{m_{dc}}{m_{bo}} = \frac{m_{bo} - m}{m_{bo}} = \frac{T_{wo} - T_w}{C - T_w} \qquad (3.98)$$

This is a linear heat–mass relationship into the still, where m_{dc} is the amount of distillate collected during day time. The parameter C is calculated from equation:

$$C = \left(\frac{dT_w}{dm/m}\right) + T_w = \frac{U_{wg}h_{rg-sky} + U_{bs}(U_{wg} + h_{rg-sky})}{c_p h_{ewg} h_{rg-sky}} h_{ev} \qquad (3.99)$$

Parameter C gives the decrease of water mass due to distillation, at water mass temperature T_w. The parameter C can be assumed to be a standard productivity which depends primary on initial water mass m_{bo}, ie, from water depth in the basin, the initial temperature T_{wo} and the temperature decrease $(T_{wo} - T_w)$. Malik and Van Vi Tran (1973) present some interesting conclusion for night operation for simple conventional passive stills:

1. The amount of distillate, by the time the brine mass cooled to temperature T_{wo}, is given as:

$$m_{dc}/m_{bo} = \Delta T_w/(C - T_{wo}) \tag{3.100}$$

2. If the water is heated further to a temperature $T_{wo} + \Delta T_w$, the amount of heat needed to heat the water to the new temperature is:

$$q = m_{bo}c_p\Delta T_w \tag{3.101}$$

3. The amount of distillate obtained by allowing the water to cool in the still from initial temperature to a final temperature $T_f = T_i$ is given by:

$$m_{dc} = \frac{q}{c_p(C - T_{wo})} \tag{3.102}$$

where

c_p	Specific enthalpy of water in the basin	$J\,kg^{-1}\,K^{-1}$
m_{dc}	Cumulative distillate output at time t	$kg\,m^{-2}$
m_{bo}	Amount of salt water in the basin at time $t = 0$	$kg\,m^{-2}$
m	Salt water mass, $m = (m_{bo} - m_{dc})$	$kg\,m^{-2}$
T_{wo}	Salt water temperature at time $t = 0$	K
T_w	Salt water temperature at time $t = t$	K
C	Systems constant, Eq. (3.99)	K
U_{wg}	Total heat transfer coefficient from water to cover	$W\,m^{-2}\,K^{-2}$
U_{bs}	Total heat transfer coefficient from bottom and sides (heat losses to the environment)	$W\,m^{-2}\,K^{-2}$

If waste heat or other auxiliary energy source is available during night operation, productivity may increase by increasing water mass temperature.

3.3.6 The Significance of the Water Depth

Many studies have been performed on the impact of water depth in the basin. For a thin layer of water in the basin, the productivity increases during day time. In peak hours, evaporation rate increases exponentially with water temperature increase. A shallow layer of water has low heat capacity and thus higher temperatures may be achieved, increasing simultaneously heat losses. Distillation stops at sunset.

Solar stills with deep water basins during day time have less productivity than thin layer water but during night time they continue to produce distillate, in decreasing flow rates, but in increasing heat losses. Large amounts of water have enough heat capacity to act as heat storage material. Night distillation continues as long as temperature difference between water and inside cover surface is enough to achieve condensation of vapor.

Tiwari and Madhuri (1987) studied the impact of water depth in the basin of solar stills and concluded that increasing the depth from 0.01 to 0.20 m productivity decreases by about 44%. The study was performed at an initial water temperature of 35°C. For an increase from 0.01 to 0.012, the distillate decreased at about 25% for temperatures of 50°C. They claim agreement with the results by Cooper (1969a), Bloemer et al. (1965b), and Tiwari and Tiwari (2006). Tripathi and Tiwari (2005) refer that water depth has important impact on heat and mass transfer coefficients. The optimum productivity at night operation is found to be at depths of 0.1−0.15 m.

Phadatare and Verma (2007) studied the influence of water depth on the internal heat and mass transfer in a plexiglass solar still. They conclude, as El-Haggar and Awn (1993) did, that the optimum output is achieved for depths of about 2.0 cm. Experience showed that to operate a simple passive solar still with thin layer water mass, care should be taken that enough water exist for complete diurnal operation. If water is not enough, stagnant conditions may be achieved.

Aybar and Assefi (2009) studied the water depth in the basin and the cover inclination impact to the distillate productivity. Their mathematical models for heat balance were solved with Fortran and the use of fourth Runge−Kutta order. Convective heat transfer equations between black surface and water are presented by the following equations:

$$h_{cblw} = 0.54 \frac{\lambda_w \, \mathrm{Ra}^{1/4}}{l_l}, \quad \text{if} \ \ \mathrm{Ra} = 10^4 - 10^7 \tag{3.103}$$

$$h_{cblw} = 0.15 \frac{\lambda_w \, \mathrm{Ra}^{1/3}}{l_l}, \quad \text{if} \ \ \mathrm{Ra} = 10^7 - 10^{11} \tag{3.104}$$

where h_{cblw} is convective heat transfer coefficient from black liner to the water surface, λ_w is water heat transfer coefficient by conduction, l a characteristic dimension, and Ra dimensionless Rayleigh number. From previous evaporative equations (Aybar, 2006) and the heat balance equations, they concluded that the best depth for optimum efficiency is the minimum possible amount of water. The amount of heat for the water evaporation is:

$$Q_{ev} = 0.027 \Delta T^{1/4} p_{sat} (1 - \phi) \tag{3.105}$$

where p_{sat} is saturation pressure, ΔT is temperature difference between water in the basin and air−humidity mixture, and ϕ is relative humidity.

For symmetric and asymmetric conventional solar stills, the relationship depth/ productivity is given by the following equation which is derived from many experimental data correlations by using the least square method. The equation is valid for solar radiation range $8-30 \, \text{MJ m}^{-2} \, \text{d}^{-1}$, inclination angles $5-45°$, and latitude $20-35°\text{N}$ (Khalifa and Hammod, 2009):

$$m_d = 3.884\text{e}^{-0.0458(d)} \quad R^2 = 0.832 \tag{3.106}$$

where $\dot{m}_d$ is daily productivity, $\text{L m}^{-2} \, \text{d}^{-1}$, and d is brine depth in the basin.

Independently from the investigation results for a passive one-stage solar still, the optimum depth is a function of solar radiation intensity, the geometry of the still, the wind velocity, and in part from the ratio of direct to diffuse incident radiation, taking in consideration that the still chamber has to be tight without any vapor losses.

3.3.7 Increasing Productivity of Simple Solar Stills

Conventional single effect solar stills have, as already has been pointed out, low productivity and low efficiency. To increase daily productivity, many trials have been made, the main of which are grouped in:

- Increasing vapor condensation rate
- Increasing solar radiation absorption and/or feed temperature
- Gaining condensation sensitive heat.

These are examined in the following sections.

For all methods that increase productivity of solar stills a special economic study has to be performed in the cases where additional devices and/or components are used, in order to find the balance between the additional construction and capital cost needed to increase productivity.

3.3.7.1 Increasing Vapor Condensation Rate

This is a procedure that can be achieved by various means as pointed out previously.

One of the methods is the use of double cover glass introducing feed water between their spaces, thus cooling the lower glass cover and consequently, increasing vapor condensation. Fig. 3.12 presents the two glass cover cooling arrangement, the feedback flow (Fig. 3.12A) and the counter count flow (Fig. 3.12B; Fath, 1998).

During the circulation of cooling seawater, a thin film of saline water flowing on the inside glass surface is created. Because the refraction index of the water film is $n = 1.33$, lower than the refraction index of glass cover ($n = 1.52$), the presence of this film would smooth transmition of the solar rays into the still. A schematic of a conventional solar still with the cooling film is shown in Fig. 3.13. The still is divided into four regions: the basin (b), the water in the basin (w), the glass cover (g), and the flowing cooling water film (f). The balance of each region is based on average temperatures. The energy balance in each region, per unit still area, assuming that the areas of glass cover and basin are equal and that the energy balance is based on the average temperatures, is described below (Abu-Hijleh, 1996; Abu-Hijleh and Mousa, 1997).

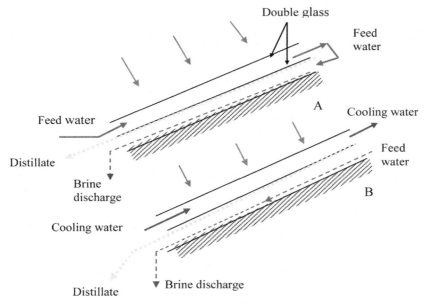

Figure 3.12 Double glass covers cooling procedure: (A) feedback flow and (B) counter count flow (Fath, 1998).

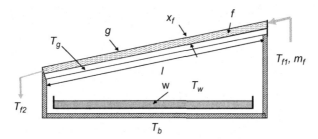

Figure 3.13 Schematic of a solar still with double glass cooling system (Abu-Hijleh and Mousa, 1997).

For the basin and for the water in the basin:

$$m_b c_{pb} \frac{dT_b}{dt} = (1 - n_g)(1 - a_g)(1 - a_w)G - q_{cwa} - q_{cbw} \tag{3.107}$$

$$m_b c_{pw} \frac{dT_w}{dt} = (1 - n_g)(1 - a_g)a_w(G) + q_{cwa} - q_{rgw} - q_{cgw} - q_{ev} - q_{cmv} \tag{3.108}$$

For glass cover and the water in the cooling film:

$$m_b c_{pg} \frac{d\overline{T}_g}{dt} = (1 - n_g)a_g(G) + q_{rwg} + q_{cwg} + q_{ev} - h_{cg}(T_g - T_f) \tag{3.109}$$

$$m_b c_{pf} \frac{dT_f}{dt} = \dot{m}_f n_f(c_{p1} T_{f1} - c_{p2} T_{f2}) + h_{cf}(T_g - T_f) - q_{cfa} - q_{rfa} - q_{ef} \tag{3.110}$$

Eqs. (3.107)–(3.110) estimate the performance of a solar still with film cooling. It is assumed that the saline water film is thin, that no radiation will be absorbed by the film, and that either there is no evaporation from the film or it is negligible.

Eq. (3.110) refers to the amount of heat that is transferred from the glass to the water film. The corresponding heat transfer equation is $\dot{q}_{cgf} = h_{cgf}(T_g - T_f)$. The term represents the heat due to evaporation from the cooling film and is calculated from the equation for heat and mass transfer over a flat plate:

$$h_{cgf} \begin{vmatrix} \dfrac{\lambda_f}{l} 0.664\, \mathrm{Re}_l^{1/2} \mathrm{Pr}_{fa}^{1/3} \\[2mm] \dfrac{\lambda_f}{l} (0.037\, \mathrm{Re}_l^{4/5} - 871) \mathrm{Pr}_{fa}^{1/3} \end{vmatrix} \quad \mathrm{Re} \le 5 \le 10^5, \quad \mathrm{Re} \ge 5 \ge 10^5 \qquad (3.111)$$

The dimensionless Reynolds number Re_l refers to laminar flow across the length of the glass cover.

The amount of heat q_{mw} needed to heat the make up water to the same temperature of the basin water is:

$$q_{mv} = \begin{vmatrix} \dot{m}_w(c_{pw} T_w - c_{p2} T_{f2}) \\[2mm] \dot{m}_f(c_{pw} T_w - c_{p2} T_{f2}) + (\dot{m}_{ev} - \dot{m}_f)(c_{pw} - c_{pa} T_a) \end{vmatrix} \quad \dot{m}_{ev} \le \dot{m}_f, \dot{m}_{ev} \ge \dot{m}_f \qquad (3.112)$$

In Eq. (3.110), the term q_{cfa} represents the heat due to evaporation from the cooling film. It is calculated from the equation for heat transfer between the water mass and flat plates as:

$$q_{cfa} = \frac{0.037 m_{air} p_{atm}}{Sc^{2/3} R} h_{fg} \frac{\dot{V}_f}{T_g \mathrm{Re}_l^{0.2}} \ln \left[\frac{p_{atm} - p_v(T_{a,wet})}{p_{atm} - p_v(T_f)} \right] \qquad (3.113)$$

where mass transfer flow rate is assumed to be constant

a_g	Coefficient of glass absorptivity	—
c_{pf}	Specific heat capacity of water film	J kg^{-1} K^{-1}
c_{pg}	Specific heat capacity of glass	J kg^{-1} K^{-1}
c_{pw}	Specific heat capacity of water at outlet	J kg^{-1} K^{-1}
c_{p1}	Specific heat capacity of cooling water at the inlet	J kg^{-1} K^{-1}
c_{p2}	Specific heat capacity of cooling water at the outlet	J kg^{-1} K^{-1}
G	Solar flux	W m^{-2}
λ_f	Thermal conductivity of water film	W m^{-1} K^{-1}
l	Length of class cover	m
M_{air}	Air molar mass	kg kmol^{-1}
m_f	Water mass per unit surface area	kg m^{-2}
m_g	Cover glass mass per unit surface area	kg m^{-2}
$\dot{m}_f$	Cooling water mass flow rate per unit width of glass	kg s^{-1} m^{-1}
n_g	Refraction coefficient of glass	—
p_{atm}	Atmospheric pressure	Pa

$P_{v(T)}$	Vapor pressure at environmental temperature T	Pa
q_{cwa}	Convection heat flow rate from water surface to environment	W m^{-2}
q_{cbw}	Convection heat flow rate from bottom surface to water surface	W m^{-2}
q_{mw}	Needed heat amount to heat the replenish water in the basin	W m^{-2}
q_{cgf}	Convection heat flow rate from glass to water film	W m^{-2}
q_{cfa}	Convection heat flow rate from water film to environment	W m^{-2}
q_{ef}	Heat due to evaporation from the cooling film	J kg^{-1}
q_{rfa}	Transferred heat from water film to the environment by radiation	W m^{-2}
R	Gas constant $= 8314$	J kmol^{-1} K^{-1}
R_l	Dimensionless number, Reynolds for laminar flow	–
Sc	Dimensionless number, Schmidt	–
T_{a-wet}	Temperature of wet atmosphere	K
$\overline{T}_f$	Mean film temperature $= (T_{f1} + T_{f2})/2$	K
T_{f1}	Cooling water inlet temperature	K
T_{f2}	Cooling water outlet temperature	K
V_f	Volumetric flow rate of culling water per unit depth	m^3 s^{-1} m^{-1}

Another way to increase output of the still is to incorporate a heat exchanger, mainly coils, inside in bottom or for asymmetrical solar stills in back side, as shown in Fig. 3.14 (Janisch and Dreschel, 1982). By this arrangement, part of the rejected brine heat is gained back.

3.3.7.2 Increasing Solar Radiation Absorption

The increase of solar radiation absorption may be achieved by using external (Figs. 3.15 to 3.17) or internal (Fig. 3.18) mirrors. Usually they reflect concentrated solar radiation to the water basin. There exist many forms of connecting mirrors in a solar still. In Fig. 3.18, the solar still is adapted to an inverted compound parabolic solar concentrator.

3.3.7.3 Increasing Feed Seawater Temperature

The increase of feed temperature and indirectly augmenting solar radiation input is also achieved by using solar collectors mainly flat plate collectors as they meet the operating temperature demand or by utilizing as feed water low temperature waste water.

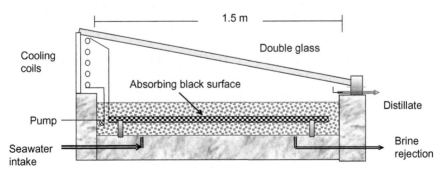

Figure 3.14 Solar still with cooling coils TU Berlin-IPAT design, erected in Porto Santo, Madeira, Portugal (Janisch and Dreschel, 1982).

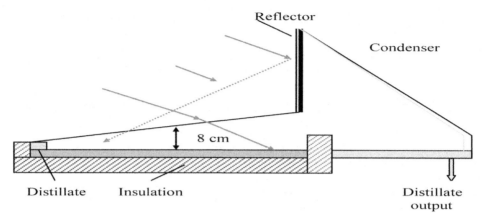

Figure 3.15 Solar still with side reflection mirror (Kabeel and El-Agouz, 2011).

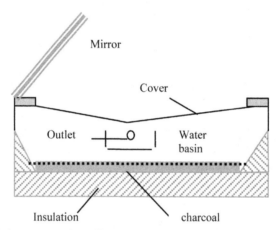

Figure 3.16 Inverted glass cover solar still with reflecting mirror.

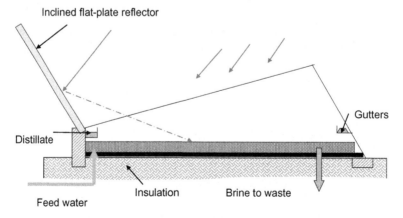

Figure 3.17 Schematic diagram of a Greek solar still with an inclined flat-plate external reflector. [Research still in National Centre for Scientific Research (NCSR) "Demokritos", Athens].

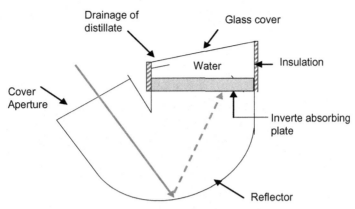

Figure 3.18 Solar still incorporated in a solar inverted absorber, asymmetric line axis compound parabolic solar concentrator (Yadav and Yadav, 2004).

Feed temperature may be increased by using waste heat as feed, by warming up feed water in solar collectors, or by recirculating part of the hot brine to warm up incoming cold feed water. This procedure is obvious that refers to active solar stills where a small pump for the circulation is necessary. The use of a water tank in higher level than the solar still may also be used for small individual solar stills taking advantage of free circulation due to height differences. Feed temperature increases as well by using coils inside the still where feed water circulates cooling the vapors, as described earlier (Fig. 3.14). The mostly used solar collector is the flat plate. An intermediate storage may help nocturnal operation.

For a solar still connected to solar flat plate collectors, the energy balance of the system is given by the following equation (Kumar, eolss):

$$C_w \frac{dT}{dt} = (1 - \rho_g)(1 - a_c)a_w(G) + \frac{F(t)Q_u}{A_w} + h_{cw-bl}(T_{bl} - T_w)$$

$$- U_{wg}(T_w - T_g) - \frac{h_{cw-bl} + U_p A_p F(t)}{A_w}(T_w - T_a)$$

(3.114)

where $F(t)$ is systems term. If $F(t) = 0$, the system is a conventional solar still. If $F(t) = 1.0$, solar collector is in operation. The useful energy Q_u needed from the collector is calculated as follows:

$$Q_u = F_R A_c [(a\tau)_c G(t) - (T_w - T_a)U_L]$$

(3.115)

where F_R is the heat removal factor of the solar collector. It is given by the equation (Duffie and Beckmann, 2006, 2013; Kalogirou, 2009, 2014):

$$F_R = \frac{\dot{m}c_p(T_a - T_c)}{A_c[G - U_L(T_c - T_a)]} = \frac{\dot{m}c_p}{A_c U_L}\left[1 - \exp\left(\frac{A_c U_L F'}{\dot{m}c_p}\right)\right]$$

(3.116)

Parameter F' is the efficiency factor of the solar collector and U_L collector overall heat losses. Singh and Tiwari (2004) present the balance of basin water mass on different terms with the collector parameters except parameter F_R:

$$Q_u + a_w(1 - a_g)A_w G(t) + h_{blw}(T_{bl} - T_w)A_{bl} = (m_w c_{pw})\frac{dT_w}{dt} + U_{wg}(T_w - T_g)A_w$$

(3.117)

where

α_g, α_w, α_c	Glass, water, and collector absorptivity, respectively	—
$(\alpha\tau)$	Collectors absorptance−transmittance product	—
A_{bl}, A_g, A_w, A_c, A_p	Black liner, glass, water, collector, and pipe surface area	m^2
c_{pw}	Water enthalpy	$J\,kg^{-1}\,K^{-1}$
d_w	Water depth in the basin	m
F_R, F'	Collectors heat removal factor and efficiency factor	
h_{blw}	Convective heat transfer coefficient from black liner to water surface	$W\,m^{-2}\,K^1$
h_{cga}	Convective heat transfer coefficient from glass to environment	$W\,m^{-2}\,K^1$
G	Incident solar radiation	$W\,m^{-2}$
m_w	Mass of water in basin $= A_w d_w \rho$	$kg\,m^{-2}$
Q_u	Useful heat of solar collector	W
T_a	Environmental temperature	K
T_{bl}	Black liner temperature	K
T_g	Cover glass temperature	K
T_w	Temperature of water in the basin	K
U_{wg}	Overall heat transfer coefficient from water surface to glass	$W\,m^{-2}\,K^{-1}$
U_{bla}	Overall heat transfer coefficient from basin bottom to environment	$W\,m^{-2}\,K^{-1}$
U_p	Overall heat losses from the pipes of the system	$W\,m^{-2}\,K^{-1}$
U_L	Overall heat losses from the solar collector	$W\,m^{-2}\,K^{-1}$

If $Q_u = 0.0$, Eqs. (3.114) and (3.115) are transformed to equations for passive solar stills.

3.3.7.4 Connected to Flat Plate Collectors and to Storage

This is another approach to nocturnal operation to increase output. Voropoulos et al. (1996), Mathioulakis and Belessiotis (2003), and Voropoulos et al. (2003) present a system of solar still, incorporated storage, and flat plate collector field. Fig. 3.19 presents a conventional, asymmetric, solar still incorporated to a storage tank. The system operates with a solar collector field or any other available external heat source which provide the storage system with warm water. The still is filled every morning

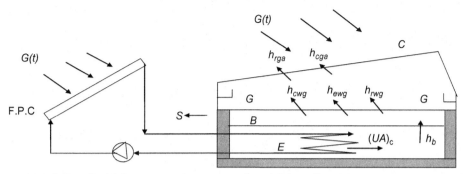

Figure 3.19 Solar still with incorporated storage tank and solar collector.

to complete diurnal losses. Circulation takes place between storage and collectors. The instantaneous energy balance between the basin and the surroundings for passive conditions is calculated as follows.

1. For day operation

The instantaneous energy balance for the solar still is presented by the modified equation (3.69b):

$$\frac{C_w}{A_w}\frac{dT_w(t)}{dt} = n_{od}k_d(t)G(t) - U_{wa}[T_w(t) - T_a(t)] + U_b[T_w(t) - T_a(t)] \qquad (3.118)$$

which is valid under the following assumptions:

a. Temperature gradient of the brine and of the transparent cover are negligible
b. Overall heat transfer coefficient U_{wa} is steady
c. Heat transfer coefficient, h_{sb}, between tank and the basin water remains steady
d. The brine mass (or brine depth) in the basin is steady during day time (this is correct when feed water circulation is automatically regulated).

In Eq. (3.118), the first term of the right part refers to the incident solar energy into the water basin, the second term refers to losses from the water mass to the environment, and finally the third term presents the energy input to the water mass in the basin from the water of the storage tank.

The energy balance within the storage tank is:

$$Q(t) = C_s\frac{dT_s(t)}{dT} + U_{wa}A_s\left[T_s(t) - T_w(t)\right] + U_sA_s[T_s(t) - T_a(t)] \qquad (3.119)$$

where C_s is the overall heat capacity of storage tank. The energy balance in the water storage tank is:

$$Q(t) + C_s\frac{dT_s(t)}{dT} = A_ck_c(t)n_oG(t) = U_cA_c[T_c(t) - T_a(t)] \qquad (3.120)$$

The last term of Eq. (3.119) presents the heat losses from the sides of the storage tank except from the upper surface which is in contact with the bottom of the still. Heat exchange between the collector fluid and tank water, considering random heat exchanger geometry and steady overall heat exchange conditions, is given by:

$$Q(t) = (UA)_{ev}(\overline{T}_c(t) - T_s(t)) \tag{3.121}$$

where subscripts c and s refer to collector and the storage tank. The general equation (3.118) for the system collector—storage is modified for day Δt_d operation:

$$C_w(T_{fin-w} - T_{int-w}) = A_w n_{od} k_d H_d - A_w U_{wg}(T_w - T_a)\Delta t + A_w h_b(\overline{T}_s - \overline{T}_w)\Delta t \tag{3.122}$$

The left part of the equation gives the overall amount of heat that is used to increase water temperature in the basin from initial T_{int-w} to final T_{fin-w} during day time period Δt_d.

Heat amount that is transferred from the storage tank to the solar still basin for the day time period is:

$$\frac{C}{BC_s} h_b A_w(\overline{T}_s - \overline{T}_w) = \int_0^{\Delta t} Q(t) - \frac{A_c n_o k_c}{B} H_d - K(\overline{T}_s - \overline{T}_a) \tag{3.123}$$

where the parameters B and K are calculated from the following equations:

$$B = 1 + C_c/C_s + A_c U_c/(UA_c) \tag{3.124}$$

$$K = \frac{(C_c/C_s)U_s A_s - A_c U_c}{1 + C_c/C_s + A_c U_c/(UA)_c} \tag{3.125}$$

$\int_0^{\Delta t} Q(t)$ is the amount of heat transferred from the collectors to the storage tank.

2. **For nocturnal operation**, the corresponding equations are

  For the solar still:

$$-\frac{C_w dT_w(t)}{A_w dt} = -U_{wg}[T_w(t) - T_a(t)] + h_{bs}[T_s(t) - T_w(t)] \tag{3.126}$$

where the term $U_{bs}[T_s(t) - T_w(t)]$ refers to the heat transfer from storage hot water to the colder basin water.

  For the storage tank:

$$-\frac{C_s dT_s(t)}{dt} = h_{bs} A_w[T_s(t) - T_w(t)] - U_s A_s[T_s(t) - T_a(t)] \tag{3.127}$$

The distillate amount produced during night time is:

$$m_{out} = A_g \frac{h_{ewg}}{h_{ev}} \frac{U_{wa}}{U_{wg}} \Delta t_n(\overline{T}_{wn} - \overline{T}_{an}) = F\Delta t_n(T_{wn} - T_{an}) \tag{3.128}$$

Finally the temperature at the end of night period before the start of day period is:

$$T_{wn-f} = \overline{T}_{an} + T_{wn-in} - \overline{T}_{an}\exp\left[\frac{U_{wg}}{(C_w/A_w) + C_s}\Delta t_n\right] \tag{3.129}$$

where

A_c, A_g, A_w	Surface area of collector, glass cover, and seawater in the basin, respectively	m^2
C_c, C_w, C_s	Heat capacity of collector, seawater in the basin and of the storage	$kJ\,kg^{-1}\,K^{-1}$
F	Systems parameter $= A_g(h_{ewg}U_{wa}/h_{ev}U_{wg})$	$kg\,s^{-1}\,K^{-1}$
h_{sb}	Heat transfer coefficient between storage water and still basin water	$W\,m^{-2}\,K^{-1}$
H_d	Daily radiation	$MJ\,m^{-2}$
k_c, k_d	Coefficient of angle of incident of solar radiation in collector and the solar still, respectively	$-$
m	Daily distillate output	$kg\,d^{-1}$
$\dot{Q}$	Instantaneous rate of energy input in collector	$W\,m^{-2}$
n_{oc}, n_{od}	Optical efficiency of the collector and the still	$-$
T_{an}	Ambient temperature during night operation	K
T_{swn}	Temperature of water in the basin during night	K
$(UA)_e$	Heat transfer coefficient in the heat exchanger	$W\,m^{-2}\,K^{-1}$
U_{wg}, U_{ga}, U_{ba}	Overall heat transfer coefficient between basin water/glass cover, glass cover/ambient, and basin/ambient	$W\,m^{-2}\,K^{-1}$
U_c, U_s	Overall heat loss coefficient in collector and storage tank	$W\,m^{-2}\,K^{-1}$
U_{sw}	Overall heat transfer coefficient from storage tank to the water	$W\,m^{-2}\,K^{-1}$

A similar system of a tubular heat exchanger incorporated into the still and an external storage tank connected to a solar collector are described by Smakdji et al. (2014). A mathematical model for the system is included.

Dashtban and Tabrizi (2011) present a solar still (Fig. 3.20A) having an incorporated storage system with phase change material (PCM) to a weir-type cascade solar still. The system is designed to increase productivity. The PCM used was paraffin wax, placed beneath the absorber plate. The setup comprises a storage tank, absorber plate, PCM reservoir, the solar still and piping. As comparison tool, an identical solar still without storage system was used. The solar still consists of a 15-step horizontal absorber. The corresponding mathematical model of the system for absorbing plate is:

$$G(t)a_{ab}\tau_g\tau_w = h_{ab-w}(T_{ab} - T_w) + (\lambda_{PCM}/s_{PVCM})(T_{ab} - T_{PCM}) + (m_{ab}c_{p-ab}/A_{ab})(dT_{ab}/dt) \tag{3.130}$$

where λ and s are thermal conductivity and thickness of PCM and τ_g, τ_w the glass and water transmissivity; m_{ab}, c_{p}, and A_{ab} are mass, specific heat, and surface area of

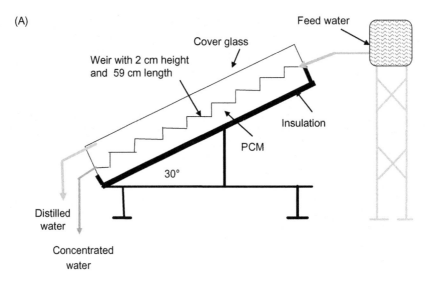

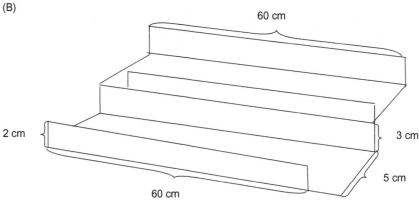

Figure 3.20 (A) Solar still incorporated phase change material thermal storage (PCM). (B) Multiple basins arrangement inclined solar stills cross-section. Water is glowing in cascading form (Dashtban and Tabrizi, 2011).

the absorbing plate, and h_{abw} is convective heat transfer coefficient from the absorbing plate to the water calculated from the expression:

$$h_{abw} = 0.54(l_w/l)(\text{Gr Pr})^{0.25} \qquad (3.131)$$

where l is characteristic dimension for a horizontal rectangular surface.

The PCM material stores the energy from the absorber plate. A small amount is lost to the surrounding via the still bottom:

$$(\lambda_{PCM}/s_{PCM})(T_{ab} - T_{PCM}) = (\lambda_{ins}/s_{ins})(T_{PCM} - T_a) + (C_{eq}/A_{ab})(dT_{PCM}/dt) \qquad (3.132)$$

where subscript *ins* refers to insulation and C_{eq} is the equivalent thermal capacity of PCM which is calculated from the enthalpy of PCM h_{PCM} (J kg^{-1}) and its specific heat capacity c_p (J kg^{-1} K^{-1}) as:

$$C_{eq} = m_{PCM}c_{ps} \quad \text{for} \quad T_{PCM} < T_m$$
$$C_{eq} = m_{PCM}h_{CPM} \quad \text{for} \quad T_m \le T_{PCM} \le T_m + \delta$$
$$C_{eq} = m_{PCM}c_{pl} \quad \text{for} \quad T_{PCM} > T_m + \delta$$

where m_{PCM} is the mass of PCM, T_m is melting temperature of the material, δ is temperature increase during change phase, and c_{ps} c_{pl} is specific heat capacity of the material in solid and liquid phases, respectively.

The corresponding equations for energy balance on the glass and salt water are:

$$G(t)a_g + h_{wg}(T_w - T_g) = h_{cga}(T_g - T_a) + (m_g c_{pg}/A_g)(dT_g/dt) \tag{3.133}$$

$$G(t)\tau_g a_w + h_{cpw}(T_p - T_w) = h_{wg}(T_w - T_g) + (m_w c_{pw}/A_w)(dT_w/dt) \tag{3.134}$$

where h_{wg}, h_{cga}, and h_{cpw} are equivalent heat transfer coefficients from water to glass, convective heat transfer coefficient from glass to ambient, and convective heat transfer coefficient from plate to water, respectively.

3.4 VARIOUS TYPES OF SOLAR STILLS

3.4.1 Wick and Multiwick Solar Stills

Many still configurations use, instead of a liner, a black porous fabric as an absorber of the solar energy. Water flows on the porous material surface and evaporates immediately by the action of solar radiation. The porous fabric may be floating on the water surface of the basin or lying on its bottom. The porous fabric sucks the water by capillary action. Usually many pieces of such fabric material are used. Fig. 3.21A and B presents a single slope and a double slope solar still (Sodha et al., 1981; Fath, 1998). Multiwick solar stills have the following advantages over conventional single effect solar stills (Tiwari and Singh, eolss).

- Water surface on the fabric can be oriented at any optimum angle to receive maximum radiation
- In installations in series there is no shading effect. The layout of plant is shown in Fig. 3.21A and B
- Any scale formation can easily brush off or easily blackened again by injecting a black dye.

Janarthranan et al. (2005) present a new type, open and closed cycle floating tilted wick solar still. An extensive mathematical analysis is presented on the evaporative heat losses and on heat transfer of the system.

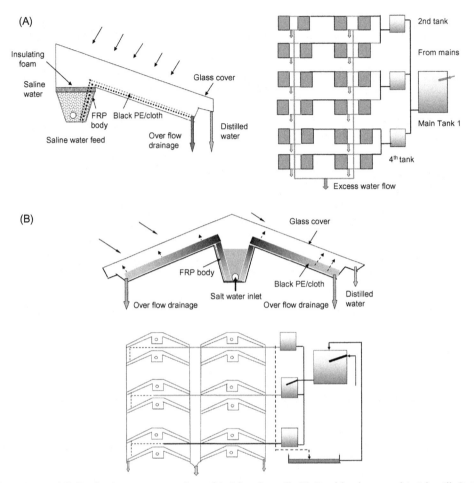

Figure 3.21 (A) Single slope asymmetric multiwick solar still. (B) Double slope multiwick still. Both of Indian design with the layout of the corresponding plants (Tiwari and Singh (eolss); Fath, 1998).

Fig. 3.22 presents a solar still in pyramided configuration having a concave basin lined with black wicks as absorbing material. The cover lies on a stainless steel frame. The author claims an efficiency of 45% and an estimated cost per liter of distillate of 0.065 \$ (Kabeel, 2009; Kabeel and El-Agouz, 2011).

The overall thermal efficiency for passive and active operation modes is presented by Singh and Tiwari (1992) and Balan et al. (2011) for double slope multiwick solar stills (Eqs. (3.75) and (3.76)).

A large variety of conventional, single, one or more effect solar stills, passive and active having various construction details exists. It is not easy to be

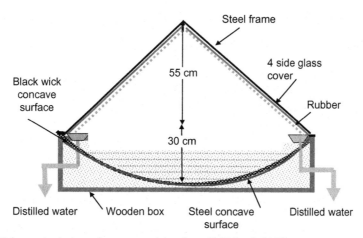

Figure 3.22 Schematic design of concave wick solar still (Kabeel, 2009).

grouped in a simple way. In general these may be classified into three general modes:

- **Horizontal type**, the basin being in horizontal position on the soil with the necessary bottom insulation
- **The inclined type**, where basin and cover are inclined, resembling flat plate collectors. The still system is supported by a metallic or other material support. Bottom is well insulated
- **The vertical type**, where basin and class cover are in vertical position, supported by a metallic structure or from other material.

A comparison of all of these different designs is not easy and reliable due to the variety of construction and operational details which in many cases are not comparable. Balan et al. (2011) present a detailed and useful review on passive solar stills and a detailed classification of the stills. They present and tabulated, for each the study performed by the authors, the aim of the study and the most essential results and remarks.

Tiwari et al. (1986) present a monthly comparative performance of various Indian designs of solar stills. They compare single and double slope (asymmetric and symmetric) fiber-reinforced plastic (FRP) stills, a double slope concrete still and asymmetric fiber-reinforced plastic (FRP) stills. Tiwari et al. (2003b) modeled passive/active solar stills by using inner glass temperature. Yadav and Tiwari (1987) studied the FRP single and double slope multiwick solar stills (Fig. 3.21A and B). They give long-term (year around) tabulated comparative data and conclude that the single slope FRP (either simple or multiwick) gives higher output in winter time and the double slope (also either simple or multiwick) has higher output in summer for Indian climatic conditions.

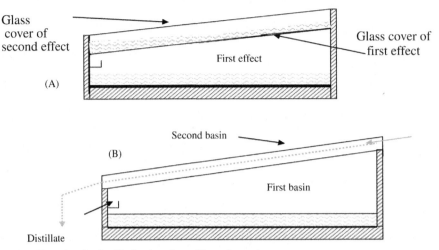

Figure 3.23 Two simple double basin solar still configurations.

3.4.2 Multiple Effect Solar Stills

Multiple effect solar stills are conventional solar still configurations, utilizing inside still chambers of two or more stages to gain as much as possible the condensation enthalpy and to take advantage of increased evaporation surface area. As for single effect stills, multiple effect stills may be of single or double slope (roof type), multiwick, of glass or plastic cover(s), etc. They can also be accommodated in horizontal position, tilted or vertical. The simplest configurations are the passive multiple effect stills. Many configurations of two or more stages, solar stills have been proposed as follows.

With stationary water in the basin (Fig. 3.23A). In this type of solar still, the basins lay one upon the other, thus gaining installation surface area. It is the simplest multiple effect solar still. The water layer in the upper basin warms up by the condensing water vapors of the first stage, coming in contact with the inside surface of the glass cover, producing additional distillate at the second basin. These types of stills have been studied by Malik et al. (1982) and by Singh and Tiwari (1992) who found about 10% productivity increase. Lobo and Aranjo (1978) proved that the upward effects have greater productivity than the bottom ones. Sodha et al. (1981) observed that the output increase is not a linear function of daily irradiance increase. They also refer an output increase of about 40% for radiation input $5.6 \, kWh \, m^{-2} \, d^{-1}$ and 35% for $5.11 \, kWh \, m^{-2} \, d^{-1}$, respectively.

For a solar still having two glass covers the space between the covers consist the second basin having a thin layer water flow. The first glass cover consist the bottom of the second basin (Fig. 3.23B). Bapeshwara et al. (1982) studied this type of double basin solar still and refer that the efficiency increases up to about 15–20% over the single basin still. Abu-Hijleh (1996) has well studied this type of still by circulating saline water at very low flow rate. He called this procedure "the film cooling method" and found that the presence of the thin water layer smoothens the transmission of the solar rays as has a refraction index $n=1.33$, lower than that of glass cover alone. Heat transfer equations and heat balance are presented.

The energy balance equations developed for single basin conventional stills can be extended to n-basin stills where intermediate basins are created by having a number of transparent glass cover set between top cover and the black liner. Assuming that surface area of the n basins is same the energy balance is (Kumar, eolss):

For the upper nth cover apply:

$$C_{gn}\frac{dT_{gn}}{dt} = [(1-\rho_g)a_{av}]_n G + U_{wgn}(T_{wn-1} - T_{gn}) - (h_{rga} + h_{cga})(T_{gn} - T_{sky}) \qquad (3.135)$$

For an intermediate cover i:

$$C_{gi}\frac{dT_{gi}}{dt} = [(1-\rho_g)a_{av}]_i G + U_{wgi}(T_{wi-1} - T_{gi}) - h_{cwbl}(T_{gi} - T_{wi}) \qquad (3.136)$$

For the water mass into the i basin:

$$C_{wi}\frac{dT_{wi}}{dt} = [(1-\rho_g)(1-a_{av})a_w]_i G + h_{cwbl}(T_{gi} - T_{wi})$$

$$- U_{wgi}(T_{wi} + T_{gi+1}) - h_{si}(T_{wi} - T_a)\frac{A_{si}}{A_b} \qquad (3.137)$$

For the black absorbing surface:

$$[(1 - \rho_g)(1 - a_{av})(1 - a_w)a_{bl}]G = h_{cwbl1}(T_{bl} - T_{w1}) + h_{bs}(T_b - T_a) \qquad (3.138)$$

where α_{tcg} is the overall absorptance of the top cover glass, ρ_{tcg} is the overall reflectance of the top cover glass, and h_{cwbl} is the convective heat transfer coefficient.

Second effect incorporated in a single solar still. A different than the previous double effect solar still is presented by Fath (1996). The single solar still has movable shutters fashion type reflectors located at its shaded sites (Fig. 3.24). The second effect cover has finned outer surface which helps keeping its surface near ambient temperature. Fath gives the heat transferred by evaporation and convection from the first effect to the second as:

$$\dot{Q}_{ev(1-2)} = \dot{Q}_{ev(1)}/(1+R_v), \quad \dot{Q}_{c(1-2)} = \dot{Q}_{c(1)}/(1+R_v) \qquad (3.139)$$

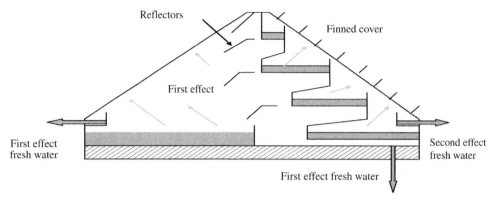

Figure 3.24 Double effect solar distillation unit (Fath, 1996).

The overall productivity is expressed by the summation of the first and second effect productivity as:

$$m_{out-T} = \sum m_{e1}(t)\Delta t + \sum m_{e2}(t)\Delta t \tag{3.140}$$

where

$\dot{Q}_{ev(1-2)}$	Heat flow rate from first to second effect by evaporation	W
$\dot{Q}_{ev(1)}$	Heat flow rate in the first effect by evaporation	W
$\dot{Q}_{c(1-2)}$	Heat flow rate from the first to second effect by convection	W
$\dot{Q}_{c(1)}$	Heat flow rate in the first by convection	W
R_V	Volume ratio, first to second effect by evaporation: $V_{ev(1)}/V_{ev(2)}$	—
m_T	Total mass collected	kg
$m_{ev(1)}, m_{ev(2)}$	Mass flow rate of 1st and 2nd effects, respectively	kg s^{-1}

He also gives the balances of the water in the basins and the glass covers as

1. First effect basin water:

$$\frac{m_{w1} C_{w1} dT_{w1}}{dt} = (\tau_g a_{w1})(G)A - Q_{ev1} - Q_{c1} - Q_{rq} - Q_{bl} \tag{3.141}$$

2. First effect glass cover:

$$\frac{m_g C_g dT_g}{dt} = a_g(G)A + Q_{eg} + Q_{cg} + Q_{r1} - Q_{air1} - Q_{sky1} \tag{3.142}$$

3. Second effect basin water:

$$\frac{m_{w2} C_{w2} dT_{w2}}{dt} = Q_{ev(1-2)} + Q_{c(1-2)} - Q_{ev2} = Q_{cgw_2} - Q_{rgw2} - Q_{b2} \tag{3.143}$$

4. Second effect glass cover:

$$\frac{m_c C_c dT_c}{dt} = Q_{ev2} + Q_{c2} + Q_{r2} - Q_{air2} - Q_{sky2} \qquad (3.144)$$

Stills with more effects. Hamdan et al. (1999) claim that multiple effect solar stills have their optimum operation efficiency when are triple effect. El-Sebaii (2005) presents an analytical model for a triple effect basin solar still based on analytical solution of energy balance equations for the various parts of the still. The hourly total productivity m_{hT} of the three effect solar still is:

$$m_{hT} = \sum_{2=1}^{3} (m_{hi}) \times 3600 \qquad (3.145)$$

The hourly productivity of each basin m_{hi} is calculated by the following equation:

$$m_{hi} = h_{hev}(T_w - T_g)/h_{ev} \qquad (3.146)$$

where h_{hev} is the evaporative heat transfer coefficient, of each effect, h_{ev} is the enthalpy of evaporation, and T_w, T_g are temperatures of water and glass of each effect, respectively. The efficiency over the day for the three effects is:

$$\eta_d(\%) = \frac{m_{Td} h_{ev}}{(A_b \sum G)}\Delta t \times 100 \qquad (3.147)$$

The corresponding temperatures of the basin and of each effect are:

$$T_b = (G_b + h_{blw} T_{w1} + U_b T_a)/(h_{blw} + U_b) \qquad (3.148)$$

$$T_{g1} = (G_{g1} + U_{wg1} T_{w1} + h_{gw1} T_{w2})/(U_{wg1} + h_{gw1}) \qquad (3.149)$$

$$T_{g2} = (G_{g2} + U_{wc2} T_{w2} + h_3 T_{w3})/(U_{wg2} + h_{gw3}) \qquad (3.150)$$

$$T_{g3} = (G_{g3} + U_{wg3} T_{w3} + h_{rgs} T_{sky} + h_{cga} T_a)/(U_{wg3} + h_{rgs} + h_{cga}) \qquad (3.151)$$

where

A_b	Basin surface area	m^2
Δt	Time interval over which solar radiation is measured	s
G_{g1}, G_{g2}, G_{g3}	Intensity of solar radiation falling on the lower (1), middle (2), and upper (3) basins of the still	W m^{-2}
h_{blw}	Heat transfer coefficient from black liner to water mass in the lower basin	W m^{-2} K^{-1}
h_{gw1}, h_{ug1}	Convective heat transfer coefficient from the water to the glass cover of the first effect and from the first effect to the water mass of the middle effect	W m^{-2} K^{-1}

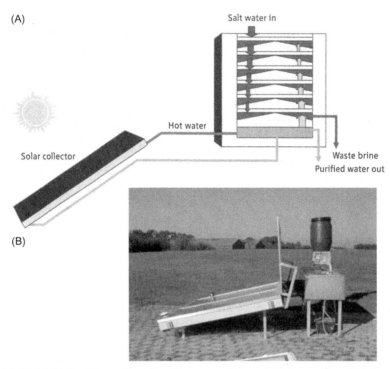

Figure 3.25 (A) Multiple effect solar still with removable trays (Schwarzer and Majumder, 2006). (B) The prototype at Solar Institute Juelich, Germany (Schwarzer et al., 2003).

h_{rgs}, h_{gw3}	Convective heat transfer coefficient from 2nd effect glass cover to the water mass of the upper effect and radiation heat transfer coefficient from the upper glass cover to the sky, respectively	W m^{-2} K^{-1}
U_b, U_{wg2}, U_{wg3}	Overall heat transfer coefficients, from the basin of the 1st effect (losses), from the water to the glass cover of the 2nd effect, and from the water to the glass cover of the upper effect, respectively	W m^{-2} K^{-1}
T_b, T_{g1}, T_{g2}, T_{g3}	Temperatures of the basin and the glass cover of the lower, middle, and upper effects, respectively	K

There exist solar stills with three or more effects, as this of Figs. 3.25 and 3.26. The solar still of Fig. 3.25 was studied at the Solar Institute Juelich, at Aachen TU, Germany and Thames Water, UK. This type of solar still where the basins lay one upon the other saves installation surface area.

The multiple effect solar still of Fig. 3.26 has a daily productivity of about 4.0 kg per hour and energy needs about 0.3 kWh per produced kg of distillate. Temperature difference between effects is 6–10 K (Khedim et al., 2004). The authors claim that

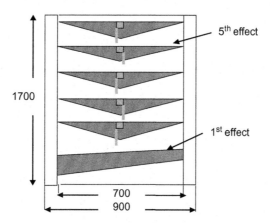

Figure 3.26 Construction principle of a multiple effect solar still with the prototype's dimensions (Khedim et al., 2004).

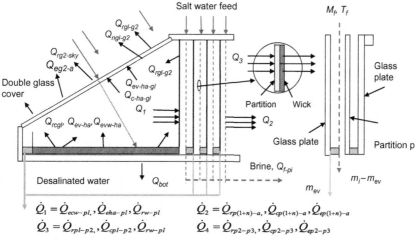

Figure 3.27 Schematic presentation of the multiple effect solar still with the heat flow diagram (Tanaka et al., 2000).

the system operates without pumps or control system. The consecutive chambers permit recovery of sensible heat thus increasing the productivity. Its efficiency is about three to four times of a conventional still. It produces 32—60 liters of distilled water for solar radiation intensity of 6—8 kWh.

Fig. 3.27 (Tanaka et al., 2000) presents a different design of a multiple effect solar still with a triangle cross-section consisting of a horizontal basin liner, a tilted double glass cover, and the vertical parallel partitions in contact with saline soaked wicks. The solar still with 5 mm diffusion gaps predicted to produce 15.4 kg m^{-2} d^{-1} water at incident solar energy of 22.4 MJ m^{-2} d^{-1}. The authors claim productivity which is 3.5 times greater than average efficiency of conventional solar stills, ie, 162% for the daily incident solar radiation of 22 MJ m^{-2} d^{-1}.

The same, as in conventional solar stills, heat transfer occurs through the gap of the single distillation still by convection, radiation, and evaporation/condensation. The proposed model takes into account the energy and mass balance for the bulk of humid air in the stills basin to exactly determine the ratio of the inner glass plate, *gl* to the first partition *pl*.

In the basin, natural convective heat and mass transfer occur from the water in the basin to the glass plate *g*1 and the partition plate *p*1 via the bulk humid air. Radiative heat transfer occurs directly between any two surfaces of the basin, g_i and p_i. The energy and mass balances are as follows.

The energy balance for basin, first partition and inner and outer glass plates, may be written as follows.

1. For the water in the stills basin:

$$\tau_g^2 a_w(G)_b + (\tau_g^2 a_w)_d(G)_d = \frac{d(C_w T_w)}{dt} + h_{ev}\dot{m}_{ev-ha} + \frac{Q_{cw-ha} + Q_{nvg1} + Q_{nvp1} + Q_{cnd}}{A_p}$$

(3.152)

2. For the first partition glass:

$$\tau_g^2 a_{p1}(G)_b \frac{\cos\phi}{\tan\Phi} + (\tau_g^2 a_{p1})_d(G)_d + h_{ev}\dot{m}_{evha-p1} + \frac{Q_{cha-p1} + Q_{nv-p1} - Q_{rp1-g1}}{A_p}$$
$$= C_p \frac{dT_{p1}}{dt} + h_{ev}\dot{m}_{evp1-p2} + \frac{Q_{con-p1} + Q_{rp1-p2} + Q_{fp1}}{A_p}$$

(3.153)

3. For the inner glass plate surface:

$$\tau_g a_g(G)_b\left(\cos\theta + \frac{\cos\phi\sin\theta}{\tan\Phi}\right) + (\tau_g a_g)_d(G)_d + h_{ev}\dot{m}_{evha-g1}$$
$$+ \frac{Q_{cha-g1} + Q_{nv-g1} + Q_{rp1-p2}}{A_g} = \frac{Q_{cg1-g2} + Q_{rg1-g2}}{A_g} + C\frac{dT_{g1}}{dt}$$

(3.154)

4. For the outer glass plates:

$$a_g(G)_b\left(\cos\theta + \frac{\cos\phi\cos\theta\sin\theta}{\tan\Phi}\right) + a_g(G)_d\frac{1+\cos\theta}{2}$$
$$+ \frac{Q_{cng1-g2} + Q_{rg1-g2}}{A_g} = \frac{Q_{cg2-a} + Q_{rg2-sky}}{A_g} + \frac{C_g dT_{g2}}{dt}$$

(3.155)

where subscripts b, d, and p refer to the direct (beam) and diffuse solar energy and the partition material of the cells, respectively. The angles ϕ and Φ represent the angle between south and the projection of the sun–earth line onto a horizontal plane, and the angle between sun ray and normal line to glass cover, respectively. Subscript ha refers to humid air.

In the distilling cells between a partitions pi and $p(i+1)$ part, $Q_{f,pi}$ of thermal energy is absorbed on the partition pi heating up the saline water fed to the wick and the remaining thermal energy is transferred to the facing partition $p(i+1)$ by conduction, radiation, and evaporation/condensation:

$$\frac{Q_{cng(i-1)-pi} + Q_{rp(i-1)-pi}}{A_p} + h_{ev}\dot{m}_{evp(i-1)-pi} = \frac{C_p dT_{pi}}{dt} + h_{ev}\dot{m}_{pi-p(i+1)}$$

$$+ \frac{Q_{cnpi-p(i+1)} + Q_{rpi-p(i+1)} + Q_{fi}}{A_p} \quad (3.156)$$

where C_g and C_p ($\mathrm{J\,m^{-2}\,K^{-1}}$) represent the thermal capacity of glass and of the partition material, per unit surface, respectively. Subscript cn denotes conduction. For the last nth effect being in contact with the environment:

$$\frac{Q_{cnp(n+1)} + Q_{rpn-p(n+1)}}{A_p} + h_{ev}\dot{m}_{evpn-p(n+1)} = \frac{C_p dT_{p(n+1)}}{dt} + h_{ev}\dot{m}_{p(n+1)-a}$$

$$+ \frac{Q_{cp(n+1)-a} + Q_{rp(n+1)-a} + Q_{fp(n+1)}}{A_p}$$

$$(3.157)$$

3.4.3 Incline Solar Stills

This type of solar still needs a support frame and their bottom of the basin is exposed to the environment. The inclined solar still of Fig. 3.28 consists of an absorber plate and a glass cover which creates a cavity. The absorber plate is covered by a black wick which distributes the water on the plate evenly, increasing the thickness of the water film. The inclination angle of the still is $30°$ with horizontal. This inclination angle permits smooth dripping of water and an incident of solar rays to normal for most hours of the day. The system produces $\sim 5.7\ \mathrm{kg\,m^{-2}\,d^{-1}}$ distilled water and about $37.70\ \mathrm{kg\,d^{-1}}$ hot water for domestic use (Aybar et al., 2005; Aybar, 2006).

The heat balance for the absorbing plate, the glass cover, and the flowing water film is given by the following equations, respectively:

$$m_p c_{p-p} \frac{dT_p}{dt} = G(\tau a) - h_{rpg}(T_p - T_g) - h_{cpw}(T_p - T_w) \quad (3.158a)$$

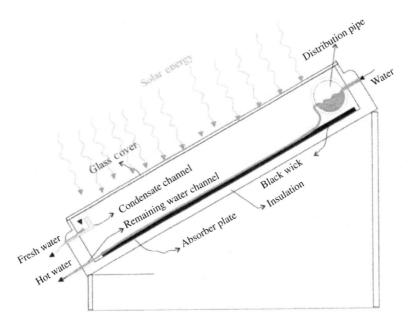

Figure 3.28 Schematic diagram of an inclined solar still (Aybar, 2006).

$$m_g c_{pg} \frac{dT_g}{dt} = Q_{rpg} + Q_{cn} - h_{rga}(T_g - T_a) - h_{cga}(T_g - T_a) \qquad (3.158b)$$

$$s c_{pw} \rho \frac{dT_{w-out}}{dt} = c_{pw}[\dot{m}_{in} T_{w-in} - \dot{m}_{out} T_{w-out}] \frac{1}{l_c} + Q_{cpw} - Q_{ev} \qquad (3.158c)$$

where l_c is the length of cavity, ie, the length of the free space of humid air and s denotes the water film thickness. Subscripts cn and p refer to conduction heat transfer and the black plate onto where the film of desalinated water flows.

A similar inclined solar still forming a flowing water film was used for the theoretical study of the falling film unit (Mousa and Abu Arabi, 2009; Abu Arabi et al., 2009). The authors present, similar to Aybar (2006), the heat balance equations for the black plate, the glass cover, and the flowing water film, respectively (Fig. 3.29).

$$m_{bl} c_{pg} \frac{dT_{bl}}{dt} = G(t) b_{bl} - q_{cbl-wf} - q_{los} \qquad (3.159)$$

$$m_{bg} c_{pg} \frac{dT_g}{dt} = G(t) b_g - q_{cwf-g} + q_{rwf-g} + q_{ewf-g} - q_{rg-sky} - q_{cg-a} \qquad (3.160)$$

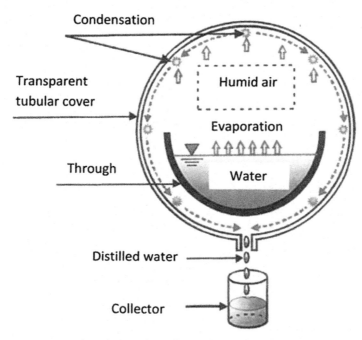

Figure 3.29 Cross-section of a tubular solar still. Condensate circulates in the inside surface of the cover. The basin for the salt water has hemispheric form (Ahsan et al., 2010a).

$$m_{uf} c_{pw} \frac{dT_{uf}}{dt} = G(t) b_{wf} - q_{cbl-uf} - \dot{m}_{uf} c_{pw} \frac{\partial T_{uf}}{\partial x} dx - q_{cuf-g} - q_{nuf-g} - q_{evuf-g} \quad (3.161)$$

The convective heat transfer between water film and glass is (Abu Arabi et al., 2009):

$$Q_{cfg} = h_{cfg}(T_{uf} - T_g) \quad (3.162)$$

Total produced desalinated water is calculated as:

$$\frac{dm}{dt} = \frac{h_{evuf-f}(T_{uf} - T_g)}{h_{ev}} \quad (3.163)$$

The amount of product water m_p (kg m^{-2} s^{-1}) for a falling film solar still may be also expressed by the mass of flowing feed water $\dot{m}_f$ (kg m^{-2} s^{-1}) (Abu Arabi et al., 2009) as:

$$\dot{m}_p = \frac{\dot{m}_f c_p (T_f - T_{out}) + (\overline{G}) A}{h_v - c_p (T_{out} - T_r)} \quad (3.164)$$

where b_g, b_{wf}, and b_{bl} are absorbance parameters given by the absorptivity of the glass cover, the water, and the black liner as:

$$b_g = (1-\rho_g)a_g, \quad b_{wf} = (1-\rho_g-\alpha_g)a_w, \quad \text{and} \quad b_{cl} = (1-\rho_g-\alpha_{wf}-\alpha_g)a_{bl}$$

Subscript wf refers to flowing water film. h_{cpw} and h_{rpg} are convection and radiative heat transfer coefficients from absorber plate to water and to glass, respectively. Temperature T_f and T_r are feed water and reference temperatures and T_{out} is temperature of outlet water, and (G) is mean radiation intensity between time interval $t_1 - t_2$.

3.4.4 Tubular Solar Stills

Further to the design shown in Fig. 3.29 tubular solar stills consist of a transparent tube having inside a suitable disk which contains the salt water. One of the first tubular solar stills was assembled of a glass tube cover of 10 cm diameter, 1.1 m length, and 0.32 cm wall thickness. A shallow black painted metal tray of 1.27 cm depth was used as basin for the salt water. The salt water rectangular disk was made of galvanized iron. This tubular active solar still was constructed for the study of the parameters that influence daily and nocturnal operation of the still (Tleimat and Howe, 1966).

Tiwari and Kumar (1988) used also glass tube as transparent cover of a tubular solar still. They present analytical expressions for the calculation of mean temperatures of salt water, glass tube surface, brine outlet for nocturnal productivity, using waste heat to preheat the salt water. They conclude that:
- Daily yield of distillate in the tubular solar still is higher than that of a conventional solar still for the same set of still and climatic parameters
- The internal heat transfer coefficient remains constant for constant inlet brine temperature in contrast with the conventional solar still for higher flow rates.

Ahsan et al. (2010) present a tubular solar still with plastic transparent cover and wicks as the absorbing material. The still may be considered as a combination of tubular and multiwick configuration. They studied two tubular modes of solar stills. The system consists of a transparent tubular cover made of polyethene sheet and for the second model a black, semicircular trough, made of galvanized iron (Fig. 3.30). The frame of the second model is assembled with six pipes and six rings made of galvanized iron that supports the plastic film. Ahsan and collaborators who studied the solar still concluded that the humid air inside the still chamber is not saturated. The authors give the following empirical equation for the hourly production of desalinated water:

$$m = 0.045 + 0.618(T_{ha}/\varphi) \tag{3.165}$$

where φ is the relative air humidity.

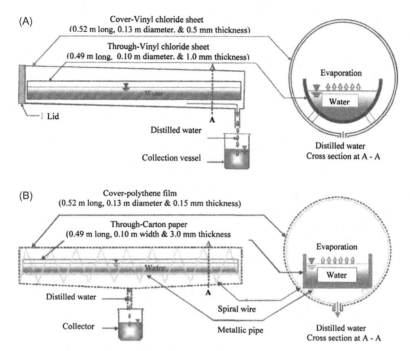

Figure 3.30 The old (upper) and new mode of the tubular solar still, with details, as presented by (Ahsan et al., 2012).

The calculation accuracy of empirical equation (3.165) was quantitatively evaluated by the root mean square deviation σ:

$$\sigma = \sqrt{\frac{1}{N}\sum_{i=1}^{N}\left(\dot{m}_{hd} - \dot{m}_{hdc}\right)^2} \qquad (3.166)$$

where $\dot{m}_{hd}$ is the hourly produced distillate flux and $\dot{m}_{hdc}$ the hourly calculated distillate flux.

For this study, ordinary polyethylene film, soft polyvinyl chloride (PVS), and Diastar (commercial name of polyolefin Durable film used in agriculture) were tested also as transparent plastic material. Diastar showed longer life time, about 5 years, but the other two plastics had only 2 years life time. The authors claim that the solar still made of Diastar plastic is lighter, cheap, simple, and durable.

Ahsan and Fukuhara (2008, 2010) present a new mass and heat transfer detailed mathematical analysis of this type of tubular solar still based on humid air properties inside the still. The model provides some new outputs for the tubular solar still such as the diurnal temperature variations, the water vapor density, and the relative humidity of the humid air. This new model, based on the heat balance of humid air and mass balance

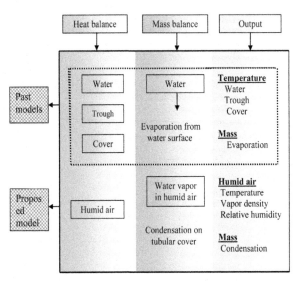

Figure 3.31 Comparison between past and proposed models for heat and mass balance in tubular solar still (Ahsan and Fukuhara, 2009).

on water vapor in the humid air, is compared with past models where heat balance is calculated between basin (trough) and cover and mass balance of water. Fig. 3.31 presents the authors comparison between the new and the past models.

The difference between evaporation and condensation rates affects the diurnal variations of vapor density and the relative humidity. The condensation flux (kg m^{-1} s^{-1}) is expressed by the evaporative mass transfer coefficient as:

$$m_c = k_{dhac}(\rho_{wha} - \rho_{vc}) \tag{3.167}$$

where k_{ew} is the evaporative mass transfer coefficient (m s^{-1}). It is derived by a dimensionless analysis developed by Ahsan and Fukuhara (2010) as:

$$k_{ev} = \frac{M_v R_v \gamma_e (T_w + T_{ha})}{2 R_g T_{ha}} \left[\frac{g \beta D_v^2 (T_w - T_c)}{\nu} \right]^{1/3} \tag{3.168}$$

Mass transfer coefficient for k_{cdha} (m s^{-1}) is given by Ahsan and Fukuhara (2009) according to filmwise condensation theory as:

$$k_{cdha} = 0.996 \gamma_c^{3/4} \left[\frac{g \rho_l (\rho_l - \rho_{vha}) \lambda_l^3 a^3 (T_{ha} - T_a)^3}{\eta_l d h_{ev}^3 (\rho_{vha} - \rho_{vc})^4} \right]^{1/4} \tag{3.169}$$

The decrease of basin water equals the evaporation rate and is presented by:

$$-m_c A_w = \rho_w \frac{dV_w}{dt} \qquad (3.170)$$

The water vapor in the humid air is calculated as:

$$\dot{m}_{ev} - m_{cd} = V_{ha} \frac{\partial \rho_{vha}}{dt} = A_w \dot{m}_{ev} - A_c \dot{m}_c \qquad (3.171)$$

where

A_w, A_c	Surface area of water in the basin (trough) and of the cover, respectively	m^2
a	Temperature difference fraction	—
β	Volumetric thermal expansion coefficient	K^{-1}
γ_c, γ_e	Coefficient of condensation and evaporation	—
D_{ev}	Molecular diffusion coefficient of vapor	$m^2\,s^{-1}$
d	Diameter	m
g	Gravitational acceleration = 9.807	$m\,s^{-1}$
h_{ev}	Vaporization enthalpy	$J\,kg^{-1}$
k_{ev}, k_{cdha}	Evaporative mass transfer coefficient from the water to the humid air and condensation mass transfer coefficient from humid air to the cover, respectively	$m\,s^{-1}$
λ	Thermal conductivity	$W\,m^{-1}\,K^{-1}$
M_v	Molecular mass of vapor = 18.016	$kg\,mol^{-1}$
m_{ev}, m_c	Evaporation and condensation fluxes, respectively	$kg\,s^{-1}$
η	Dynamic viscosity	$kg\,m^{-1}\,s^{-1}$
R	Universal gas constant = 8315	$J\,kmol^{-1}\,K^{-1}$
R_v	Water vapor gas constant = 461.5	$J\,kg^{-1}\,K^{-1}$
ρ_{vu}, ρ_{vha}, ρ_l, ρ_{vc}	Density of vapor, vapor in humid air, saturated liquid water, and of vapor at the cover	$kg\,m^{-3}$
T_w, T_c, T_{ha}, T_a	Temperature of water, the cover, humid air, and ambient temperature, respectively	K
V	Volume of water in the basin	m^3

In addition, Ahsan et al. (2012) present a detail analysis of the design and fabrication of the old type (Fig. 3.30A) and the new, improved, tubular solar still (Fig. 3.30B). The frame of the improved tubular still consists of two galvanized iron pipes and a galvanized iron wire arranged in spiral shape. The cover material used was a cheap Diastral polythene film, which reduced cost and weight of the new model by about 92%. The authors present a detailed cost and water production analysis of models, the old one and the improved new configuration.

Zeng et al. (2013) designed, constructed, and studied a series of single effect, two effect, and three effect tubular solar stills by using solar and waste heat energy. Fig. 3.32 presents the cross-section of a two effect tubular solar still.

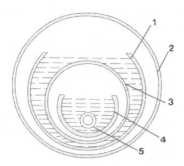

Figure 3.32 Cross-section of a two effect tubular solar still: (1) 2nd effect, (2) tubular shell, (3) 2nd effect inlet, (4) 1st effect inlet, and (5) desalted water outlet of 2nd effect (Zeng et al., 2013).

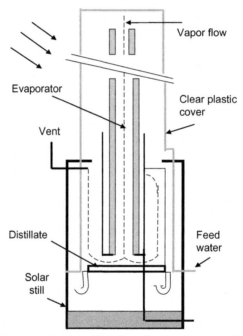

Figure 3.33 One of the first vertical solar stills using a feedback system (Correy, 1975).

3.4.5 Vertical Solar Stills

A vertical, multiple effect, solar still was proposed first by Dunkle (1961). A vertical microporous evaporator, solar distillation unit (Fig. 3.33) was designed by a company in the United States, which also proposed various modes of the vertical still for ground with moist soil and/or ground water underneath and a floating system. The system was designed for agricultural purposes (Correy, 1975). Later

Figure 3.34 Photograph of the vertical solar still erected at Algeria (Boukar and Harmin, 2007).

Kiatsiriroat et al. (1987) presented an analytical model to predict performance of a vertical, multiple effect still connected to flat plate collector, as a heat generator. The schematic of Fig. 3.34 presents a vertical solar distillation unit (Boukar and Harmin, 2004, 2005). The device consists of a head tank for brackish water, a vertical chamber, the solar still covered by a spongy absorbing material and a metallic support. The still is oriented in the direction at which the highest average solar radiation is obtained. Fig. 3.35 presents a photograph of the vertical solar still erected in Adrar, Algeria (latitude 27°53′N).

3.4.6 Other Types of Solar Stills—Solar Plants

A huge amount on references concerning solar stills exist describing all possible configurations of stills including theory and mathematical models which are limited, in most of the cases, to each special still design. Every design has its advantages and limitations. A description of all these studies would be unnecessary but the papers of the first investigators are given here (Loef et al., 1969; Telkes, 1953; Morse and Read, 1968; Cooper, 1969a,b), which reported almost all necessary equations for heat and mass transfer and related mathematical models for all possible operation schemes of conventional stills.

No any special paper has been issued about operation, performance, efficiencies, etc. of large-scale practical application, of solar still plants and only few of the studies

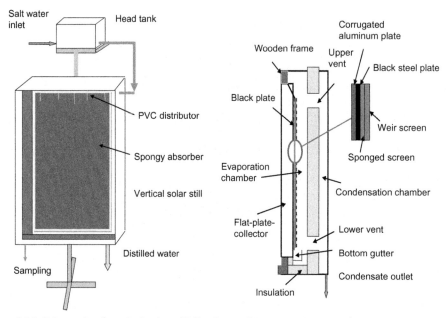

Figure 3.35 Schematic of vertical solar still (Boukar and Harmin, 2004; 2005).

include long-term operation data, ie, yearly data, either as modular units for individual use or as units connected together to form solar distillation plants. No information is also given about economic evaluation, ie, about real cost of water resulting by the increased capital cost to achieve thermal energy reuse or efficiency increase in connection to higher distillate output, except for very few cases. This may depend on the fact that by using local material and labor, economics differ considerably from place to place. In any case, the ratio of capital cost to the cost of water output may be useful information.

Frankly to construct, erect, and operate a solar distillation unit with unskilled personnel, for remote, poor communities, no mathematical calculations and models are necessary. Common sense and carefully fitting the individual components of the still are enough. Operating experience will help to optimize operation parameters as feed flow rate, etc. Mathematical models proved by long-term tests on site, ie, round the year are necessary for solar distillation data in order to be reliable.

Data collected for few days or weeks are not suitable for the design, optimization and/or construction of commercial solar stills in order to predict their output and achieve the optimum performance and efficiency.

Here some passive and active solar still plants having different configurations are presented, most of them not any more in operation, from the first ones to the latest in solar still bibliography starting from the simplest one. It is not possible to

Figure 3.36 The Aquamate, inflatable, simple solar still produces 0.5–2.0 liters per day fresh water. The design serves about 40 years, individuals and military worldwide. *Landfall navigation. Available from: http://www.landfallnavigation.com.*

describe here all studied solar still configurations. We selected some that according to our opinion are representative or played a significant role to the erection of solar distillation plants. Some have been already mentioned into the corresponding part of the text. The Report of Battelle Memorial Institute (Talbert et al., 1970) presents detailed information on erected solar distillation plants. For economic and engineering evaluation of solar distillation for small communities, the MacGill University Report (Lawand, 1968) is a useful tool. Figs. 3.36–3.43 present various such designs.

3.5 ECONOMIC EVALUATION OF SOLAR DISTILLATION

An economic comparison of various types of solar stills has many risks. The material, the geometry, and local prices differ considerably not only from place to place but also, in some occasions, from one site to another of the same region. In small, poor communities especially in remote regions, the main target is for a cheap as possible construction independently of life time. Cost of operation is also minimal in comparison with that for conventional desalination processes. The total capital cost investment per unit of still production capacity, coupled with the interest charges on this investment and

Figure 3.37 Pyramid type (called the Water Pyramid) solar still of Aqua-Aero, Holland has been installed in Gambia, Africa, and is a stand-alone water factory. It produces 2–3 liters fresh water per m³ of pyramid space and collects rain water about 1500 liters per m² ground area (Nitzsche, 2006).

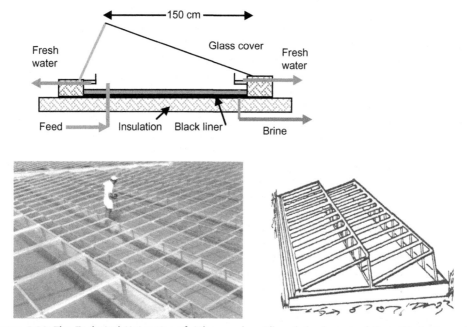

Figure 3.38 The Technical University of Athens solar still and the layout of the stills in the solar distillation plants erected in the Greek Islands. Photograph of the Nissiros Island solar distillation plant (Delyannis and Piperoglou, 1967b, 1968 *Photograph E. Delyannis).*

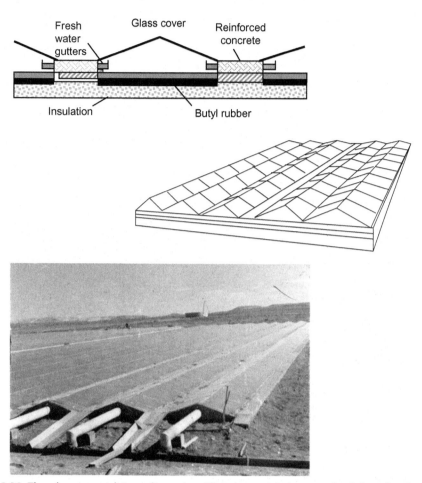

Figure 3.39 The glass-covered Australian solar still. Layout and photograph of the solar distillation plant at Central Australia Cooper and Read, 1974, Morse, 1968 *(Photograph by A. Delyannis)*.

the amortization rate depending upon the useful life of the system are the primary and controlling elements in the cost of solar distillation system.

Water cost produced in solar distillation plants depends mainly, as referred previously, on capital cost, operation and maintenance costs and from energy consumption by pumps and control system. This can be calculated from the following relations formed by a United Nations Committee (United Nations, 1971; Delyannis and Delyannis, 1985; Goosens et al., 2000) as:

$$C = \frac{10I(\overline{IA} + M\overline{R} + T\overline{I}) + 1000(K_c + s)}{A(Y_D - Y_R)} \qquad (3.172)$$

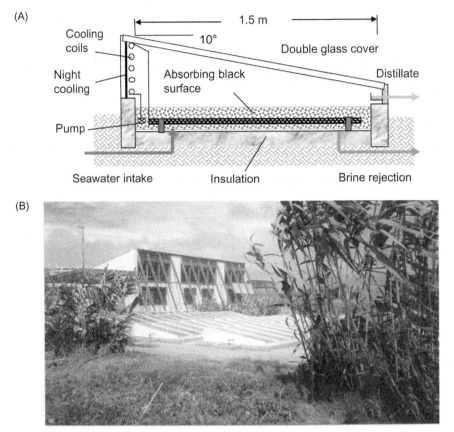

Figure 3.40 The TU Berlin-IPAT solar still (plant in the Island of Porto Santo, Madeira, Portugal) (Janisch and Dreschel, 1982).

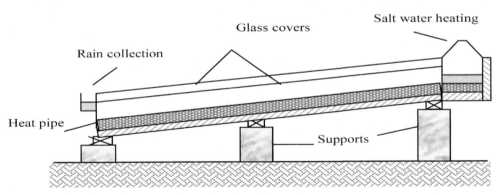

Figure 3.41 Inclined solar still with incorporated heat pipe onto the rain collector system (Janisch and Dreschel, 1982).

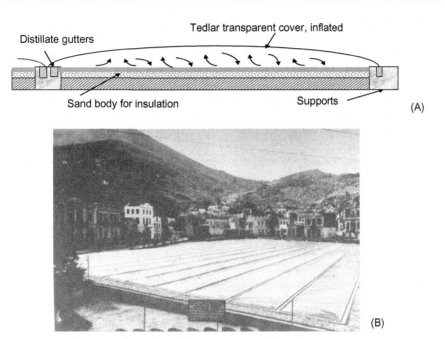

Figure 3.42 (A) The inflated plastic (tedlar) transparent cover solar system designed by Edlin (1965). (B) Solar distillation plants were installed in the Greek Aegean islands and in Cabo-Verde (Eckstrom, 1965, 1966). *(Photograph by E. Delyannis).*

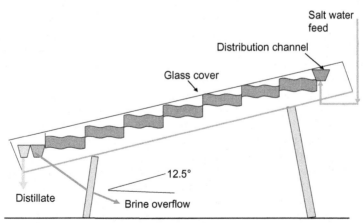

Figure 3.43 The SUNDIALS, Australia, commercial, tilted solar still called solar water purifier (Ward, 2003).

The annual interest and amortization rate is:

$$\overline{IA} = r\left[1 + \frac{1}{(1 + r(100)^n - 1)}\right] \qquad (3.173)$$

where

C	Cost of water	€ or US $
c	Operating and labor wage	€, US $, per man-hours
I	Total capital investment	€, US $
$\overline{IA}$	Annual interest and amortization rate	%
$\overline{MR}$	Annual maintenance and repair labor and material	% of investment
O	Annual operating labor	man-hours
r	Annual interest rate	%
S	Total constant cost salt water supply	€, US $
$\overline{TI}$	Annual taxes and insurance charges	% of investment
Y_D	Annual unit yield of distilled water	$(L\ m^{-2})$
Y_R	Annual unit yield of collected rain water	$(L\ m^{-2})$

A standardized procedure for costing, applicable to all designs of solar stills and all countries, was recommended by a Committee (Delyannis and Howe, 1971) organized following the Solar Energy International Conference in Melbourne, Australia, 1971. The issued Report by the Committee is summarized in tables. Table 3.4 includes the parameters that affect construction cost and Table 3.5 the parameters that influence the cost of produced water.

Singh and Tiwari (1995) and Tiwari and Singh (eolss) developed a cost estimation procedure based on yearly life time which can be applied to single, two effect and to multiple wick solar stills. The equations are formulated on Indian monetary system (Rupees), although they may be applicable to any other currency.

Table 3.4 Cost of Construction of Solar Distillation Systems

1. Construction of distillation units include:
   Construction of basin
   Construction of the cover
2. Land and site preparation for the system includes:
   Cost of land
   Suitable preparation of the land
   Fencing of the systems field
3. Auxiliaries include:
   Piping, pipes for salt water and fresh water circulation
   Storage of salt water and of produced fresh water
   Any other investment expenses
4. Total investment cost should be reported either per square meter of evaporating surface or per square meter of cover projected area

Table 3.5 Cost of Produced Distilled Water

1. Power of pumping, kg of fuel or kWh
2. Cost of raw water, if purchased
3. Pretreatment cost, if necessary
4. Maintenance cost of material
5. Labor for operation of the system and labor for repair and maintenance
6. Amortization of capital cost
7. Taxes and insurance, when applicable
8. Annual operating charges should be computed per m^3 of product water

$$C_{alcc} = C_{i,i}f_{crf} + M_{mc}f_{crf} S_{sv} F_{sff} + C_{fc}f_{crf} SS_s \qquad (3.174)$$

The parameters f_{crf} of capital recovery factor (CRF), M_{mc} maintenance cost, and F_{sff} the sinking fund factor (SFF) are evaluated from the following equations:

$$f_{crf} = CRF = [r(1+r)^n]/[(1+r)^n - 1] \qquad (3.175)$$

$$F_{sff} = SFF = [r]/[(1+r^n) - 1] \qquad (3.176)$$

$$M_{mc} = C_{i,i}R[(1+p)^{n-1} - (1\_r)^{n-1}]/(p-r)(1+r^{n-2}) \qquad (3.177)$$

The payback period, in years, at the time at which the first cost and annual expenses with compounded interest equals total energy cost savings is:

$$n_{pb} = \ln \frac{[(E - RC_{i,i})/(r-e)] - \ln[(E - RC_{i,i})/(r-p) - C_{i,i}]}{\ln[(1+r)/(1+p)]} \qquad (3.178)$$

where

C_{fc}	Fixed cost including installation and electricity	€
$C_{i,i}$	Initial capital investment, present value	€
C_{alcc}	Annualized life cycle cost	€
E	Yearly saved energy	€
e	Energy inflation rate	%
n	Number of useful years where the system operated satisfactory	—
p	Rate of increase in maintenance	%
r	Rate of interest	%
R	Initial maintenance cost at the end of the year	€
SS_s	Subsidized value	€
S_s	Scrap value	€

Kamal et al. (1999) performed an investigation for the estimation of cost of water produced by solar powered membrane distillation and by solar stills. They refer that the last 15 years the trend has been away from solar distillation plants. This applies almost for the next 12 years up today. They conclude that the contributing factors for this reason are:

- Cost of the large amount land area needed
- No money to be made from solar stills
- Too much confidence is given from researchers
- The technology was not seen by potential users as off-the-shelf equipment.

For this investigation as source a 315 kg d^{-1} water production solar distillation unit in the Grenadines was used. The comparison to membrane distillation unit of same capacity favored the solar still.

3.6 CONCLUSION

Conventional solar stills are very useful piece of equipment for the production of desalinated water from brackish or seawater for individual or for small communities use, especially for remote regions. As mainly are addressed to poor communities it is more economical to construct solar stills and/or small plants with local available materials and take care for proper operation and maintenance, if necessary.

Despite the numerous designs and means to increase productivity, the final results do not differ considerably between them in such a way that justifies a complicated design that increases the economy of the system. Unfortunately there exist not enough data from large solar distillation plants. It is obvious that the operation and the data collected from their operation may differ considerably from the data collected from a single solar still.

In general, solar stills have some advantages and also important disadvantages.

A. Advantages of solar stills

- They can be constructed with locally available, not expensive materials easily, as simple equipment for the production of fresh water
- They do not need special skilled personnel for the assembly, operation, and/or maintenance
- They can incorporate rain catchment canals to increase productivity, at low additional cost.

B. Disadvantages of solar stills

- They operate at low efficiency, about 30% or in best conditions at about 40−50%
- They need high initial capital cost which is counterbalanced by lower operational cost
- Solar distillation plants need large installation surface areas. If it cannot provided free the land price adds a considerable amount to the initial capital cost

- They are vulnerable to all extreme weather conditions, such as wind, storms, and hale, especially the plastic covered stills. They need strong foundations and stills have to be anchored to the earth
- Winds leave on the transparent cover a dust layer of sand or soil, dried leaves, and other contaminants decreasing considerably radiation penetration and consequently absorption, thus decreasing their efficiency. In these cases, a part of produced water is used to clean up the surfaces. This restricts the installation to protected sites.

Special care must be taken to avoid formation of algae and scale on the black absorbing liner by stagnant conditions.

REFERENCES

Abu Arabi, M., Mousa, H., Abdelrahman, R., 2009. Solar desalination unit with falling film. Desalin. Water Treat. 3, 58−63.

Abu-Hijleh, B.A.K., 1996. Enchanted solar still performance using water film cooling of glass cover. Desalination. 109, 235−244.

Abu-Hijleh, B.A.K., Mousa, H.A., 1997. Water film cooling over the glass cover of a solar still including evaporation effects. Energy. 22 (1), 43−48.

Aristotle, 1962. Book I, Chapter 9. 127/183 .Meteorologica. Harvard Univ. Press, Cambridge, MA

Ahsan, A., Fukuhara, T., 2008. Evaporative mass transfer in tubular solar still. J. Hydrosci. Hydraulic Eng. 26 (2), 15−25.

Ahsan, A., Fukuhara, T., 2009. Condensation mass transfer in unsaturated humid air inside tubular solar still. Annu. J. Hydraulic Eng. JSCE. 53, 97−102.

Ahsan, A., Fukuhara, T., 2010. Mass and heat transfer model of tubular solar still. Solar Energy. 84, 1147−1156.

Ahsan, A., Shafiur Islam, K.M., Fukuhara, T., Ghazali, A.H., 2010. Experimental study on evaporation, condensation and production of a new tubular solar still. Desalination. 260, 172−179.

Ahsan, A., Imleaz, M., Rahman, A., Yusuf, B., Fukuhara, T., 2012. Design, fabrication and performance of an improved solar still. Desalination. 292, 105−112.

Aybar, H.S., 2006. Mathematical modelling of inclined solar water distillation system. Desalination. 190 (1−3), 63−70.

Aybar, H.S., Assefi, H., 2009. Simulation of a solar still to investigate water depth and glass angles. Desalin. Water Technol. 7, 35−40.

Aybar, H.S., Egelioglu, F., Aticol, U., 2005. An experimental study on solar water distillation system. Desalination. 180 (1−3), 285−289.

Balan, R., Chandrasekaran, J., Shabmugan, S., Janaaarthanan, B., Kumar, S., 2011. Review on passive solar distillation. Desalin. Water Treat. 28, 217−238.

Bapeshwara, V.S.V., Singh, V., Tiwari, G.N., 1982. Transient analysis of double basin solar stills. Energy Convers. Manage. 23 (2), 83−90.

Belessiotis, V., (eolss). Solar Collectors. Available online: http://www.desware.net/Sample-Chapters/D06/D10-005.pdf.

Belessiotis, V., Delyannis, E., 2000. The history of renewable energies for water desalination. Desalination. 128, 147−159.

Belessiotis, V., Voropoulos, K., Delyannis, E., 1996. Experimental and theoretical method for the determination of daily output of a solar still−input−output method. Desalination. 100, 99−104.

Birket, J., 1984. A brief illustrated history of desalination, from Bible to 1940. Desalination. 50, 17−91.

Birket, J., (eolss). The History of Desalination Before Large-scale Use. Available online: http://www.desware.net/Sample-Chapters/D01/01-003.pdf.

Bittel, A., 1959. Zur Geschichte multiplikativer Trennverfahren (on Tales of multiple separation methods). Chemie-Ingenieur-Technik. 31 (6), 365—424.

Bloemer, J.W., Collins, R.A., Eibling, J.A., 1961. Field evaluation of solar sea water stills. Office of Saline Water (OSW), Res. Develop. Progr. Rept. No. 50.

Bloemer, J.V., Collins, R.A., Eibling, J.A., OSW Res. Develop. Progress Rept. No. 190 (1965a).

Bloemer, J.W., Collins, R.A., Eibling, J.A., Loef, G.O.G., 1965b. A practical basin type solar still. Solar Energy. 9, 197—203.

Bloemer, J.V., Collins, R.A., Eibling, J.A., OSW Res. Develop. Progress Rept. No. 19 (1966).

Bloemer, J.V., Collins, R.A., Eibling, J.A., 1970. Advances in Chemistry Series No. 28. American Chemical Society, pp. 166—177.

Boukar, M., Harmin, A., 2004. Parametric study of a vertical solar still under desert climatic conditions. Desalination. 168, 21—28.

Boukar, M., Harmin, A., 2005. Performance evaluation of a one-sided vertical solar still tested in the desert of Algeria. Desalination. 183, 113—126.

Boukar, M., Harmin, A., 2007. Design parameters and preliminary experimental investigation of an indirect vertical solar still. Desalination. 203, 444—454.

Brace Research Institute, 1993. The case for family-sized desalination system—methodology and questionnaire. Technical Report No. T-177.

Briegel, Z., 1918. Zur Entsalzung des Meerwassers bei Aristotles (The desalination of seawater by Aristotle). Chemiker Zeitung. 42, 302.

Cooper, P.I., 1969a. The absorption of radiation in solar stills. Solar Energy. 12, 333—346.

Cooper, P.I., 1969b. Digital simulation of transient solar still processes. Solar Energy. 12, 313—331.

Cooper, P.I., 1973. The maximum efficiency of single-effect solar stills. Solar Energy. 15, 205—217.

Cooper, P.I., Read, W.R.W., 1974. Design philosophy and operating experience for Australian solar stills. Solar Energy. 16, 1—8.

Correy, J.P., 1975. Vertical solar distillation. Solar Energy. 17, 375—378.

Dashtban, M., Tabrizi, F.F., 2011. Thermal analysis of a weir-type cascade solar still integrated with PCM storage. Desalination. 279, 415—422.

Delyannis, A., Howe, E.D., 1971. Report of Working Party on recommended procedure for costing of solar stills. CSIRO, Division of Mechanical Engineering, Internal Report No. 77, pp. 17—26, Melbourne.

Delyannis, A., Delyannis, E., 1974. Gmelin Handbuch der Anorganische Chemie, Sauerstoff, *Anhangband*, "Water Desalting", 2.17, 339 pp. Solar Energy as Heat Source. Springer Verlag, Berlin

Delyannis, A., Piperoglou, E., 1967a. Handbook of Saline Water Conversion Bibliography, Vol. 1, Antiquity to 1940, 121 pp.

Delyannis, A., Piperoglou, E., 1967b. Solar distillation developments in Greece. Sun at Work. First Quarter, 14—18.

Delyannis, A., Piperoglou, E., 1968. The Patmos solar distillation plant. Technical Note. Solar Energy. 12, 113—115.

Delyannis, E., 2003. Historic background of desalination and renewable energies. Solar Energy. 75, 357—366.

Delyannis, E., Belessiotis, V., 2004. Solar Water Desalination, Encyclopedia of Energy, Vol. 5. Elsevier Publishing Co, Oxford, UK, pp. 685—694.

Delyannis, E.E., Delyannis, A., 1985. Economics of solar stills. Desalination. 52, 167—176.

Duffie, J.A., Beckmann, W.A., 2006. Solar Engineering Thermal Processes. John Wiley and Sons, Inc., Hoboken, New Jersey, 908 pp., 4th Ed., 2013.

Dunkle, R.V., 1961. Solar water distillation, the roof type still multiple effect diffusion, Int. Development in Heat Transfer. Proc. ASME, Int. Heat Transfer. Part V, 895—902.

Eckstrom, R., 1965. Design and construction of the Symi still. Sun at Work. 10 (1), 6—8.

Eckstrom, R., 1966. New solar still on the Greek Islands. Sun at Work. 11 (2), 13.

Edlin, F.E., Air supported solar still. US Patent, 3.174.915, March 23, 1965.

Edlin, F.E., Willauer, D.E., 1961. Plastic films for solar energy applications. In: UN Conference, Rome, August 1961, Session III, 35/5/33, pp. 1—28.

El-Haggar, S.M., Awn, A.A., 1993. Optimum conditions for solar still and its use for a greenhouse using the nutrient film technique. Desalination. 94, 55—68.

El-Nashar, A., Delyannis, E., (eolss). A Short Historical Review in Desalination by Renewable Energies. Available online: http://www.desware.net/Sample-Chapters/D06/D10-002.pdf.

El-Sebaii, A.A., 2004. Effect of wind speed on active and passive solar stills. Energy Conversion & Management. 45, 1187—1204.

El-Sebaii, A.A., 2005. Thermal performance for a triple-basin still. Desalination. 174 (1), 23—37.

Fath, H.E.S., 1996. High performance of a simple design two effect solar distillation unit. Desalination. 107, 223—233.

Fath, H.E.S., 1998. Solar distillation—a promising alternative for water provision with free energy. Simple technology and clean environment. Desalination. 116, 45—56.

Grange, S.W., The transmissivity of selected glasses and plastics to solar radiation. Master of Science Thesis, University of California, May 1966.

Goosen, M.F.A., Sablani, S.S., Shayya, W.H., Paton, C., Al-Hinai, H., 2000. Thermodynamic and economic consideration in solar desalination. Desalination. 129, 63—89.

Hamdan, M.A., Mousa, A.M., Jabran, B.A., 1999. Performance of a solar still under Jordanian climate. Energy Convers. Manage. 40, 495—503.

Harding, J., Apparatus for solar distillation. Proceedings Institution Civil Engineers, London, 73 (1883), pp. 284—288.

Howe, E.D., 1968a. Pacific water systems using combined solar still and rain collection. Solar Energy. 10 (4), 178—181.

Howe, E.D., 1968b. Measurement and control in solar distillation plants. Desalination. 59, 307—320.

Janarthanan, B., Chandrasekaran, J., Kumar, S., 2005. Evaporative heat loss and heat transfer for open-and closed-cycle systems of a floating tilted wick solar still. Desalination. 18 (1—3), 291—305.

Janisch V., Dreschel, H. Solare Meerwasserentsalzung, (Seawater desalination), 1982, Gesellschaft fuer Technische zusammenarbeit (GTZ), Escborn, Germany, Projekt- No 77.2163.2-01.100.

Jurges, W., Gesunheit Ingenieur, 19 (1924), (in McAdams W. C. (1954), heat Transfer, 3rd Edition. McGraw Hill, New York.

Kabeel, A.E., 2009. Performance of solar still with concave wick evaporation surface. Energy. 34, 1504—1509.

Kabeel, A.E., El-Agouz, S.A., 2011. Review of researches and development on solar stills. Desalination. 276, 1—12.

Kalogirou, S.A., 2005. Seawater desalination using renewable emery sources. Prog. Energy Combust. Sci. 31, 242—281.

Kalogirou, S.A., 2014. Solar Energy Engineering, Processes and Systems. 1rst Edition 2009: 2nd Edition Academic Press, Elsevier, London, pp. 776—840.

Kamal, M.R., Simandle, J., Ayoup, J., 1999. Cost comparison of water produced from solar powered distillation and solar stills. Desalin. Water Reuse. 9, 74—75.

Khalifa, A.J.N., Hamood, A.M., 2009. Performance correlations for basin type solar stills. Desalination. 249, 24—28.

Khedim, A., Schwarzer, K., Faber, C., Mueller, C., 2004. Production décentralisée de l' eau potable á l' énergie solaire (Production of decentralized potable water by solar energy). Desalination. 168, 13—20.

Kiatsiriroat, T., Bhattaccharya, S.C., Wibulswas, P., 1987. Performance analysis of a multiple effect vertical still with flat plate solar collector. Solar Wind Technol. 4 (4), 451—457.

Kumar, A., (eolss). Configuration, Theoretical Analysis and Performance of Simple Solar Stills. Available online: http://www.desware.net/Sample-Chapters/D06/D10-018.pdf.

Lawand, T.A., 1968. Engineering and economic evaluation of solar desalination for small communities. Brace Research Inst., McGill Univ., Technical Report. No MT-6; pp. 262.

Lawand, T.A., 1975. Systems for solar distillation. In: Proceedings of International Conference on Appropriate Technologies for Semiarid Areas: Wind and Solar Energy for Water Supply, September 15—20, Berlin (West), pp. 201—250.

Lawand, T.A., Ayoub, J., (eolss). Materials for Construction of Solar Stills. Available online: http://www.desware.net/Sample-Chapters/D06/D10-021.pdf.

Lobo, P.C., Aranjo, S.R.D., 1978. Design of a simple multi-effect basin type solar still. In: Proc. Int. Solar Energy Congress, New Delhi, pp. 2026—2032.

Loef, G.O.G., Eibling, J.A., Bloemer, J.W., 1961. Energy balances in solar distillers. AIChE. J. 7, 641—647.

Loef, G.O.G., Eibling, J.A., Bloeme, J.W., 1969. Energy balance in solar distillers. AIChE. J. 7 (4), 641—649.

Malik, M.A.S., Tran, V.V., 1973. A simplified mathematical model for predicting the nocturnal output of a solar still. Solar Energy. 14, 571—585.

Malik, M.A., Tiwari, G.N., Sodha, M.S., 1982. Solar Distillation. Pergamon Press, Oxford.

Mathioulakis, E., Belessiotis, V., 2003. Integration of solar still in a multi-source, multi-use environment. Solar Energy. 75, 403—411.

Mathioulakis, E., Voropoulos, K., Belessiotis, V., 1999. Modelling and prediction of long-term performance of solar stills. Desalination. 122, 85—93.

Morse, R.N., 1968. The construction and installation of solar stills in Australia. Desalination. 5, 82—89.

Morse, R.N, Read, W.R.W., 1968. A rational basis for the engineering development of a solar still. Solar Energy. 12, 5—17.

Mousa, H., Abu Arabi, A., 2009. Theoretical study of water desalination by a falling film unit. Desalin. Water Treat. 12, 331—336.

Nebbia, G., Nebia-Menozzi, G., 1966. Aspeti storici della disalazione delle acque salmastre. Acqua Industr. 41—42, 3—20.

Nebbia, G., Nebia-Menozzi, G., 1967. A short history of water desalination. In: Proceedings of International Symposium, Milano, April 1966, pp. 129—172.

Nitzsche, M., 2006. Gambia tests solar distillation plant. Desalin. Water Reuse. 16 (2), 45.

Papanicolaou, E, Voropoulos, K, Belessiotis, V., 2002. Natural convective heat transfer in an asymmetric, greenhouse type solar still. Effect of inclination. Num Heat Transfer A. 42 (8), 853—880.

Phadatare, MK, Verma, S.K, 2007. Influence of water depth on internal heat and mass transfer in a plastic solar still. Desalination. 217 (1—3), 267—275.

Phadatare, M.K., Verma, S.K., 2009. Effect of cover material on heat and mass transfer coefficients in a plastic still. Desalin. Water Treat. 2, 248—253.

Rowly, F.B., Eckley, W.A., 1932. Surface coefficients as affected by direction of wind. Trans. A.S.H.V.E. 38, 35—45.

Rowly, F.B., Algren, A.B., Blackshaw, J.L., 1931. Surface conductance as affected by air velocity, temperature and character surface. Trans. A.S.H.V.E. 36, 501—508.

Schwarzer, K., Majumder, D., 2006. UK—Germany collaboration on solar still. Desalin. Water Reuse. 16 (1), 44—45.

Schwarzer, K., Vieira, M.E., Müller, C., Lehmann, H., Coutino, L., 2003. RIO 3—World Climate & Energy Event, December 1—5, Rio de Janeiro, Brazil.

Setodeh, N., Rahimi, R., Ameri, A., 2011. Modelling and determination of heat transfer voefficient in a basin solar still using CFD. Desalination. 268, 103—110.

Sharma, V.B., Mullic, S.C., 1991. Estimation of heat transfer coefficients, the upward heat flow and evaporation in a solar still. Trans. ASME, J. Solar Energy Eng. 113, 36—41.

Sharma, V.B., Mullic, S.C., 1993. Calculation of hourly output of a solar still. Trans. ASME, J. Solar Energy Eng. 115, 231—236.

SI, 2008. The International System of Units, NIST Special Publication 330, B.N. Taylor, A. Thompson, Eds.

Singh, A.K., Tiwari, G.N., 1992. Performance of a thermal evaporation of a multi effect distillation system. Heat Recovery CHP. 12 (5), 445—450.

Singh, A.K., Tiwari, G.N., 1995. Thermoeconomic analysis of various solar stills based on monthly performance, technical note. Int. J. Ambient Energy. 16 (2), 89—94.

Singh, A.K., Tiwari, G.N., 2004. Monthly performance of passive and active solar stills for different Indian climatic conditions. Desalination. 168, 145—150.

Smakdji, N., Kaabi, A., Lips, B., 2014. Optimization and modeling of a solar still with heat storage. Desalin. Water Treat. 52 (7—9), 1761—1769.

Sodha, M.S., Kumar, A., Singh, U., Tiwari, G.N., 1980. Transient analysis of solar still. Energy Convers. Manage. 20 (3), 191—198.

Sodha, M.S., Kumar, A., Tiwari, G.N., Tyagi, R.C., 1981. Simple multiwick solar still—analysis and performance. Solar Energy. 26 (2), 127−131.

Swinbank, W.C., 1963. Long wave radiation from clear skies. Quarter. J. Royal Meteorol. Service. Vol. 69.

Talbet, S.G., Eibling, J.A., Loef, G.O.G., 1970. Manual on Solar Distillation of Saline Water. Battelle Memorial Institute, Columbus Laboratories, 279 pp.

Tanaka, H., Nosoko, T., Nagata, T., 2000. A highly productivity basin-type-multi-effect coupled solar still. Desalination. 130, 279−293.

Telkes, M., 1953. Fresh water from seawater by solar distillation. Ind. Eng. Chem. 45 (5), 1105−1114.

Telkes, M., 1955. Solar stills. In: Proc. World Symposium on Applied Solar Energy, Phoenix, AZ, November 1−5, 1954. Johnson Reprint Corp., New York.

Tiwari, A.K., Tiwari, G.N., 2005a. Effect of the condensing cover slope on internal heat and mass transfer in distillation-an indoor simulation. Desalination. 180 (1−3), 73−88.

Tiwari A.K., Tiwari G.N., 2005b. An optimization of slope of condensing cover of solar still for maximum yield in summer climatic conditions. ISES Congress, Orlando, FL, USA, August 6−12, 2005. Paper No. 1146.

Tiwari, A.K., Tiwari, G.N., 2006. Effect of water depth on heat and mass transfer coefficients in a passive solar still: in summer climatic conditions. Desalination. 195 (1−3), 78−94.

Tiwari, G.N., 2002. Solar Energy, Fundamentals, Design, Modelling and Application. Alpha Science International Ltd., Pangbourne, England.

Tiwari, G.N., Kumar, A., 1988. Nocturnal production bu tubular solar stills using waste heat to preheat brine. Desalination. 69 (3), 309−318.

Tiwari, G.N., Madhuri, 1987. Effect of water depth on daily yield of the still. Desalination. 61, 67−75.

Tiwari G.N., Singh A.K., (eolss). Solar Distillation. Available online: http://www.desware.net/Sample-Chapters/D01/E6-106-18.pdf.

Tiwari, G.N., Mukherjee, K., Ashoh, K.R., Yadav, Y.P., 1986. Comparison of various designs of solar stills. Desalination. 60, 191−202.

Tiwari, G.N., Singh, A.K., Tripathi, R., 2003a. Present status of solar distillation. Solar Energy. 75, 367−373.

Tiwari, G.N., Shukla, S.K., Singh, P., 2003b. Computer modelling of passive/active solar stills by using inner glass temperature. Desalination. 154 (2), 197−206.

Tleimat, V.W., Howe, E.W., 1966. Nocturnal production of solar distillers. Solar Energy. 10 (2), 61−66.

Tleimat, V.W., Howe, E.W., 1969. Comparison of plastic and glass covers for solar distillers. Solar Energy. 12, 293−404.

Tripathi, R., Tiwari, G.N., 2005. Effect of water depth on internal mass and heat transfer for active distillation. Desalination. 173 (2), 187−200.

United Nations, Department of Economic and Social Affairs, Solar Distillation, V. Baum, A, A, Delyannis, J. AS. Duffie. E. D. Howe, G. O. G. Loef, R. N. Morse, H. Tabor, U.N. ST/ECA/121, 1971.

Von Lippman, E., 1910. Chemisches und Alchemisches aus Aristoteles (Chemistry and Alchemistry from Aristotle). Chemiker Zeitung. 2, 233−300.

Von Lippman, E., 1911. Die Entsalzung des Meerwassers bei Aristoteles (The desalination of sea water by Aristotle). Chemiker Zeitung. 35, 629−630.

Voropoulos, K., 2003. An integrated solar still-thermal storage tank (in Greek). Dissertation. 270 pp.

Voropoulos, K., Delyannis, E., Belessiotis, V., 1996. Thermo-hydraulic simulation of a solar distillation system under pseudo steady state conditions. Desalination. 107, 45−51.

Voropoulos, K., Mathioulakis, E., Belessiotis, V., 2000. Transport phenomena and dynamic modelling in greenhouse type solar still. Desalination. 129 (3), 273−281.

Voropoulos, K., Mathioulakis, E., Belessiotis, V., 2003. Solar stills with solar collectors and storage tank-analytical simulation and experimental variations of energy behaviour. Solar Energy. 75 (3), 199−206.

Ward, J., 2003. A plastic solar water purifier with high output. Solar Energy. 75, 433−437.

Watmuff, J.H., Charters, W.W.S., Proctor, D., 1977. Solar and wind induced external coefficients—solar collectors. Technical Note, COMPLES, Revue International d-Heliotechnique, 12th Semester, 56.

Wheeler, N.W., Evans, W.W., 1879. US P., 102.633.

Yadav, Y.P., Tiwari, G.N., 1987. Monthly comparative performance of solar still coupled to various designs. Desalination. 67, 565–572.

Yadav, Y.P., Yadav, S.K., 2004. Parametric studies on the transient performance of a high temperature solar distillation system. Desalination. 170, 251–262.

Zeng, H., Chang, Z., Chen, Z., Sie, G.X., Wang, H., 2013. Experimental investigation and performance analysis on a group of multi-effect tubular solar desalination devices. Desalination. 311, 62–68.

EXTENDED RECENT REVIEWS ON SOLAR STILLS

Ayoub, G.M., Malaeb, L., 2012. Developments in solar still desalination systems: a critical review. Crit. Rev. Environ. Sci. Technol. J.. Available from: http://dx.doi.org/10.1080/10643389.2011.

Balan, R., Chandrasekaran, J., Shabmugan, S., Janaaarthanan, B., Kumar, S., 2011. Review on passive solar distillation. Desalin. Water Treat. 28, 217–238.

Kabeel, A.E., El-Agouz, S.A., 2011. Review researches and developments on solar stills. Desalination. 276, 1–12.

Kalogirou, S.A., 2005. Seawater desalination using renewable energy sources. Prog. Energy Combust. Sci. 31, 242–281.

Kaushal, A., Varum, 2010. Solar stills: a review. Renew. Sustainable Energy Rev. 14, 446–453.

Mathioulakis, E., Belessiotis, V., Delyannis, E., 2007. Desalination by using alternative energy: review and state-of-the-art. Desalination. 203, 346–365.

Velmurugan, V., Sithar, K., 2011. Performance analysis of solar still on various factor affecting the productivity—a review. Renew. Sustainable Energy Rev. 15, 1294–1304.

CHAPTER FOUR

Membrane Distillation

4.1 INTRODUCTION

The membrane distillation (MD) is a relative new and promising technology for brackish and seawater desalination. It is a technology of small to medium desalinated water supplies, operating at atmospheric pressure and temperatures below 100°C and therefore is suitable for solar energy supply.

MD is a nonisothermal membrane separation process, in which the water vapor molecules are transferred through a microporous hydrophobic membrane. Driving force is the partial pressure difference across the membrane that is created by the temperature difference between the two sides of the membrane. This is an attractive alternative solution for medium desalinated water supplies, giving high quality distillate and the possibility of operating at low temperatures. Compared to other membrane separation processes, MD operates at low pressures, lower than the pressures of the reverse osmosis (RO). Moreover, the capability of utilizing solar thermal energy or waste heat from other processes makes MD an energy and cost-efficient and environmental friendly process.

One of the most crucial aspects of the MD is to have at disposal membranes with desirable properties for the better performance of the process. The membrane does not substantially takes part in the separation process, but acts as a physical barrier between the two phases preventing the penetration of the liquid phase into the membrane pores and simultaneously provides the liquid–gas interface for mass and heat transfer. In MD process, membranes with low thermal conductivity are used, but with porosity and other physicochemical properties that cannot meet the requirements for high performance of the process, they were originally intended for other separation processes (microfiltration and nanofiltration). An ideal MD membrane should exhibit high permeability (or low membrane mass transfer resistance), high liquid entry pressure of water (LEP$_w$, ie, hydrophobicity), and low thermal conductivity. Moreover, it should have good thermal stability and excellent chemical resistance to feed solution. In particular, the essential characteristics that a membrane for the MD process should have are as follows (Khayet, 2011; Harryson et al., 1986):

- The membrane must be porous and to be comprised of a single layer or multi layers, wherein at least one of the layers should be made of a hydrophobic material
- No capillary condensation should take place inside the membrane pores

Thermal Solar Desalination

- The membrane should not alter the vapor—liquid equilibrium of each component involved in the process
- LEP_w must be as high as possible in order not to allow pore wetting to occur. High LEP_w can be achieved by using materials with low surface energy and small maximum pore size
- The membrane pore size may be from several nanometers to few micrometers (100 nm—1.0 µm). The pore size distribution should be as narrow as possible, so as the liquid solution does not penetrate the membrane pores
- The membrane tortuosity should be small, since it is inversely proportional to the vapor mass flux
- The porosity must be high in the limits of 70—80%. High porosity increases the membrane permeability and provides larger space for evaporation. In spite of that, higher porosity leads to decrease of membrane mechanical strength and stability, which tends to form cracks and to be compressed
- The membrane should have an optimized value of thickness, since thin membranes exhibit high permeability, while by increasing thickness the energy losses are minimized
- The thermal conductivity of the membrane should be as low as possible, so that thermal losses due to heat transfer by conduction through the membrane is minimal
- The membrane should be resistant to organic fouling
- The membrane must have good thermal stability and excellent chemical resistance to various feed solutions.

4.1.1 A Short Historical Introduction

MD is derived from its sister method "pervaporation," which was applied for the first time by Cover in 1917. The MD process started as an application in the 1960s. In early 1963, the first US patent application filed by Bodell (1968) named the method as "membrane distillation using silicone rubber membrane." Later, the first traced report on the method is given by Van-Haut and Henderyckx (1967), although they referred two earlier publications on water vapor permeability through the pores of a flat sheet membrane. In 1967, a new US patent was issued by Weyl (1967) on "the recovery of demineralized water from saline water," who proposed also the use of hydrophobic porous membranes. In the late 1960s, Findley published the results of his work on vaporization through porous membrane and later in a second article he presented a basic theoretical study and the experimental results of direct contact membrane distillation (DCMD), using a variety of membrane materials (Findley, 1967, 1969). Findley concluded that "if low cost, high temperature, long-life membranes with desirable characteristics can be obtained, then MD process could become a successful method for water desalination." At that time, the interest in this process has been faded quickly, due partly to the observed lower MD production compared to the RO method. With the advent of new membrane manufacturing techniques in the early 1980s, interest in MD was renewed, as became available membranes with porosities as high as 80% and thicknesses as low as 50 μm.

In May 1986, in Rome, at the "Workshop on Membrane Distillation," mainly to standardize the relevant terminology, different names were proposed for that method such as transmembrane distillation, thermopervaporation, pervaporation, and membrane evaporation (Franken and Rippenger, 1988; Smolders and Franken, 1989). Eventually, the name "Membrane Distillation" was established as the most suitable for the process. The term "membrane distillation" arises from the similarity of the process to conventional distillation, since both methods rely on liquid–vapor equilibrium as a basis for separation.

In addition to the name of the process, in that Workshop some characteristics that the method should have were defined:
- The membrane must be porous
- The membrane should not be wetted by the liquid phase of the process
- No capillary condensation should take place inside the membrane pores
- During the process, the membrane must not alter the vapor–liquid equilibrium of each component
- At least one side of the membrane should be in direct contact with the process liquid (mainly brackish water or seawater)
- For each component, the driving force of the process is the partial pressure difference of the vapor phase.

4.2 TERMINOLOGY

In this section, a short description of the specific terminology involved in the MD process is provided, which is different from the two processes, the solar distillation and the humidification–dehumidification, previously presented. Those specific terms include the key vocabulary used in the international literature related with scientific, technical, and commercial aspects of the membrane desalination processes. General membrane features, characteristic properties, and performance parameters of both the membranes and the MD process are presented as comprehensively as possible in the English language. The terminology is taken from the published articles and reports (Smolders and Franken, 1989; Boddeker, 1990; Koros et al., 1996; Gekas, 1988).

Back flushing (BF) is the temporary reversal of the direction of the permeate flow. It is achieved by applying pressure in the opposite direction of the flow, so that the feed stream enters the module through the permeate side.

Composite membranes (CM) are membranes made from composite materials (organic and inorganic), usually prepared by the deposition of a thin polymer film on a porous substrate.

Concentration factor (CF) refers to the degree of increasing the concentration of a component in a membrane operation. It is expressed as the ratio of the concentration of the component in the retentate stream (C_R) to the concentration of the same component in the feed stream (C_F) and is given as:

$$CF = \frac{C_R}{C_F} \tag{4.1}$$

Concentration polarization coefficient (CPC). The concentration polarization is a frequent phenomenon that occurs in the pressure-driven membrane processes. It is the development of a concentration gradient of the retained components i in a boundary layer near the membrane surface. This is due to the accumulation of the retained compounds near the membrane—liquid interface, which leads to the decrease of the partial pressure difference and consequently to the reduction of the mass flux of the components (here the water vapor) passing through the membrane. This phenomenon is expressed by the CPC, which is the ratio of the concentration difference of the retained components in the feed—membrane interface and in the permeate stream ($C^i_{f,m}, C^i_{m,p}$) to the concentration difference of the retained components in the feed and solution and in the permeate stream ($C_{f,i}, C_{p,i}$) and is given as follows (Lawson and Lloyd, 1997):

$$CPC = \frac{C^i_{f,m} - C^i_{m,p}}{C_{f,i} - C_{p,i}} \tag{4.2}$$

Cross-flow (CF). Mode of a membrane module operation, where the feed stream and the permeate stream flow parallel to the membrane surface either in cocurrent flow or in countercurrent flow.

Diffusion (D) refers to a fundamental, irreversible process, where the concentration difference within a system is reduced simultaneously with the flux of the components from the highest to the lowest concentration. The diffusion is continued until dynamic equilibrium occurs between the molecules. The forces acting in the diffusion are the partial pressure difference of the components. Diffusion is expressed by the diffusion coefficient or diffusivity, $D \, [=] \, \text{m}^2 \, \text{s}^{-1}$.

Dynamic membranes (DM) are membranes in which an active layer is formed on the membrane surface by the deposition of substances contained in the fluid to be treated.

Evaporation efficiency (EE) is a parameter that characterizes the efficiency of an MD operation. EE is defined as the ratio of the heat needed for evaporation to the total heat input in the membrane module. Besides evaporation, a certain heat transfer due to conduction also takes place. Therefore, EE is always lower than unity.

Flux (J_i) refers to the density of the flow rate of moles, mass, or volume of a specified component (here the vapor) passing per unit time through a unit of membrane surface area. The flux can be described as:

- Mass flux, J_m [=] $kg\ m^{-2}\ s^{-1}$
- Molar flux, J_{mol} [=] $kmol\ m^{-2}\ s^{-1}$
- Volume flux, J_v [=] $m^3\ m^{-2}\ s^{-1}$.

Fouling. The deposition of various materials, mainly microorganisms or organic substances, which form a thin film deposited on the external surface of the membrane, as its pores, leading to a decrease of the mass flux of the permeating components.

Liquid entry pressure of water (LEP$_w$) is the minimum pressure that must be applied onto pure water before it penetrates into the nonwetted (dry) membrane pores. The wetting pressure, as sometimes is called, is a function of the membrane pore size and the membrane hydrophobicity.

Membrane is commonly regarded as a thin physical barrier that allows the selective transport of some mass species or mixture molecules, due to a certain driving force, such as the pressure gradient, the concentration gradient, or the chemical potential gradient.

Membrane module (MM) refers to an independent process assembly, the smallest practical unit, containing one or more membranes and supporting structures.

Membrane distillation (MD) is a distillation process, in which the liquid and gas phases are separated by a porous hydrophobic membrane, the pores of which must not be wetted by the liquid phase.

Membrane performance (MP). MP is characterized by the flux of a component through the membrane and the selectivity (or the separation factor).

Membrane thickness (δ). An important membrane parameter, as it gives information on both the mechanical strength of the membrane and the expected flux of the component passing through the membrane.

Permeability (P_i) is defined as the flux of a component (here the vapor) through the membrane per unit driving force per unit membrane thickness. The permeability is a function of permeant concentration and permeant diffusivity in the membrane. It is referred to every liquid or gas component involved in a fluid mixture and it is expressed as:

- Solvent permeability, P_v [=] $m\ s^{-1}\ Pa^{-1}$
- Solute permeability, P_s [=] $kmol\ m^{-2}\ s^{-1}\ Pa^{-1}$
- Gas permeability, P_g [=] $Nm^3\ m^{-2}\ s^{-1}\ Pa^{-1}$.

Permeance (Q_i) is the transport flux of a component through the membrane per unit driving force. It is defined as kmol m^{-2} s^{-1} Pa^{-1}.

Permeate (P) is the portion of the gas mixture or liquid solution from the feed stream that passes through the membrane. In the case of MD process, the term "distillate" is also used to describe the permeate.

Pore size (d_p or r_p) is related with the diameter (d_p) or the radius (r_p) or with equivalent values for irregular cross-sections of membrane pores. Membrane pores are usually nonuniform of different size and morphology and with abrupt changes in direction.

Porosity (ε) is considered the mass of a material, which contains finely passages (pores) that allow the fluid flow through them. It is defined as the ratio of the volume of the pores to the total volume of the membrane (that is to say the volume of the solid mass of the membrane and the pores). Pore volume is considered as the gas volume trapped inside the membrane pores. Porosity can be expressed by the densities of the membrane, ρ_m, and the material, ρ_p, as:

$$\varepsilon = 1 - \frac{\rho_m}{\rho_p} \tag{4.3}$$

Process efficiency (PE) is a parameter that characterizes the efficiency of the MD process. It is defined as the ratio of the heat which contributes to the evaporation of the distillate to the total heat input to the process. In case that the process only consists of the membrane module, then PE is equal to EE.

Recovery (R) is regarded as the ratio of the amount of a specific component A collected as a useful product to the amount of the same component entering the process.

Retentate (R) is the portion of the gas mixture or liquid solution from the feed stream that has been depleted of penetrants and leaves the system without passing through the membrane. In RO as well as in MD process, this term is referred mainly to the brine.

Retention and retention coefficient (RC) are referred to the membrane ability to hinder a component to pass through its pores. It is expressed by RC, which is defined as the degree of separation of a certain component from the solution, through the membrane, under defined operating conditions. RC is given by the following equation:

$$R = 1 - \frac{C_P}{C_R} \tag{4.4}$$

where C_P and C_R are the concentrations of the component at the permeate and retentate streams, respectively.

Selectivity (S) is the ratio of the permeabilities of two components A and B that pass through the membrane.

Separation factor (SF) is defined as the ratio of the concentrations of two components A and B in the permeate stream relative to the concentrations of these components in the retentate stream. In MD process, the separation factor is used in case that the permeating components are volatile compounds (VOCs).

Temperature polarization coefficient (TPC). Temperature polarization (TP) is a known phenomenon that occurs in the MD process. It is expressed with TPC, which cannot be measured directly, but it can be determined only when the hydrodynamic conditions on both sides of the membrane are known. Even if that coefficient is physically not correct, it is used very often in the scientific literature. The value of TPC is defined as the ratio of the temperature difference between the evaporation and condensation surface $(T_{f,m}, T_{m,p})$ to the temperature difference between the feed and permeate stream (T_f, T_P) and is given as:

$$TPC = \frac{T_{f,m} - T_{m,p}}{T_f - T_p} \tag{4.5}$$

Tortuosity (τ) is defined as the ratio of the length of the actual path that the fluid travels inside the porous material to the shortest path length in the direction of the flow. It reflects the random orientation of the network of membrane pores.

Vapor mass transfer coefficient (K_m) is expressed as the ratio of the vapor flux through the membrane to the partial pressure difference between the two sides of the membrane. It is an important parameter that depends on the characteristic properties of membrane, such as the porosity, the pore size, the tortuosity, and the membrane thickness, as the temperature operation range. It is usually referred in literature as membrane mass transfer coefficient or permeance.

4.3 MEMBRANE DISTILLATION

4.3.1 The Process

MD is a relative simple hybrid process, which is a combination of the thermal distillation and the membrane technology. It is, that is to say, a thermal membrane separation process (Fig. 4.1), in which water vapor molecules and/or volatile compounds are transferred from a hot aqueous solution (usually saline water) through a microporous hydrophobic membrane because of the partial pressure difference created due to the temperature difference on both sides of the membrane (Lawson and Lloyd, 1997).

The operation of the MD process is based on the principle of the liquid–vapor equilibrium (Fig. 4.2). The hot aqueous solution flows towards the hot membrane surface, with which is in direct contact, releasing vapor. The hydrophobic nature of

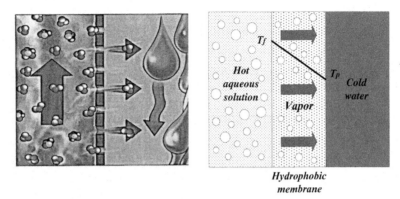

Figure 4.1 Schematic illustration of membrane distillation process.

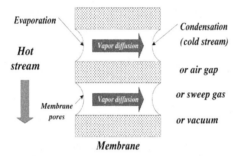

Figure 4.2 Principle of operation of MD process.

membrane prevents the penetration of the liquid phase into the membrane pores, because of the surface tension forces, but allows the passage of the vapor molecules through it. As the surface tension of the aqueous phase is higher than the critical surface tension of the polymeric film of the membrane, it withholds the liquid solution so as not to enter into the membrane pores. As a result, a liquid–vapor interface is created in the membrane pores entrance, wherein the liquid phase is separated from the gas phase. Since the process is nonisothermal, the temperature difference across the membrane creates a partial pressure gradient, which causes firstly the water vapor molecules and the volatile components to start to evaporate in the liquid–vapor interface at the entrance of the membrane pores, and secondly the formed vapor to diffuse in the membrane. As the vapor pressure difference is maintained between the two membrane sides, the vapor molecules are transferred through the membrane pores from the high partial pressure area (hot membrane surface) to the low partial pressure area (cold membrane surface), and is finally condensed either internally downstream of the membrane or externally outside the membrane module (Lawson and Lloyd, 1997; Curcio and Drioli, 2005; El-Bourawi et al., 2006).

4.3.2 Process Configurations

The MD process, depending on the condensation and the vapor recovery and the application of driving force, can be classified into four different configurations (El-Bourawi et al., 2006; Alkhudhiri et al., 2012):

- **Direct Contact Membrane Distillation (DCMD)**: In this technique (Fig. 4.3), the membrane is in direct contact with the hot aqueous feed solution and the cold stream of water on both sides. The partial pressure difference, created due to the temperature difference in the two membrane sides, forces the water molecules to evaporate at the hot feed stream—membrane interface. The formed vapor is passed through the membrane pores and is condensed in the membrane—cold stream interface in the membrane module. DCMD is the simplest MD configuration and is widely applied in the desalination process (Martinez-Diaz and Florido-Diez, 2001) and in the concentration of aqueous solutions (Drioli et al., 1987). Its main drawback is the heat loss due to the conduction.

- **Air Gap Membrane Distillation (AGMD)**: It is a variation of MD (Fig. 4.4), wherein an air gap is interposed between the membrane and a condensation surface, which is cooled by a cold water stream. In this case, the vapor molecules penetrate through the membrane pores and the air gap region and they finally condense in the cold surface inside the membrane module. Although the design of this configuration has the benefit of the reduced heat lost by conduction, an additional resistance to mass transfer is created, which is considered as a drawback. AGMD is a suitable method for seawater desalination (Kubota et al., 1988) and for the removal of VOCs from aqueous solutions (Banat and Simandl, 1996).

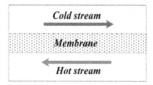

Figure 4.3 Direct contact membrane distillation.

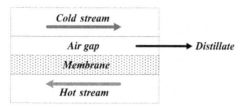

Figure 4.4 Air gap membrane distillation.

- **Sweep Gas Membrane Distillation (SGMD)**: This configuration is based on the use of cold inert gas (eg, nitrogen or air), which acts as a carrier of the vapor formed from the permeate side of the membrane in a separate chamber outside of the membrane module, where the condensation takes place (Fig. 4.5). SGMD is useful for removing VOCs from aqueous solutions (Duan et al., 2001). The main disadvantage of this technique is that a large condenser is required, as the small volume of the formed vapor is diffused in a large volume of sweep gas.
- **Vacuum Membrane Distillation (VMD)**: In this method, the driving force is the vacuum applied at the permeate side of the membrane by a vacuum pump. As the applied vacuum pressure is lower than the saturation vapor pressure, the released vapor is sucked from the membrane pores and is condensed in a separate condenser outside of the membrane module (Fig. 4.6). The VMD configuration is mainly used for the separation of aqueous volatile solutions (Sarti et al., 1993). Its main advantage is the negligible heat loss due to conduction.

4.3.3 Advantages, Disadvantages, and Applications

The MD process constitutes an attractive alternative solution to the conventional desalination techniques, due to the production of high purity distillate and the possibility of operating at lower temperatures. Compared to other membrane technologies (reverse osmosis, nanofiltration, and electrodialysis), MD operates at relative low pressures leading to low equipment cost and increased process safety. The mechanical properties of the membranes used are more flexible than those of the other separation processes and their large available surface allows high concentration of vapor molecules. Significant advantage of MD is the greater separation factor of VOCs, as well as

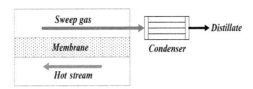

Figure 4.5 Sweep gas membrane distillation.

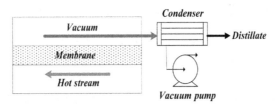

Figure 4.6 Vacuum membrane distillation.

the high, in percentage, 100% theoretical, rejection of ions, macromolecules, colloidal cells, and nonvolatile components (eg, salts) (Al-Hinai and Drioli, 2008). The capability of utilizing solar thermal energy or even waste heat from other processes makes MD an energy and cost-efficient and environmental friendly process (Koschikowski et al., 2003; Ding et al., 2005).

Additional benefits of MD process may be considered (Hsu et al., 2002): (1) the reduced vapor space compared to the conventional distillation, (2) the absence of limitations caused by the osmotic pressure effects, (3) the feed water does not require extensive chemical pretreatment, (4) the negligible organic fouling, (5) the compactness of process units as well as their simple and fast assembly and installation, and (6) the low energy cost.

Shneider et al. (1988) report also that:
- All the component parts usually consist of plastic material and as a result there are no corrosion problems caused by seawater
- The desalinated water is free of particles and bacteria
- The process depends only on the pressure and consequently on the temperature developed on both membrane sides
- The diffusion distance of the vapor molecules between the evaporation and condensation is extremely short due to that the membrane is very compact and thin (m^2/m^3)
- For systems with heat recovery, the performance factor can reach up to 8.

In spite of the large number of advantages that the MD process presents, it has not been fully commercialized due to some disadvantages such as:
- The low flux of the produced vapor compared to the other membrane separation processes
- The high sensitivity of the produced vapor flux in the feed concentration and temperature conditions due to the concentration and temperature polarization (TP) phenomena
- The risk of membrane wetting from the liquid phase of the feed aqueous solution
- The high membrane cost
- The low thermal energy performance
- The high-energy consumption compared to the reverse osmosis.

The MD process is widely used for the production of high water purity from the desalination of brackish and seawater (Martinez-Diaz and Florido-Diez, 2001; Kubota et al., 1988; Drioli et al., 2002) and has been successfully applied to the concentration of ions, colloids, and other nonvolatile aqueous solutions (Drioli et al., 1987; Gryta et al., 2001). In recent years, MD has been proposed as a separation technique of azeotropic aqueous mixtures, such as the alcohol—water mixtures (Udriot et al., 1994; Banat and Simandl, 1993), removal of trace volatile organic compounds (VOCs) (Banat and Simandl, 1996; Duan et al., 2001), or extraction of other organic compounds from water (Sarti et al., 1993), as well as concentrating acids (Tomaszewska et al., 1995; Khayet et al., 2004). Its applications have been also extended to food processing for the concentration of juices and milk (Nene et al., 2009), to biomedicine for the removal of water from blood (Sakai et al., 1986), to textile industry for the removal of dyes (Calabro et al., 1991), to wastewater and radioactive solutions treatment (Zolotarev et al., 1994; Zakrzewska-Trznadel et al., 1999), and to many other areas, where high temperature processes cause fluid degradation.

4.4 MASS AND HEAT TRANSFER

4.4.1 Direct Contact Membrane Distillation

4.4.1.1 Mass Transfer

In DCMD, the membrane is in direct contact with the hot aqueous feed solution and the cold stream on both sides. The mass transfer in that configuration, as illustrated in Fig. 4.7, involves the following steps:

- Transport of water vapor molecules from the hot feed stream to the hot feed stream—membrane interface through the feed boundary layer
- Diffusion of water vapor molecules through membrane pores and their convection from the hot feed stream—membrane interface to the membrane—cold stream interface
- Transport of water vapor molecules from the membrane—cold stream interface to the cold water stream through the cold stream boundary layer.

The third step is not taken into account, since only the water vapor molecules are transferred to the permeate stream (their mole fraction is equal to 1).

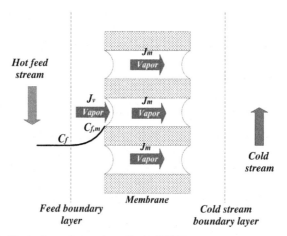

Figure 4.7 Schematic illustration of mass transfer in DCMD.

4.4.1.1.1 Mass Transfer Through the Feed Boundary Layer

The boundary layer resistance to mass transfer can be ignored when a single component, here the pure water, is considered to be present to the feed stream. On the contrary, when more than one components are present to the aqueous feed solution, then a concentration gradient is created in the feed area.

In the case that the aqueous feed solution contains nonvolatile components (eg, salt), an additional boundary layer is formed adjacent to the membrane surface and due to the permeation of water vapor through the membrane pores, the solute concentration at the feed–membrane interface ($C_{f,m}^s$) becomes higher than that at the bulk feed (C_f^s) with time as long as the separation process is taking place (Fig. 4.7). According to Eq. (4.2) and taking into account that the retained components at the feed–membrane interface are non-VOCs, ie, $C_{P,i} = 0$, then the CPC term that is used to quantify the mass transport resistance within the concentration boundary layer is (Martinez-Diez and Vazquez-Gonzaalez, 1999):

$$CPC = \frac{C_{f,m}^s}{C_f^s} \tag{4.6}$$

Applying mass balance across the feed boundary layer, based on the film theory (Bird et al., 2002), a relationship is derived between the mass flux of water vapor that passes through the membrane, J_v (kg m^{-2} s^{-1}), the solute mass transfer coefficient, $k_{f,s}$ (m s^{-1}), as well as the solute concentration at the feed stream, C_f^s (mol m^{-3}):

$$C_{f,m}^s = C_f^s \, \exp\left(\frac{J_v}{\rho_f k_{f,s}}\right) \tag{4.7}$$

where ρ_f (kg m^{-3}) is the density of the aqueous feed solution.

The solute mass transfer coefficient can be estimated from Sherwood number via an appropriate dimensionless empirical correlation for mass transfer, which is derived employing an analogy between the mass and heat transfer. This involves substitution of Nusselt number for the Sherwood number and Schmidt number for Prandtl number. The empiric Sherwood correlations for mass transfer are presented in Table 4.1 and they have the following form:

$$Sh = a Re^b Sc^c \tag{4.8}$$

where Sh, Re, and Sc are the dimensionless numbers Sherwood, Reynolds, and Schmidt, respectively, which are expressed as:

$$Sh = \frac{k_{f,s} d_h}{D_{AB}}, \quad Re = \frac{u_f d_h \rho_f}{\mu_f}, \quad Sc = \frac{\mu_f}{\rho_f D_{AB}} \tag{4.9}$$

where d_h (m) the hydraulic diameter, D_{AB} (m s^{-2}) the binary diffusion coefficient in the liquid phase, u_f (m s^{-2}) the fluid flow velocity in the hot feed stream, and μ_f (kg m^{-1} s^{-1}) the solution viscosity.

Table 4.1 Empirical Sherwood Correlations for Mass Transfer for Various Membrane Modules of Different Geometry (Chiam and Sarbatly, 2013)

Empirical Sherwood Correlations	Description
$$\text{Sh} = 1.86 \left(\text{Re Sc} \frac{d_h}{L} \right)^{1/3}$$	Laminar flow (Re < 2100): For flat sheet and tubular membrane modules
$$\text{Sh} = 3.66 + \frac{0.0668 \left(\text{Re Sc } d_h/L \right)}{1 + 0.045 \left(\text{Re Sc } d_h/L \right)^{2/3}}$$	Laminar flow (Re < 2100): For hollow fiber membrane modules
$$\text{Sh} = 0.116(\text{Re}^{2/3} - 125)\text{Sc}^{1/3} \left(1 + \left(\frac{d_h}{L} \right)^{2/3} \right)$$	Transition flow (2100 < Re < 10,000): For flat sheet membrane modules
$\text{Sh} = 0.023\text{Re}^{0.8}\text{Sc}^n$ $n = 0.4$ for heating, $n = 0.3$ for cooling	Turbulent flow (Re > 2100): For flat sheet and tubular membrane modules
$\text{Sh} = 0.34\text{Re}^{0.75}\text{Sc}^{0.33}$	Turbulent flow (Re > 2100): For flat sheet membrane modules

4.4.1.1.2 Mass Transfer Through the Membrane

Numerous studies for transport of gases and vapors through porous media have been performed as presented in the literature reports (Schofield et al., 1990a). Various theoretical models, based on the kinetic theory of gases, have been developed with the aim to predict MP for each configuration of the MD process (Present, 1958; Khayet, 2011; Phattaranawik et al., 2003).

The mass transfer in MD process occurs by convection and diffusion of the permeating compounds through the membrane pores. The different types of mechanisms that have been proposed for the mass transport through an MD membrane are: the Knudsen diffusion, the ordinary molecular diffusion, the viscous flow (or Poiseuille flow), and/or the combination between them. The Knudsen diffusion takes place when the membrane pore size is small compared to the mean free path of the molecules, resulting in collisions between molecules and membrane pore walls. In the viscous flow, the gas molecules act as a continuous fluid driven by a pressure gradient between the two sides of the membrane. The molecular diffusion occurs when the diffusing molecules move corresponding to each other under the influence of the concentration gradient created across the membrane.

In MD process, two mass transfer models are widely used: the Dusty gas model (DGM) (Mason and Malinauskas, 1983) and Schofield's model (Schofield et al., 1990b). DGM is a known general model that assumes the porous membrane is an array of dust particles held fixed in space. These particles, according to the classic kinetic theory of gases, are assumed to be giant molecules in the interactions between gas and membrane surface. This model combines the above mass transfer mechanisms, neglecting the surface diffusion, and is expressed with the following set of equations:

$$\frac{J_i^D}{D_{ie}^k} + \sum_{j=1\neq i}^{n} \frac{P_j J_i^D - P_i J_j^D}{D_{ije}^0} = -\frac{1}{RT}\nabla P_i \tag{4.10}$$

$$J_i^{vis} = -\frac{\varepsilon r^2 P_i}{8RT\tau\mu_i}\nabla P \tag{4.11}$$

with

$$D_{ie}^k = \frac{2\varepsilon r_p}{3\tau}\sqrt{\frac{8RT}{\pi M_i}} \quad \text{and} \quad D_{ije}^0 = \frac{\varepsilon}{\tau}PD_{ij}^0 \tag{4.12}$$

where J_i^D (kg m^{-2} s^{-1}) is the diffusive flux of the transported component inside the membrane, J_i^{vis} (kg m^{-2} s^{-1}) is the viscous flux, $J_i(=J_i^D+J_i^{vis})$ (kg m^{-2} s^{-1}) is the total flux, D_{ie}^k (m^2 s^{-1}) is the Knudsen diffusion coefficient, D_{ije}^0 (Pa m^2 s^{-1}) is the effective diffusion coefficient, D_{ij}^0 (m^2 s^{-1}) is the molecular diffusion coefficient, P_i (Pa) is the partial pressure of the component, ε, τ, r_p (μm) are the porosity, the tortuosity, and the pore size of the membrane, respectively, and μ_i, M_i are correspondingly the viscosity (Pa s) and the molecular weight (kg mol^{-1}) of the transported component.

It must be pointed out that DGM was originally developed for isothermal systems under relatively small thermal gradients by assuming an average value of temperature across the membrane.

Regardless of which model is used, the mass transfer through the membrane is often described by Darcy's law, whereby the mass flux of the permeating component (here the water vapor) is proportional to the partial vapor pressure difference across the membrane:

$$J_{v,m} = C_{mem}\Delta P_i = C_{mem}(P_{f,m} - P_{m,p}) \tag{4.13}$$

where C_{mem} (kg m^{-2} s^{-1} Pa^{-1}) is the membrane mass transfer coefficient, $P_{f,m}, P_{m,p}$ (Pa) is the partial vapor pressure in the corresponding interfaces of feed–membrane and membrane–permeate streams. For diluted aqueous solutions (eg, brackish water desalination) and when the temperature difference across the membrane is less than 10°C, Eq. (4.13) can be written in terms of temperature difference (Schofield et al., 1987):

$$J_{v,m} = C_{mem}\frac{dP}{dT}(T_{f,m} - T_{m,p}) \tag{4.14}$$

The term dP/dT represents the relationship between the partial vapor pressure and the temperature, which can be expressed by the Clausius–Clapeyron equation, as follows:

$$\left(\frac{dP}{dT}\right)_{T_m} = \left(\frac{\Delta H_v}{RT_m^2}\right)P_v^{sat}(T_m) \tag{4.15}$$

where ΔH_v (J kg^{-1}) is the latent heat of evaporation and P_v^{sat} (Pa) is the saturation water vapor pressure, estimated from Antoine equation in an average membrane temperature, T_m.

However, Schofield et al. (1987) adapted Eq. (4.14) for more concentrated solutions, as follows:

$$J_{v,m} = C_{mem}\frac{dP}{dT}(T_{f,m} - T_{m,p} - \Delta T_{th}) \tag{4.16}$$

with

$$\Delta T_{th} = \frac{RT^2}{M_w\Delta H_v}\frac{x_{f,m} - x_{p,m}}{1 - x_m} \tag{4.17}$$

where ΔT_{th} (K) is the threshold temperature, $x_{f,m}, x_{p,m}, x_m$ are the corresponding mole fractions of the solute at the hot and cold membrane surfaces, as well as inside the membrane.

For the real aqueous solutions, nonideal, the partial vapor pressure at the hot feed stream–membrane interface is defined as:

$$P_{f,m}(T_{f,m}, x) = x_{f,m}^w\gamma_{water}(x, T_{f,m})P_v^{sat}(T_{f,m}) \tag{4.18}$$

where $x_{f,m}^w$ and γ_{water} are the mole fraction and the activity coefficient of water at the hot feed stream–membrane interface.

When the single transported component through the membrane is water vapor, the partial vapor pressure can be estimated from Raoult's law by assuming ideal aqueous solutions:

$$P_{f,m}(T_{f,m}, x) = x_{f,m}^w P_v^{sat}(T_{f,m}) \tag{4.19}$$

In the case that dissolved species of non-VOCs (eg, salt) are present at the feed side, then the reduction in the partial vapor pressure is described as:

$$P_{f,m} = (1 - x_{f,m}^{salt})P_v^{sat}(T_{f,m}) \tag{4.20}$$

where $x_{f,m}^{salt}$ and $P_v^{sat}(T_{f,m})$ are the mole fraction of salt and the saturation water vapor pressure (Pa) at the feed–membrane interface, respectively.

The membrane mass transfer coefficient is mainly function of the structural membrane properties (porosity, pore size, tortuosity, and membrane thickness), the physicochemical properties of the transported vapor through the membrane, as well as the operating conditions (temperature and pressure). This coefficient depends on the type of the mass transfer mechanism that dominates in the membrane pores, which is related with the collisions of the diffusing molecules with each other or with the membrane pore walls. The quantity that determines the governing mechanism in the membrane is the Knudsen number (Kn), which is defined as:

$$\text{Kn} = \frac{\lambda_i}{d_p} \qquad (4.21)$$

where λ_i (μm) and d_p (μm) are the mean free path traveled by the transported molecule and the pore size, respectively. The parameter λ_i is given as:

$$\lambda_i = \frac{k_B T_m}{\sqrt{2}\pi P_m \sigma_i^2} \qquad (4.22)$$

where k_B is the Boltzmann constant (1.38×10^{-23} J K^{-1}), T_m and P_m are the average temperature (K) and pressure (Pa) in the membrane, respectively, and $\sigma_i(\text{Å})$ is the collision diameter of the molecule (2.61 Å for water vapor).

For Kn > 1 or $d_p < \lambda_i$, the mean free path of the transported vapor molecules is larger than the diameter of the membrane pores, which means that the molecule–pore wall collisions are dominant over molecule–molecule collisions. In that case, the Knudsen diffusion is responsible for the mass transfer through the membrane pores and the membrane mass transfer coefficient is defined as:

$$C_{mem}^{Kn} = \frac{2}{3}\frac{\varepsilon r_p}{\tau \delta_m}\left(\frac{8M_i}{\pi R T_m}\right)^{1/2} \qquad (4.23)$$

where ε, r_p, τ, and δ_m are the porosity, the pore size (μm), the tortuosity, and the thickness (μm) of the membrane, respectively; M_i (kg mol^{-1}) is the molecular weight of the permeating component; R is the gas constant (m^3 Pa mol^{-1} K^{-1}); and T_m (K) is the average temperature in the membrane.

When Kn < 0.01 or $d_p > 100\lambda_i$, the mean free path is very small relatively to the membrane pore size and the molecule–molecule collisions are prevailing over the molecule–pore wall collisions. The viscous flow is the dominant mass transfer mechanism in the membrane and the membrane mass transfer coefficient is given by the following relationship:

$$C_{mem}^{vis} = \frac{1}{8\mu_i}\frac{\varepsilon r_p^2}{\tau \delta_m}\frac{M_i P_m}{R T_m} \qquad (4.24)$$

where μ_i (Pa s) is the viscosity of the permeating component in the gas phase and P_m (Pa) is the average pressure in the membrane.

In case that air is trapped inside the membrane pores, the resistance in mass transfer becomes essential. Therefore, if Kn < 0.01, the molecular diffusion is the prevailing mass transfer mechanism due to the presence of a stagnant air film inside the membrane pores. The membrane mass transfer coefficient of molecular diffusion is:

$$C_{mem}^{mol} = \frac{\varepsilon}{\tau \delta_m} \frac{M_i}{RT_m} \frac{P_{tot} D_{i-air}}{P_{air}} \tag{4.25}$$

where P_{tot} (Pa) is the total membrane pressure, equal to the partial pressure of the air (P_{air}) and the water vapor and D_{i-air} (m^2 s^{-1}) is the diffusion coefficient of the component i to the air.

In the transition region, $0.01 < Kn < 1$ or $\lambda < d_p < 100\lambda$, both collisions between molecules and molecules—pore walls collisions are taking place. If there is also a stagnant air film inside the membrane pores, then a combination of a Knudsen diffusion—molecular diffusion—viscous flow is used to describe the mass transfer in the membrane. The membrane mass transfer coefficient is written as:

$$C_{mem}^{Kn-vis-mol} = \left\{ \left(\frac{2}{3} \frac{\varepsilon r_p}{\tau \delta_m} \left(\frac{8M_i}{\pi RT_m} \right)^{1/2} \right)^{-1} + \left(\frac{\varepsilon}{\tau \delta_m} \frac{M_i}{RT_m} \frac{P_{tot} D_{i-air}}{P_{air}} \right)^{-1} \right\}^{-1} + \frac{1}{8\mu_i} \frac{\varepsilon r_p^2}{\tau \delta_m} \frac{M_i P_m}{RT_m} \tag{4.26}$$

4.4.1.2 Heat Transfer

MD is a phase change process, in which the heat transfer occurs simultaneously with the mass transfer from the hot feed stream to the cold stream. The two main mechanisms governing the heat transfer in MD process are the latent heat due to evaporation and condensation of water and the heat transfer by conduction across the membrane. In the DCMD configuration, the heat transfer is described by the following steps:

- Heat transfer from the hot feed stream to the hot feed stream—membrane interface through the thermal feed boundary layer
- Heat transport through the membrane
- Heat transfer from the membrane—cold stream interface to the cold stream through the thermal boundary layer of the cold stream.

4.4.1.2.1 Heat Transfer Through the Thermal Boundary Layer

The heat transfer across the boundary layer in the liquid feed phase is often the rate limiting step for high permeability membranes, because a large quantity of heat must be supplied to the hot feed stream—membrane interface to evaporate the liquid solution. Since evaporation occurs at the hot membrane surface and condensation at the other membrane surface, the temperature difference, created across the membrane, causes the formation of thermal boundary layers adjacent to the two membrane surfaces. The thermal boundary layer at the feed side creates a resistance in heat transfer, as a result the temperature at the hot membrane surface, $T_{f,m}$, is lower than that at hot feed stream, T_f (Fig. 4.8). This phenomenon is called temperature polarization. The temperature polarization coefficient (TPC), term that is used to quantify the magnitude of the boundary layer resistance over the total heat transfer resistance, is defined as:

$$\tau = \frac{T_{f,m} - T_{m,c}}{T_f - T_c} \tag{4.27}$$

where $T_{f,m}$, $T_{m,c}$ (K) are the temperatures at hot and cold membrane surfaces and T_f, T_c (K) the temperatures at the hot feed and cold stream, respectively.

TPC is an indirect index of MD process efficiency, with its value approaching unity for well-designed systems that are mass transfer limited. However, for heat transfer limited systems (poor-designed systems), TPC value is very low. Schofield et al. (1987) report that in satisfactory designed MD modules, TPC values range between 0.4 and 0.7. In order to reduce the impact of TP, several efforts have been attempted such as the use of spacers (Schofield et al., 1987).

The heat is transferred from the hot feed stream to the hot membrane interface through the thermal boundary layer, due to the convection, $q_{f,conv}$, and the mass transfer, q_f^m (Dufour effect). The heat flux through the feed boundary layer, Q_{fbl} (W m^{-2}), is given by the following relationship:

$$Q_{fbl} = q_{f,conv} + q_f^m = h_f(T_f - T_{f,m}) + J_{v,m}H_{L,f}(T_{avg,f}) \tag{4.28}$$

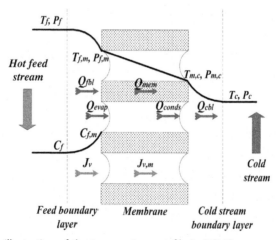

Figure 4.8 Schematic illustration of the temperature profile in DCMD.

where h_f (W m^{-2}K^{-1}) is the heat transfer coefficient in the feed boundary layer, J_v (kg m^{-2} s^{-1}) is the mass flux of vapor, $H_{L,f}$ (J kg^{-1}) is the enthalpy of the liquid solution, and $T_f, T_{f,m}, T_{avg,f}$ (K) are the temperatures in the hot feed stream, in the hot membrane interface, and in an intermediate point between the feed and the hot membrane interface, respectively.

Phattaranawik and Jiraratananon (2001) showed for the DCMD configuration that the effect of mass transfer on heat transfer (Dufour effect) at the thermal boundary layer at the feed is higher than this at the boundary layer of the cold stream. The maximum values of q_f^m and q_c^m were 13% and 4%, respectively, at 368K. q_f^m can be considered negligible at low temperatures, while q_c^m can be neglected under all conditions.

The heat transfer coefficient in the boundary layer of the feed stream (as also of the cold stream) can be estimated from Nusselt number via existing empirical correlations for heat transfer. The empirical Nusselt correlations for heat transfer are presented in Table 4.2 and they have the following form:

$$\text{Nu} = a\text{Re}^b\text{Pr}^c \tag{4.29}$$

where Nu, Re, and Pr are the dimensionless Nusselt, Reynolds, and Prandtl numbers, respectively, which are expressed as:

$$\text{Nu} = \frac{h_f d_h}{k_{f,L}}, \quad \text{Re} = \frac{u_f d_h}{\mu_f}, \quad \text{Pr} = \frac{C_{p,f}\mu_f}{k_{f,L}} \tag{4.30}$$

where $k_{f,L}$ is the thermal conductivity (W m^{-1} K^{-1}) and $C_{p,f}$ is the specific heat capacity (J kg^{-1} K^{-1}) of the liquid feed solution.

Table 4.2 Empirical Nusselt Correlations for Heat Transfer in Various Membrane Modules of Different Geometries (Chiam and Sarbatly, 2013)

Nusselt Empirical Correlations	Description
$\text{Nu} = 1.86\left(\text{Re Pr}\frac{d_h}{L}\right)^{1/3}$	Laminar flow (Re < 2100): For flat sheet and tubular membrane modules
$\text{Nu} = 3.66 + \dfrac{0.0668\left(\text{Re Pr } d_h/L\right)}{1+0.045\left(\text{Re Pr } d_h/L\right)^{2/3}}$	Laminar flow (Re < 2100): For hollow fiber membrane modules
$\text{Nu} = 0.116\,(\text{Re}^{2/3} - 125)\text{Pr}^{1/3}\left(1 + \left(\frac{d_h}{L}\right)^{2/3}\right)$	Transition flow (2100 < Re < 10,000): For flat sheet membrane modules
$\text{Nu} = 0.023\text{Re}^{0.8}\text{Pr}^n$ $n = 0.4$ for heating, $n = 0.3$ for cooling	Turbulent flow (Re > 2100): For flat sheet and tubular membrane modules
$\text{Nu} = 0.023\text{Re}^{0.8}\text{Pr}^{1/3}\left(1 + \frac{6d_h}{L}\right)$	Turbulent flow (Re > 2100): For flat sheet and hollow fiber membrane modules

4.4.1.2.2 Heat Transfer Through the Membrane

In the microporous membrane, the heat is transported simultaneously with the mass flux through the membrane pores. The main mechanisms that govern the heat transfer across the membrane is the heat by conduction through the membrane material and its pores, q_{cond}, the heat due to the mass transfer, q_m^m, and the latent heat of evaporation, q_v. The heat flux across the membrane, Q_{mem} (W m^{-2}), considering linear temperature distribution, is given by the following relationship:

$$Q_{mem} = q_{cond.} + q_m^f + q_v = \frac{k_m}{\delta_m}(T_{f,m} - T_{m,c}) + J_{v,m}C_{p,v}(T_{f,m} - T_{m,c}) + J_{v,m}\Delta H_v\{T_{f,m}\}$$

(4.31)

where k_m (W m^{-1} K^{-1}) is the thermal conductivity coefficient of the membrane, J_v (kg m^{-2} s^{-1}) is the mass flux of the vapor through the membrane, $C_{p,v}$ (J kg^{-1} K^{-1}) is the specific thermal capacity of the vapor, and $\Delta H_v\{T_m\}$ (J kg^{-1}) is the latent heat of evaporation at the hot membrane interface.

However, recent theoretical studies for DCMD consider nonlinear temperature distribution across the membrane and non-isenthalpic flow of water vapor molecules (Phattaranawik and Jiraratananon, 2001; Phattaranawik et al., 2003; Bahmanyar et al., 2012). According to the above assumptions, the heat flux across the membrane is written as follows:

$$Q_{mem} = -k_m\frac{dT}{dx} + J_{v,m}H_v\{T_{f,m}\}$$

(4.32)

with

$$H_v\{T\} = H_v\{T_0\} + C_{p,v}(T - T_0)$$

(4.33)

where H_v is the enthalpy of the vapor molecules at the hot membrane interface and $H_v\{T_0\}$ is the heat of evaporation at reference temperature T_0.

It is worth to be noted that the amount of heat transferred by conduction through the membrane and its gas-filled pores is considered as a heat loss and should be minimized so as to reduce the effect of TP and increase the MD process efficiency. According to the investigation of Fane et al. (1990), the heat loss by conduction represents 20–50% of the total heat transferred in MD.

The thermal conductivity coefficient of the membrane, k_m, can be calculated via the thermal conductivities of the solid and gas phases by using various models (Phattaranawik et al., 2003). The "Isostrain" model is the known common expression that is widely used in the MD literature:

$$k_m = k_g\varepsilon + (1 - \varepsilon)k_s$$

(4.34)

where k_g and k_s (W m^{-1} K^{-1}) are the thermal conductivities of the diffusing gas in the membrane pores and the membrane material, respectively.

The polymeric materials used to fabricate the MD membranes, such as polypropylene (PP), polytetrafluoroethylene (PTFE), polyvinylidene fluoride (PVDF), and
polyethylene (PE), have thermal conductivities that range from 0.15 to
0.51 W m^{-1} K^{-1}. The thermal conductivities of water vapor and air range from
0.019 to 0.024 W m^{-1} K^{-1} and 0.025 to 0.032 W m^{-1} K^{-1}, respectively.

4.4.2 Air Gap Membrane Distillation

4.4.2.1 Mass Transfer

AGMD came from a Swedish patent as a new configuration of MD method (Jonsson
et al., 1985). This configuration is characterized by the presence of an air gap between
the membrane surface and a cooled condensing surface in order to reduce the heat
losses due to the conduction through the membrane. The mass transfer in AGMD,
as illustrated in Fig. 4.9, occurs: (1) from the hot feed stream to the hot membrane
interface through the feed boundary layer, (2) through the membrane pores, and (3)
from the less hot membrane interface to the surface of the condensing film through
the air gap.

The mass transfer through the feed boundary layer can be described by the film
theory (see Section 4.4.1.1.1).

The transport of the vapor molecules through the membrane pores is related with
the membrane morphology and the existence of the trapped air in the membrane
pores. For membranes with small pore size ($d_p < 0.45$ µm), the vapor mass flux across

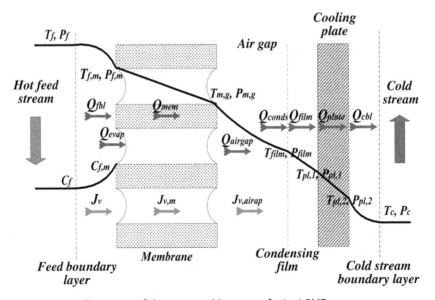

Figure 4.9 Schematic illustration of the mass and heat transfer in AGMD.

the membrane can be written as a combination of Knudsen diffusion and molecular diffusion:

$$J_{v,m} = \left\{ \left(\frac{2}{3} \frac{\varepsilon r_p}{\tau \delta_m} \left(\frac{8M_i}{\pi R T_m} \right)^{1/2} \right)^{-1} + \left(\frac{\varepsilon}{\tau \delta_m} \frac{M_i}{R T_m} \frac{P_{tot} D_{i-air}}{P_{air}} \right)^{-1} \right\}^{-1} (P_{f,m} - P_{m,g}) \quad (4.35)$$

In case that the membrane pore size is relatively large, ie, $d_p > 0.45 \, \mu m$, the vapor mass flux can be described satisfactorily by the molecular diffusion mechanism:

$$J_{v,m} = \frac{\varepsilon}{\tau \delta_m} \frac{M_i}{R T_m} \frac{P_{tot} D_{i-air}}{P_{air}} (P_{f,m} - P_{m,g}) \quad (4.36)$$

Moreover, from DGM a simplified equation is derived, considering only molecular diffusion through the gas-filled membrane pores (Lawson and Lloyd, 1997).

$$J_{v,m} = \frac{D_{ije}^0}{T^b} \frac{T^{b-1}}{\delta_m R |P_{air}|_{lm}} (P_{f,m} - P_{m,g}) \quad (4.37)$$

where $|P_{air}|_{lm}$ is the log mean air pressure in the two membrane sides. Eq. (4.37) is applied only to MD process when the water is the sole diffusing component in the membrane ($b = 2.234$).

The vapor molecules, since penetrate the membrane, are transferred through the space between the membrane interface and the condensing film surface. In this region, a stagnant air film is considered to exist due to the low solubility of air in the water. In this case, the Stefan diffusion is used to describe the mass transfer through the air gap (Bird et al., 2002).

$$J_{v,airgap} = -\frac{c D_{i-air}}{1 - y_i} \frac{dy_i}{dz} \quad (4.38)$$

The Stefan equation was solved by Kimura and Nakao (1987):

$$J_{v,airgap} = -\frac{c D_{i-air}}{1 - y_i} \ln \left(\frac{1 - y_{i,film}}{1 - y_{i,m}} \right) \quad (4.39)$$

where $y_{i,m}$ and $y_{i,film}$ are the mole fraction of the permeating component in the membrane interface and the condensing film surface, respectively, and δ_{gap} (μm) is the air gap thickness.

Banat and Simandl (1998) used Stefan diffusion and binary type relations (ie, Fick's equation of molecular diffusion) to describe the multicomponent mass transfer in AGMD systems. In this case, the mass flux through the air gap region can be written in terms of pressure:

$$J_{v,airgap} = \frac{P_{tot} D_{i-air}}{R T_g \delta_{gap} |P_{air}|_{ln}} (P_{m,g} - P_{film}) \quad (4.40)$$

where T_g is the mean temperature (K), $|P_{air}|_{lm}$ is the log mean air pressure in the air gap region, and $P_{m,g}$ and P_{film} (Pa) are the partial vapor pressures in the less hot membrane surface and in the condensing film surface, respectively.

Taking into account the porosity, the tortuosity, and the thickness of the membrane, the vapor mass flux through the membrane and the air gap is expressed as:

$$J_v = \frac{\varepsilon P_{tot} D_{i-air}}{R T_{avg} (\delta_{gap} + \tau \delta_m) |P_{air}|_{lm}} (P_{f,m} - P_{film}) \tag{4.41}$$

Another expression of the vapor mass flux through the membrane and the stagnant air film, in an average operating temperature T_a, is given by Liu et al. (1998):

$$J_v = \frac{T_f - T_c}{a T_a^{-2.1} + \beta} \tag{4.42}$$

where α and β are parameters that can be determined experimentally.

4.4.2.2 Heat Transfer

In AGMD configuration, the heat transfer (Fig. 4.9) takes place in the following zones:

- From the hot feed stream to the feed—membrane interface through the feed boundary layer
- Through the membrane pores
- Through the air gap
- In the condensing film
- In the cooling plate
- From the cooling plate to the cold stream through the boundary layer of the cold stream.

The total heat is transported from the hot feed stream to the hot membrane interface through the feed boundary layer by convection and due to the mass transfer (Dufour effect). The heat flux across the boundary layer, Q_{fbl} (W m^{-2}), is defined as:

$$Q_{fbl} = h_f (T_f - T_{f,m}) + J_{v,m} H_{L,f} (T_{avg,f}) \tag{4.43}$$

The transferred heat from the hot feed stream to the entrance of membrane pores is consumed by the latent heat of evaporation due to the movement of vapor molecules through the membrane (sensible heat) and by conduction through the solid phase of the membrane material and its pores. The total heat flux through the membrane, Q_{mem} (W m^{-2}), is expressed as:

$$Q_{fbl} - J_v \Delta H_v \left\{ T_{f,m} \right\} = Q_{mem} = \frac{k_m}{\delta_m} (T_{f,m} - T_{m,g}) + J_{v,m} C_{p,v} (T_{f,m} - T_{m,g}) \tag{4.44}$$

The heat transported from the membrane—air gap interface to the condensing film surface, through the air gap region, is affected by conduction or natural convection and by the transfer of vapor molecules through the air gap region (sensible heat). The natural convection occurs due to the temperature difference $\left(T_{m,g} - T_{film}\right)$ between the two surfaces and is depended on the Rayleigh number (Ra) (Bergman et al., 2011):

$$Ra = \frac{g\beta(T_{m,g} - T_{film})\delta_{gap}^3}{\nu_a \alpha_a} \tag{4.45}$$

where g, β, ν_a, and α_a are the gravitational acceleration (m s^{-2}), the thermal expansion coefficient (K^{-1}), the kinematic viscosity (Pa s), and the thermal conductivity of the air (m^2 s^{-1}), respectively. For $\left(T_{m,g} - T_{film}\right) = 40°C$ and $\delta_{gap} = 3$ mm (typical values for an AGMD system), the Ra number is equal to 85. For Ra < 1000, the natural convection is considered negligible relative to the heat conduction across the air gap (MacGregor and Emery, 1969).

The heat flux through the air gap, Q_{gap} (W m^{-2}), is given by the following equation:

$$Q_{gap} = \frac{k_{air}}{\delta_{gap}}(T_{m,g} - T_{film}) + J_{v,airgap} C_{p,v}(T_{m,g} - T_{film}) \tag{4.46}$$

where k_{air} (W m^{-1} K^{-1}) is the thermal conductivity of the air and T_{film} (K) is the temperature at the condensing film surface.

As the water vapor is condensed on the cooling plate, a condensing film is formed at the top of the plate and flows downward under the influence of gravity. The heat from the air gap region is conducted to the condensing film and a small part of the heat is rejected as heat of condensation. The heat flux in the condensing film, Q_{film} (W m^{-2}), is defined as:

$$Q_{gap} + J_v\Delta H_v\left\{T_{film}\right\} = Q_{film} = \frac{k_{lw}}{\delta_{film}}(T_{film} - T_{pl,1}) \tag{4.47}$$

where k_{lw} (W m^{-1} K^{-1}) is the thermal conductivity of the liquid water in the condensing film surface, δ_{film} (μm) is the film thickness, and $T_{pl,1}$ (K) is the temperature in the film—cooling plate interface.

The heat transfer coefficient of the condensing film, h_{film}, can be estimated by the relationship for laminar film condensation on a vertical plate (Bergman et al., 2011):

$$h_{film} = \frac{k_{lw}}{\delta_{film}} = 0.943\left(\frac{g\rho_{lw}(\rho_{lw} - \rho_v)k_{lw}^3 \Delta H_v(T_{film})}{4\mu_{lw}(T_{film} - T_{pl,1})L_{plate}}\right)^{1/4} \tag{4.48}$$

where ρ_{lw} and μ_{lw} are the density (kg m^{-3}) and the viscosity (Pa s) of the liquid water, respectively, ρ_v is the density of the vapor, and L_{plate} (m) is the length of the plate.

The heat from the condensing film is transferred in the cooling plate by conduction. The heat flux through the plate, Q_{pl} (W m^{-2}), can be written as:

$$Q_{pl} = \frac{k_{pl}}{\delta_{pl}}(T_{pl,1} - T_{pl,2}) \qquad (4.49)$$

where k_{pl} and δ_{pl} are the thermal conductivity of the plate material (W m^{-1} K^{-1}) and the thickness of the plate (μm), respectively, and $T_{pl,2}$ (K) is the temperature at the cooling plate—cold stream interface.

The heat transport through the boundary layer of the cold stream is governed by convection and the heat flux in this side is given by:

$$Q_{cbl} = h_c(T_{pl,2} - T_c) \qquad (4.50)$$

where h_c (W m^{-2} K^{-1}) is the heat transfer coefficient of the cold boundary layer and T_c (K) is the average temperature in the cold stream of water.

4.4.3 Sweep Gas Membrane Distillation

SGMD is an alternative configuration of MD process that is mainly used for the removal of VOCs from aqueous solutions. Its operation lies to the use of a cold inert gas, usually air, to sweep the produced vapor at the permeate membrane side and to lead it in an external condenser. One of the first studies presented for the application of SGMD in water desalination was that of Basini et al. (1987), who used flat sheet and tubular membranes, studied experimentally the effect of the inlet temperature and flow rate of the hot aqueous solution and the cold air, respectively, at the evaporation rate. In parallel, with the development of a mathematic model they validated their experimental results.

The mass and heat transfer in the SGMD configuration, as in DCMD, occurs (1) from the hot feed stream to the feed—membrane interface through the boundary layer, (2) through the membrane pores, and (3) from the membrane—cold stream interface through the boundary layer to the cold air stream (Fig. 4.10).

In the feed boundary layer, the mass transfer can be described according to the film theory, where the vapor mass flux, J_v, is related with the solute mass transfer coefficient, $k_{f,s}$, as well as with the solute concentration at the feed stream, C_f^s, and at the feed—membrane interface, $C_{f,m}^s$:

$$J_v = \rho_v k_{f,s} \ln\left(\frac{C_{f,m}^s}{C_f^s}\right) \qquad (4.51)$$

The mass flux through the membrane pores can be explained in the frameworks of different mass transfer mechanisms such as the Knudsen diffusion, the molecular diffusion, the viscous flow (Poiseuille flow), and/or the combination between them.

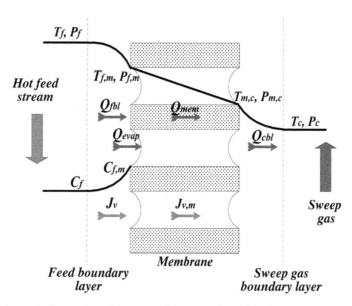

Figure 4.10 Schematic illustration of mass and heat transfer in SGMD.

Khayet et al. (2000) analyzed the nature of the mass transfer in SGMD by using two commercial membranes. They determined, with a theoretical model, the mass flux through the membrane and by comparing the experimental results with the theoretical predictions, they concluded that the transport of vapor molecules through the membrane pores takes place via a combination of Knudsen diffusion and molecular diffusion. In this case, the mass flux can be written according to the following equation:

$$J_v = \frac{M_i}{RT_m\delta_m} = \left(\frac{1}{D_K} + \frac{P_{air}}{D_M}\right)^{-1}(P_{f,m} - P_{m,c}) \qquad (4.52)$$

with

$$D_K = \frac{2}{3}\frac{\varepsilon r_p}{\tau}\left(\frac{8RT_m}{\pi M_i}\right)^{1/2} \text{ and } D_M = \frac{\varepsilon}{\tau}P_{tot}D_{i-air} \qquad (4.53)$$

where D_K and D_M (m^2 s^{-1}) are the coefficients of the Knudsen and molecular diffusion, respectively.

The water vapor partial pressure in the membrane—cold air stream interface is given as a function of the total pressure, P_{tot}, and the humidity ratio, ω:

$$P_{m,c} = \frac{\omega P_{tot}}{\omega + 0.622} \qquad (4.54)$$

The humidity ratio, ω, varies along the membrane module length and its value is not known. However, its value can be related with the cold air flow rate, $\dot{m}_{air}$ (kg s^{-1}), and with the value of the humidity at the module inlet, ω_{in}:

$$\omega = \omega_{in} + \frac{J_{v,m}A_m}{\dot{m}_{air}} \tag{4.55}$$

where A_m (m) is the membrane area.

The main mechanisms that govern the heat transfer in the SGMD process is the heat by conduction and the latent heat of evaporation. The heat from the hot feed stream is transferred to the hot membrane interface through the boundary layer by convection and due to the mass transfer (Dufour effect). The heat flux across the boundary layer, Q_{fbl} (W m^{-2}), is defined as:

$$Q_{fbl} = h_f(T_f - T_{f,m}) + J_{v,m}H_{L,f}(T_{avg,f}) \tag{4.56}$$

The transferred heat from the hot feed stream to the feed–membrane interface is consumed by the latent heat of evaporation, due to the transport of vapor molecules through the membrane (sensible heat) and by conduction through the solid phase of the membrane material and its pores. The heat flux through the membrane, Q_{mem} (W m^{-2}), is given by the following relationship:

$$Q_{fbl} - J_v\Delta H_v\left\{T_{f,m}\right\} = Q_{mem} = \frac{k_m}{\delta_m}(T_{f,m} - T_{m,c}) + J_{v,m}C_{p,v}(T_{f,m} - T_{m,c}) \tag{4.57}$$

The heat from the cold membrane surface is transported to the cold air stream through the boundary layer by convection. The heat flux through the boundary layer of the cold stream, Q_{cbl} (W m^{-2}), is expressed as:

$$Q_{cbl} = h_c(T_{m,c} - T_c) \tag{4.58}$$

4.4.4 Vacuum Membrane Distillation (VMD)

VMD constitutes a configuration of MD process, in which a vacuum is applied at the permeate side of the membrane by a vacuum pump. As the liquid aqueous solution is vaporized at the hot feed stream–membrane interface, the vapor molecules formed are diffused in the membrane pores and they are transported at the permeate side of the membrane, wherein under the influence of the vacuum pressure they flow in a separate chamber, where they condense.

The VMD method had previously classified for some time in the pervaporation process. The difference of these two processes lies in the role of the membrane. In the pervaporation, dense and compact membranes are used, where the feed aqueous solution is dissolved and diffused in the membrane pores (solution–diffusion). On the contrary, in the VMD the membranes are microporous and supply the liquid–vapor interface, which defines the separation conditions of the components. The VMD process has been proposed for the production of high purity water from brackish water and seawater for the extraction of organic compounds from water as well as for the selective removal of VOCs from aqueous solutions (Mericq et al., 2009; Sarti et al., 1993; Bandini et al., 1997).

The mass and heat transfer in VMD configuration, as is illustrated in Fig. 4.11, occurs in the following zones:

- From the hot feed stream to the hot feed stream—membrane interface through the feed boundary layer
- Through the membrane pores
- From the membrane—permeate stream through the permeate boundary layer to the permeate stream.

The low vacuum pressure applied at the permeate side of the membrane (which is lower than the saturation pressure of vapor in the hot membrane surface) prevents the formation of the boundary layer at the permeate stream. Therefore, the resistance of the permeate boundary layer to the mass transfer can be neglected.

The mass transfer through the liquid phase at the feed boundary layer can be described by the film theory (see Section 4.4.1.1.1). In the case that at the feed aqueous solution, there are VOCs and due to their permeation through the membrane pores, their concentration at the feed—membrane interface becomes smaller than that of the feed side. The mass flux through the boundary layer, J_v, is defined as:

$$J_v = k_f c_f \ln \left(\frac{x_{i,f,m} - J_i/J_v}{x_{i,f} - J_i/J_v} \right)$$

(4.59)

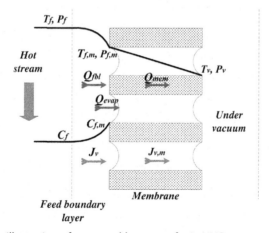

Figure 4.11 Schematic illustration of mass and heat transfer in VMD.

where c_f (mol s^{-1}) is the concentration of the feed aqueous solution, $x_{i,f,m}$ and $x_{i,f}$ are the mole fraction of the permeating component at the feed–membrane interface and at the feed, respectively, and J_i (kg m^{-2} s^{-1}) is the mass flux of the permeating component.

The mass transfer through the membrane occurs by convection and diffusion of the vapor molecules in the membrane pores. The main mechanisms that govern the mass flux of the vapor is the Knudsen diffusion, the molecular diffusion, and the viscous flow. Since in the VMD a low vacuum pressure is applied, the partial pressure of the noncondensable gases (here the air) becomes very small compared to that of the water vapor molecules. In such a case, the molecular diffusion mechanism is not playing a significant role and its resistance to the mass transfer can be neglected, since it is proportional to the partial pressure of the air in the membrane pores.

In most applications of the VMD configuration, where the pressure at the permeate side is relatively low and the membrane pore size is much smaller than the mean free path of the vapor molecules, the Knudsen diffusion is the controlling mass transfer mechanism in the membrane pores. The mass flux of the permeating components is proportional to the partial pressure difference across the membrane:

$$J_v = \frac{K_m}{\sqrt{M_i}}(P_{f,m} - P_v) \tag{4.60}$$

with

$$K_m = \frac{2}{3} \frac{\varepsilon r_p}{\tau \delta_m R T_m} \left(\frac{8RT_m}{\pi}\right)^{1/2} \tag{4.61}$$

where K_m (mol$^{1/2}$ s m^{-1} kg$^{-1/2}$) is the Knudsen diffusion coefficient (or membrane permeability) and P_v (Pa) is the vacuum pressure at the permeate side of the membrane.

In case that the total pressure gradient is maintained across the membrane and the membrane pore size is comparable to the mean free path, then the contribution of the viscous flow should be taken into account. The mass flux of the vapor molecules is expressed as:

$$J_{v,m} = \left(\frac{2}{3} \frac{\varepsilon r_p}{\tau \delta_m} \left(\frac{8M_i}{\pi R T_m}\right)^{1/2} + \frac{1}{8\mu_i} \frac{\varepsilon r_p^2}{\tau \delta_m} \frac{M_i P_m}{R T_m}\right)(P_{f,m} - P_V) \tag{4.62}$$

On the other hand, if the pore diameter of the membrane is greater than the mean free path, then the vapor transport is determined only by the viscous flow. In the Poiseuille's flow regime, the mass flux of the vapor is calculated by the following equation:

$$J_{v,m} = \left(\frac{1}{8\mu_i} \frac{\varepsilon r_p^2}{\tau \delta_m} \frac{M_i P_m}{R T_m}\right)(P_{f,m} - P_V) \tag{4.63}$$

Lawson and Lloyd (1996) applied both contributions, Knudsen diffusion and viscous flow, in their mathematical model and they calculated the importance of the Poiseuille's flow relative to Knudsen diffusion as follows:

$$\xi = 0.2\frac{r_p P_m}{\mu_i u} \tag{4.64}$$

where u (m s^{-1}) is the mean molecular speed of the water.

In the VMD process, the total heat is transferred from the hot feed stream through the boundary layer to the hot membrane interface by convection and due to the mass transfer (Dufour effect). The heat flux across the boundary layer, Q_{fbl} (W m^{-2}), is defined as:

$$Q_{fbl} = h_f(T_f - T_{f,m}) + J_{v,m}H_{L,f}(T_{avg,f}) \tag{4.65}$$

The heat in the feed–membrane interface is consumed by the latent heat of evaporation and is transferred across the membrane by conduction through the solid phase of the membrane material and its pores and due to the transport of the vapor molecules through the membrane (sensible heat). The heat flux across the membrane, Q_{mem} (W m^{-2}), can be written by the following relationship:

$$Q_{fbl} - J_v\Delta H_v\left\{T_{f,m}\right\} = Q_{mem} = \frac{k_m}{\delta_m}(T_{f,m} - T_{m,c}) + J_{v,m}C_{p,v}(T_{f,m} - T_{m,c}) \tag{4.66}$$

4.4.5 Performance Parameters of MD Process

In MD, there is no specified equation to describe the performance of the process. On the contrary, there are several parameters, such as the gained output ratio (GOR) and the performance ratio (PR), which are applied many times with different factors by various researchers.

The thermal efficiency of the process, η_{th}, is given as the ratio of the latent heat of evaporation, q_{ev}, to the total heat input to the system per unit mass of distillate, q_{tot} (Alklaibi and Lior, 2004):

$$\eta_{th} = \frac{q_{ev}}{q_{tot}} = \frac{q_{ev}}{(q_{ev} + q_{cond})} \tag{4.67}$$

where the total heat, q_{tot}, is equal to the latent heat of evaporation and to the heat lost by conduction through the membrane, q_{cond} (W m^{-2}).

Koschikowski et al. (2003) define GOR as follows:

$$\text{GOR} = \frac{\dot{m}_d\Delta H_v}{q_{in}} = \eta_{th}\frac{T_{c,out} - T_{c,in}}{T_{f,out} - T_{f,in}} \tag{4.68}$$

The thermal efficiency, η_{th}, is expressed by:

$$\eta_{th} = \frac{\dot{m}_d \Delta H_v}{\dot{m}_f C_p (T_{c,out} - T_{c,in})} \qquad (4.69)$$

where $\dot{m}_d$ and $\dot{m}_f$ (kg m^{-2} s^{-1}) are the mass flow rate of the distillate and the aqueous feed solution, and $T_{f,out}$, $T_{f,in}$, $T_{c,out}$, $T_{c,in}$ (K) are the outlet and inlet temperatures at the hot feed and cold stream, respectively.

PR has already been mentioned in terminology (see chapter: "Solar Distillation— Solar Stills," Humidification—Dehumidification), but in the case of MD system with heat recovery it can be calculated by the following relationship (Koschikowski et al., 2003):

$$PR = \frac{\dot{m}_d}{q_{in}} \qquad (4.70)$$

The input heat to the system, q_{in}, is given as:

$$q_{in} = \dot{m}_f C_p (T_{f,in} - T_{f,out}) \qquad (4.71)$$

4.5 CHARACTERISTICS OF MD CONFIGURATIONS

The four basic configurations of MD process present some operational differences between them. DCMD and AGMD are the most studied technologies in laboratory scale, while a few pilot units have been installed without been fully commercialized yet. In general, for the four different configurations the following are mentioned:

- DCMD is simplest to operate and requires the minimum equipment. It is suitable for applications where the major permeate compound is the water vapor. However, in this configuration the evaporation and condensation surfaces are very close to each other, their distance is equal to membrane thickness, as a result the resistances to mass and heat transfer are minimal. Due to this small distance, small temperature differences across the membrane are created and therefore small driving force and mass flux of vapor. Generally, DCMD presents high heat losses by conduction through the membrane. Apart from water desalination, it is used in the concentration of food and other aqueous solutions
- On the contrary, AGMD, where an air gap is interposed to increase the distance between the evaporation and condensation surfaces, presents higher temperature differences across the membrane and therefore higher mass flux of vapor than DCMD. In AGMD, the heat losses are smaller than the previous method. This configuration is used for seawater desalination and for the removal of VOCs from aqueous solutions

- SGMD is the less studied MD configuration. Its main characteristic is the small heat losses by conduction through the membrane and the small resistance to mass transfer. Its drawback is the use of large quantities of air relative to the small quantities of the permeate vapor through the membrane, which cause problems to the external condensation system
- Finally, VMD exhibits higher mass flux of vapor than DCMD and AGMD and the heat losses through the membrane are negligible. Nevertheless, the membrane wetting is most likely to occur due to the large pressure difference at the cold membrane surface. The VMD configuration is used mainly for the separation of aqueous VOCs, but recently intensive studies are carried out for its application in desalination. Moreover, it has been applied in combination with RO to concentrate the volume of the brine discharged from the RO unit (Mericq et al., 2010).

4.6 HEAT RECOVERY

In an MD system, the energy efficiency is very important, since the investment and maintenance costs are enormous. Therefore, the system design has to focus on a very good recovery function to minimize the need for thermal energy.

Heat recovery can be achieved by an external heat exchanger or by an internal heat recovery function, where the feed water is used as coolant for the condenser channel (Koschikowski et al., 2003).

The principle of the internal heat recovery function is presented in Fig. 4.12. The hot water stream is directed along the membrane and as passes the evaporator channel is cooled down. The vapor that passes through the membrane is condensed on a condensing film, where it transfers the heat into the cold water stream, by conduction and by sensible heat of condensation. The feed water as passes through the condensation channel receives the heat of evaporation, by condensation, and is heated. A part of the heat of evaporation can be recovered by a suitable heat exchanger.

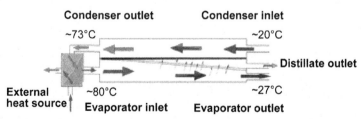

Figure 4.12 Principle of the internal heat recovery function in a membrane unit (Koschikowski et al., 2003).

According to Alklaibi (2008), the economy of an autonomous MD system can be assessed mainly with three parameters: the specific energy required for heating, the specific power required for the circulating, and the specific membrane area.

The specific energy required for heating, q_{ht}, is defined as:

$$q_{ht} = \frac{\Delta H_v}{\text{PR}} \qquad (4.72)$$

where PR is the performance ratio.

In Fig. 4.13, a heat recovery system with corresponding typical operating temperature is shown, wherein PR is equal to 4. For the process to be competitive and comparable to other processes, the specific energy required for heating should not exceed 300 kJ kg^{-1}, the typical heating cost for MSF.

The total specific energy required for circulation, e_{circ} (W kg^{-1}), over 1 membrane length is given by the following equation:

$$e_{circ} = \frac{1}{\dot{m}_{v,l}} \int_{0}^{l} E_{circ,\Delta l} dl \qquad (4.73)$$

where $E_{circ,\Delta l}$ (W) is the energy required for pumping over segment Δl of membrane length and is a function of the pressure drop and the volumetric flow rate in both sides of the membrane.

$$E_{circ,\Delta l} = \frac{\Delta p_{h,\Delta l} \dot{V}_{h,x} + \Delta p_{c,\Delta l} \dot{V}_{c,x}}{\eta_p} \qquad (4.74)$$

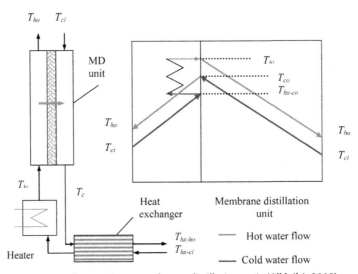

Figure 4.13 Heat recovery diagram in a membrane distillation unit (Alklaibi, 2008).

where $\Delta p_{h,\Delta l}$, $\Delta p_{c,\Delta l}$ and $\dot{V}_{h,x}$, $\dot{V}_{c,x}$ are the pressure drop (Pa) and the volumetric flow rate (m^3 s^{-1}) at l of membrane length of the hot feed and cold stream, respectively, and η_p is the pumping efficiency.

The specific membrane area, A_m, of membrane with length l is expressed as:

$$A_m = \frac{\int_0^l \frac{\dot{m}_{f,l}}{u_{h,l}\rho_{h,l}d_h} \, dl}{\int_0^l d\dot{m}_{v,l}} \tag{4.75}$$

with feed mass flow rate in the inlet of the evaporator channel:

$$\dot{m}_{f,l} = \frac{\dot{m}_v \Delta H_v}{PRC_p(T_{h,l} - T_{hx,co})} \tag{4.76}$$

where $\dot{m}_v$ is the mass flow rate of the permeate (kg s^{-1}) and $T_{hx,co}$ is the temperature of the feed solution at the outlet of the heat exchanger (K).

The effectiveness of the heat exchanger, ε_{hl}, can be calculated from the difference of the achieved temperatures:

$$\varepsilon_{hl} = \frac{T_{hx,co} - T_{hx,ci}}{T_{co} - T_{hx,ci}} \tag{4.77}$$

4.7 SOLAR POWERED MEMBRANE DISTILLATION

MD as a phase change process requires significant quantities of energy to achieve separation. The coupling of MD with renewable energy sources, such as wind, solar, and geothermal energy, or with low-grade waste heat, makes MD an economic and energy efficient process for seawater desalination.

The solar energy as an alternative energy source can be used as a thermal energy to preheat the aqueous feed solution through a solar collector. Hogan et al. (1991) were among the first that coupled the solar energy with the MD technique. They described a pilot solar powered membrane distillation (SPMD) installation producing 50 kg of desalinated water per day using a solar collector of 3 m^2 collecting surface area. Their system, which was tested in Sydney (Australia), consisted of a hollow fiber membrane module and a heat recovery exchanger for reducing the capital costs (Fig. 4.14). The SPMD unit was found to be technically feasible and compatible with this energy source.

Later, Bier and Plantikow (1995) investigated the operation of another SPMD unit, with the difference that an AGMD module was used instead of a DCMD, which was tested by Hogan et al. (1991). The heat recovery function was integrated in a spiral wound membrane module. However, the additional mass transfer resistance

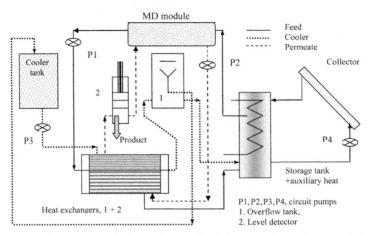

Figure 4.14 Layout of a solar powered membrane distillation installation (Hogan et al., 1991).

created by the air gap resulted in a dramatic reduction of the water vapor mass flux. Koschikowski et al. (2003) used a similar membrane module as used by Bier and Plantikow (1995) in their study of an SPMD pilot plant. According to their calculations, without heat storage, the plant can distill 150 L of water per day in the summer in a southern country. On the other hand, Banat et al. (2002) integrated an MD module with a solar still to produce portable water from a simulated seawater. In their investigations, the solar still was used for both seawater heating and portable water production. Their experimental results showed that the contribution of the solar still in the distillate production was no more than 20% of the total flux production.

Kullab et al. (2005) presented a simulation study based on laboratory experimental data for a high-performance MD system of 8.5 m^3 h^{-1} desalinated water production. The produced water has TDS < 10 ppm. The simulation layout of the plant, as is given in Fig. 4.15, has the following characteristics:

- Collector area: 3330 m^2
- Thermal energy consumption: 150 kWh m^{-2}
- Electrical consumption: 0.5−0.7 kWh m^{-2}.

Since 2004, within the frame of European funded projects (SMADES, MEDESOL, MEDIRAS), several compact systems have been installed with a nominal daily capacity equal to 100 L distillated water in different countries: Pozo Izquierdo and Tenerife (Gran Canary), Alexandria (Egypt), Irbid (Jordan), and Freiburg (Germany).

Under the umbrella of SMADES project, Banat et al. (2007a) designed and studied a stand-alone solar-driven MD system (compact SMADES), which was tested to generate a stable water supply in different remote areas (Jordan, Morocco, and Egypt).

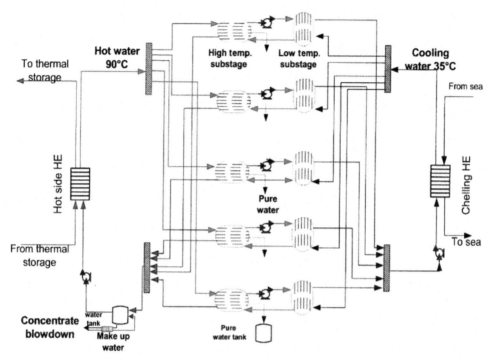

Figure 4.15 Simulation layout of a solar-driven MD system (Kullab et al., 2005).

The results from that compact system were used for designing and scaling up a "large SMADES" system, which was assembled in Aqaba, Jordan (Banat et al., 2007b). The power supply in that pilot unit can be given both by solar collectors and PV panels. The feed water was real seawater from the Red Sea. The unit can produce 120 L of desalinated water per day, ie, $19 \text{ L m}^{-2} \text{ d}^{-1}$, with specific energy consumption in the range of $200-300 \text{ kWh m}^{-3}$, which is claimed to be less than that of conventional solar distillation in the range of 640 kWh m^{-3}.

Recently among others, a small scale solar-driven MD system has been installed in Gran Canary, Spain (Koschikowski et al., 2009). It consisted of four compact units of total distillate output of $1-1.5 \text{ m}^3 \text{ d}^{-1}$ with temperatures at the evaporator inlet of $60-85°C$ and GOR $3-6$. The heat required for heating is supplied by high-performance flat plate solar collectors, while PV panels supplied the energy for the pumps operation. Moreover, another three compact systems have been installed in Tunisia and Tenerife and two more two-loop systems have been installed in Gran Canary and Panthelleria (Italy). The latter is powered by a hybrid system of solar energy and waste heat from diesel engines with nominal production up to $5 \text{ m}^3 \text{ d}^{-1}$ (Cipollina et al., 2011).

Fath et al. (2008) claimed that the advantages of stand-alone MD plants are the following:

- The operating temperatures from 60°C to 90°C coincide with those of high-performance solar collectors, while there is the possibility of waste heat utilization
- Contrary to RO process, specific pretreatment of feed water is not required
- The membranes used in the MD process are not sensitive to organic fouling as the RO membranes
- The MD modules operate at atmospheric pressure, contrary to RO modules that operate under pressure
- The feed water salinity has no influence to the process performance, while in the RO process the salinity determines the magnitude of the applied pressure and therefore the systems performance
- The process produces very high purity distillate having electrical conductivity in the range of $2-10 \, \mu S \, cm^{-1}$.

Ding et al. (2005) described the dynamic behavior of a batch-operated SPMD pilot plant. The system consisted of a feed tank, an external heat exchanger, and evacuated tube solar energy collectors. The enthalpy balance for the feed tank is:

$$\frac{d(m_f C_p T_{f,in})}{dt} = m_s C_p(T_s - T_{f,in}) - n m_f C_p(T_{f,in} - T_{f,out}) \qquad (4.78)$$

where m_f and m_s are the mass of water in the feed tank and in the collector (kg), respectively; T_s, $T_{f,in}$, and $T_{f,out}$ are the temperature at the collector and the inlet and outlet of the hot feed stream (K), respectively; and n is the number of channels or membrane layers.

For the evacuated tube solar collectors:

$$\frac{d(m_s C_p T_s)}{dt} = A_c \alpha \lambda G - A_\alpha U_{LT}(T_s - T_a) - m_s C_p(T_s - T_{f,in}) \qquad (4.79)$$

where A_c and A_a are the corresponding area for the water heating in the collector and for the absorbed solar radiation (m^2), α and λ are the solar absorptance and transmissivity of the collector, G and U_{LT} are the daily solar radiation (kJ m^{-2} d^{-1}) and the heat loss coefficient of the collector (W m^{-2} K^{-1}), respectively.

The analysis of the system has shown that heat recovery via an external heat exchanger is not only possible, but even effective, and an economical way to intensify the SPMD process.

A simulation model of a solar-driven AGMD unit with a storage tank for heating water and a heat exchanger was presented by Chang et al. (2011, 2012). In this model (Fig. 4.16), the solar absorber is simulated with a thermostat, which supplies the

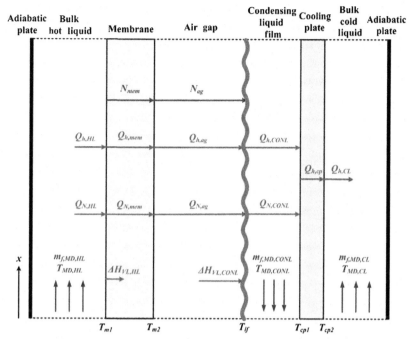

Figure 4.16 Schematic illustration of mass and heat transfer in AGMD system (Chang et al., 2011, 2012).

thermal energy according to the intensity of solar radiation, I (W m^{-2}). The mass balance equations for the hot feed and cold stream are written as:

$$\frac{d\dot{m}_f}{dx} = -J_m M w_w L_{MD} \tag{4.80}$$

$$\frac{d\dot{m}_c}{dx} = -J_{gap} M w_w L_{MD} \tag{4.81}$$

where $\dot{m}_f$ and $\dot{m}_c$ are the mass flow rates in the hot feed and cold stream (kg s^{-1}), J_m and J_{gap} are the water vapor mass flux through the membrane and the air gap (kg m^{-2} s^{-1}), respectively, and L_{MD} is the membrane module length (m).

The energy balances in the hot feed and cold stream, taking into account the convective heat transfer across the boundaries ($Q_{h,f}$, $Q_{h,c}$) and the sensible heat transfer ($Q_{m,f}$) are given by the following equations:

$$\frac{dT_f}{dt} = -L_m \left[\frac{\dot{m}_f}{M_f} \frac{dT_f}{dx} + \frac{W_m}{M_f C_{p,f}} (Q_{h,f} + Q_{m,f}) \right] \tag{4.82}$$

$$\frac{dT_c}{dt} = -L_m \left[\frac{\dot{m}_c}{M_c} \frac{dT_c}{dx} + \frac{W_m}{M_c C_{p,c}} Q_{h,c} \right] \tag{4.83}$$

where T_f and T_c are the temperatures at the hot feed and cold stream (K), L_m and W_m are the membrane length and width (m), and M_f and M_c are the fluid mass (kg) at the hot feed and cold stream, respectively.

The balance of the thermal energy that is produced and provided to the fluid of the solar collector is given by:

$$\frac{dT_s}{dt} = -L_s \frac{\dot{m}_s}{M_s} \frac{dT_s}{dx} + \frac{A_s I(t)}{M_s C_{p,s}} \tag{4.84}$$

where T_s, $\dot{m}_s$, and M_s are the temperature (K), the mass flow rate (kg s^{-1}), and the mass (kg) of the fluid in the collector.

The fluid inside the thermal storage tank is assumed to be in plug flow, the temperature variation of the fluid is one dimensional. Thus:

$$\frac{dT_{st}}{dt} = -H_{st} \frac{\dot{m}_{st}}{M_{st}} \frac{dT_{st}}{dx} \tag{4.85}$$

where T_{st}, $\dot{m}_{st}$, M_{st}, and H_{st} are the temperature (K), the mass flow rate (kg s^{-1}), the mass (kg) of the fluid in the storage tank, and the storage tank height (m), respectively.

The energy balances for the countercurrent flow fluids are:

$$\frac{dT_{hx,f}}{dt} = -L_{hx} \frac{\dot{m}_{hx,f}}{M_{hx,f}} \frac{dT_{hx,f}}{dx} - \frac{A_{hx} U_{hx}}{M_{hx,f} C_{p,hx,f}} (T_{hx,f} - T_{hx,c}) \tag{4.86}$$

$$\frac{dT_{hx,c}}{dt} = -L_{hx} \frac{\dot{m}_{hx,c}}{M_{hx,c}} \frac{dT_{hx,c}}{dx} + \frac{A_{hx} U_{hx}}{M_{hx,c} C_{p,hx,c}} (T_{hx,f} - T_{hx,c}) \tag{4.87}$$

where $T_{hx,f}$, $T_{hx,c}$, $\dot{m}_{hx,f}$, $\dot{m}_{hx,c}$, and $M_{hx,f}$, $M_{hx,c}$ are the temperature (K), the mass flow rate (kg s^{-1}), and the mass of the hot and cold fluids in the exchanger, respectively, and L_{hx} is the heat exchanger length (m).

This mathematic model, which was validated by experimental data, provides information for operation strategies for sunny and cloudy days.

For the better performance of MD process, several multistage systems have been designed and tested. A multistage air gap system (AGMD) was developed both with hollow fiber and plate and frame configuration (Hanemaaijer et al., 2004, 2010). This unit has been labeled MEMSTIL and pilot plants have been installed by Keppel Seghers Company in Singapore and by Eon in Rotterdam, with characteristic capacities of 80 and 50 m^3 d^{-1} (Meindersma et al., 2006; Hanemaaijer et al., 2006). The nominal performances of the units were 25−50 m^3 per day per module for water production, recovery around 50%, and thermal energy consumption ranging from 80 to 240 MJ m^{-3}.

A hybrid system, combination of multiple effect distillation and VMD, was developed by MEMSYS Company. This unit consists of four evaporation—condensation effects with recovery of condensation heat from the previous effect. The effects are connected in series and every one of those operates at lower temperatures and pressures from the previous one (Fig. 4.17). The MEMSYS pilot unit operated with solar energy has been installed in Singapore (Qin et al., 2010), while an autonomous MEMSYS system combining solar thermal collectors and solar photovoltaic (PV) recently installed in Saudi Arabia (Chafid et al., 2014). Technical data provided by the company website indicate GOR values 2—4, electrical energy consumption ranging from 0.75 to 1.75 kWh m^{-3}, thermal energy requirement 175—350 kWh m^{-3}, hot stream temperature between 60°C and 100°C, and cooling water temperature below 40°C.

Summers and Lienhard (2013) presented a novel solar-powered AGMD model, which employs direct heating of membrane by forwarding solar rays using Fresnel mirrors (Fig. 4.18). The membrane material, like PTFE, can be coated by a hydrophilic polymer, such as polycarbonate or cellulose acetate, which absorbs the solar

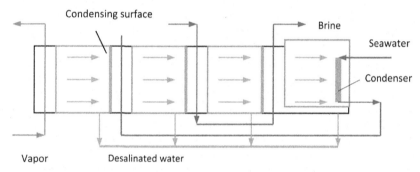

Figure 4.17 Hybrid system, combination of multiple effect distillation and VMD (Qin et al., 2010).

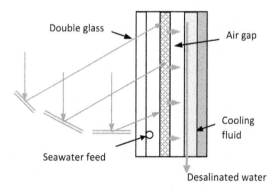

Figure 4.18 AGMD system with direct heating of membrane by solar energy using Fresnel mirrors (Summers and Lienhard, 2013).

radiation. The solar radiation component enters at the membrane according to the following equation:

$$m_f dh_f = -[J_m(h_{ev} + h_{f,m} - h_f) + q_m] \, dA + S dA \tag{4.88}$$

where h_f, h_{ev}, and $h_{f,m}$ are the enthalpy at the hot feed stream, the evaporation enthalpy, and the enthalpy at the hot membrane surface (J kg^{-1}), respectively; S is the solar radiation flux absorbed by the membrane (W m^{-2}); and J_m is the water vapor flux through the membrane (kg m^{-2} s^{-1}).

GOR can be expressed by the incident solar radiation subtracting the thermal losses, q_{loss}:

$$GOR = \frac{\dot{m}_v \Delta H_v}{(S - q_{loss})A} \tag{4.89}$$

Moreover, the coupling of the VMD systems with solar energy was studied. Zrelli et al. (2014) optimized the geometric configuration of a helically coiled fiber, which was placed in the absorber of a parabolic concentrator. The performance of helical and linear fibers was compared and a mathematic model was developed studying the influence of fibers configuration in the flux of produced water.

A mathematic model for an autonomous SPMD plant was developed by Frikha et al. (2014). The system consists of a field of solar collectors, the membrane module, and a heat exchanger. Each of the three system units was modeled and a computer program was established for the simulation and the study of the whole system. Additionally, an experimental comparison of the membranes performance in DCMD and VMD configurations for different operation conditions was given by Koo et al. (2013).

4.7.1 Technoeconomic Performance of Solar-Powered Systems

For solar-driven MD systems, the thermal recovery ratio, TRR, is defined as the ratio of the heat that is theoretically required to distill an amount of distillate to the total used energy in the system (Banat et al., 2007b).

$$TRR = \frac{0.69 \, V_{prod}}{I A_s} \tag{4.90}$$

The solar collector efficiency is:

$$\eta_s = \frac{\dot{m}_f C_p (T_{s,in} - T_{s,out})}{I A_s} \tag{4.91}$$

where V_{prod} is the produced water volume (m^3), I is the intensity of incident solar radiation (W m^{-2}), $\dot{m}_f$ is the mass flow rate of the hot feed water (kg s^{-1}), A_s is the collectors surface (m^2), and $T_{s,in}$ and $T_{s,out}$ are the temperatures at the inlet and outlet of the collector (K), respectively.

Saffarini et al. (2012) define GOR as:

$$\text{GOR} = \frac{\dot{m}_v \Delta H_v}{q_{in}} \qquad (4.92)$$

where $\dot{m}_v$ is the distillate mass flow rate (kg s^{-1}), ΔH_v is the latent heat of evaporation (J kg^{-1}), and q_{in} is the external heat flow rate in the system, which can be calculated as:

$$q_{in} = \dot{m}_f C_p (T_{f,in} - T_{heat,in}) \qquad (4.93)$$

where $T_{f,in}$ and $T_{heat,in}$ are the inlet temperatures at the hot feed stream and at the solar collector (K), respectively.

Also, GOR can be expressed as:

$$\text{GOR} = \frac{\dot{m}_v}{\dot{m}_f} \frac{\Delta H_v}{C_p (T_{f,in} - T_{heat,in})} \qquad (4.94)$$

The system performance ratio is then given as:

$$\text{PR} = \frac{\dot{m}_v}{q_{in}} = \frac{\dot{m}_v}{\dot{m}_f C_p (T_{f,in} - T_{heat,in})} \qquad (4.95)$$

Saffarini et al. (2012) present a timeline table of SPMD systems and their general features. In Table 4.3, the solar-driven MD plants, until that date, together with some of their characteristics are listed.

Table 4.3 Solar-Powered Membrane Distillation Plants and General Features (Saffarini et al., 2012)

Year	Location	MD Configuration	Energy Source
1991	Australia	DCMD	Solar collector
1995	Australia	AGMD	Evacuated tube collector
1997	Tokyo, Japan	—	Solar collector—PV
1999	Texas, USA	AGMD	Solar pond—Electrical grid
1999	Jordan	DCMD	Solar still
2003	Freiburg, Germany	AGMD	Flat plate collector—Electrical grid
2004	Texas, USA	AGMD	Solar pond—Electrical grid
2007	Irbid, Jordan	AGMD	Flat plate collector—PV
2007	Aqaba, Jordan	AGMD	Flat plate collector—PV
2008	Alexandria, Egypt	AGMD	Flat plate collector—PV
2008	Mexico	DCMD	Concentrated parabolic collector—Grid
2009	Almeria, Spain	AGMD	Concentrated parabolic collector—Grid
2009	Hangzhou, China	VMD	High-performance collector—Grid
2009	Nevada, USA	DCMD	Solar pond
2009	Gran Canaria, Spain	AGMD	Flat plate collector—PV
2011	Singapore	VMD	Flat plate collector—PV

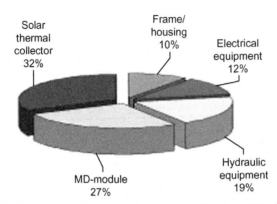

Figure 4.19 Schematic illustration of the costs of a compact MD plant (Wieghaus et al., 2008).

In addition, Wieghaus et al. (2008) gave a distribution of the costs of a compact MD plant (Fig. 4.19) indicating that the price of the produced desalinated water is 0.01 € per liter (10 € per cubic meter).

4.8 MEMBRANE'S CHARACTERISTIC PROPERTIES

4.8.1 Membrane Permeability

Membrane permeability is defined as the mass flux of the vapor molecules through the membrane divided by the partial pressure difference developed across the membrane and the membrane thickness. This property depends on the membrane parameters such as porosity, pore size, tortuosity, and membrane thickness. The relationship between the flux through the membrane and the membrane different parameters is given by:

$$J \propto \frac{d_p \varepsilon}{\tau \delta_m} \tag{4.96}$$

where d_p, ε, τ, and δ_m are the pore size (pore diameter), the porosity, the tortuosity, and the thickness of the membrane, respectively.

Eq. (4.96) illustrates the importance, in terms of flux through the membrane, of maximizing the membrane pore size and porosity, while minimizing the transport length through the membrane ($\tau \delta_m$). Therefore, to achieve high permeability, the membrane pore size and porosity must be as large as possible and its surface layer as thin as possible. However, there must be an optimization of membrane thickness, since thinner membranes favor the mass transport, while thicker membranes achieve minimal heat losses by conduction through the membrane.

4.8.2 Liquid Entry Pressure (LEP)

LEP is the minimum hydrostatic pressure that must be applied into the feed liquid solution before it exceeds the hydrophobic forces of the membrane and penetrates into the membrane pores. The value of LEP is characteristic of each membrane and depends on the maximum pore size and the membrane hydrophobicity. According to Franken et al. (1987), LEP can be estimated by the following expression:

$$LEP = \Delta P = P_f - P_c = -\frac{B\gamma_l \cos\theta}{d_p^{max}} \tag{4.97}$$

where P_f and P_c are the hydrostatic pressure at the hot feed stream and cold stream, respectively, γ_l is the liquid solution surface tension, B is the geometric pore coefficient (equal to 1 for cylindrical pores), θ is the contact angle, and d_p^{max} is the maximum pore size.

According to the above relationship, high values of LEP are achieved by using membrane material with high hydrophobicity and a small maximum pore size. However, as the maximum pore size decreases, the mean pore size of the membrane decreases and therefore the membrane permeability becomes low.

With regard to surface tension of the liquid solution, Zhang et al. (2010) studied the impact of salt concentration (NaCl) on the water surface tension and the following equation is given:

$$\gamma_{l,new} = \gamma_l + 1.467 C_f \tag{4.98}$$

where γ_l is the pure water surface tension at 25°C ($=72$ mN m^{-1}).

4.9 MEMBRANE MODULES

In the MD process, a relevant large variety of membrane modules with different geometry, such as flat sheet (plate and frame, spiral wound) and tubular (capillary and hollow fiber), have been designed and tested. The design of MD modules must permit high values of flow rate at the hot feed and cold stream and low pressure drop along the module. Due to the fact that the process is nonisothermal, the MD module should provide a guarantee of a good heat recovery function and thermal stability.

The most part of laboratory scale modules is designed for use with flat sheet membranes because they are versatile, simple, and can be easily removed from their modules for cleaning, examination, and replacement.

In plate and frame membrane module, the membranes, the porous support plates, and the spacers are stacked between two end plates. The support plate separates the flat sheet membranes to provide a channel for the permeate flow. The feed solution flows across the surface of each membrane sheet in the stack (Fig. 4.20A).

In spiral wound membrane module, the membrane, the spacer, and the porous support are enveloped and wound around a perforated central permeate collection tube. The feed solution flows axially in the spacers channel and the permeate, that passes through the membrane, flows spirally to reach the central collection tube (Fig. 4.20B). The spiral wound module has high packing density and low cost compared with the plate and frame module. However, the last one is easier in cleaning and membrane replacement as well as having a high resistance to fouling.

The tubular membrane modules have a shell-and-tube configuration. The membranes are tube shaped and are inserted between two cylindrical chambers (Fig. 4.20C). These modules are more attractive from a commercial stand point,

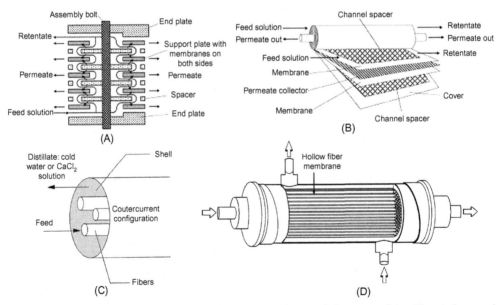

Figure 4.20 Membrane modules as used in MD: (A) plate and frame module, (B) spiral wound module, and (C and D) tubular, hollow fiber, and capillary modules (Chiam and Sarbatly, 2013; Mar-Camacho et al., 2013).

because they have less tendency to organic fouling, are easier to clean, and they have high active surface area. Even if they are more productive compared to flat sheet membranes, they have lower packing density and higher operating cost.

The hollow fiber membrane module can contain thousands of hollow fiber membranes. The membranes in that module are bundled (diameter below 1 mm) and sealed inside the shell-and-tube unit (Fig. 4.20C and D). The feed solution flows through the lumen side of the fibers and the permeate is collected on the outside of the fibers (configuration inside/out) or vice versa (configuration outside/in). This module has very high packing density and low cost, while it presents low resistance to fouling and difficulty in cleaning and in maintenance.

The capillary membrane modules are similar to hollow fiber membrane modules. The diameter of the membranes is 1–3 mm and the membranes can be configured in an inside/out or outside/in shell-and-tube module, as shown in Fig. 4.20D. Compared with the hollow fiber membrane module, they are easier in cleaning and the distribution of the capillaries is more uniform. However, they have low packing density.

4.10 MEMBRANE TYPES
4.10.1 Commercial Membranes

One of the essential requirements for the MD membranes, as mentioned before, is the hydrophobicity, that is to say the membrane pores not be wetted by the liquid phase of the aqueous feed solution. Therefore, microporous hydrophobic membranes made from polymeric materials with low surface energy, such as PTFE, PP, PVDF, and PE, are commercially available in different forms (flat sheet, tubular, hollow fiber, and capillary) and they are widely used in MD laboratory experiments. However, as shown by their characteristics which are given by the manufacturers (Table 4.4), these membranes are developed for the processes of microfiltration and nanofiltration.

PTFE is an ideal material for MD membrane manufacturing, since it has the lowest surface energy. It is a highly crystalline polymer with excellent thermal stability and chemical resistance. Nevertheless, it presents difficulty in processing to prepare membranes. The hydrophobic PTFE membranes are normally produced using sintering—extraction method or melt—extraction—stretching method.

The porous PP membranes are prepared either by molten—extraction technique followed by stretching or by thermal phase separation process with dissolution of polymer at high temperatures in less common solvents. The performance of these membranes is relevantly low due to their symmetric structure and their moderate thermal stability at elevated temperatures.

Table 4.4 Characteristic Properties of Commercially Available Membranes that Typically Used in the MD Process

Manufacturer	Commercial Name	Membrane Type	Thickness, δ (μm)	Pore Size, d_p (μm)	Porosity, ε (%)
Gelman Inst. Co.	TF200	PTFE/PP[a]	178	0.20	80
	TF450	PTFE/PP[a]	178	0.45	80
	TF1000	PTFE/PP[a]	178	1.00	80
Millipore	Durapore	PVDF	140	0.45	75
	GVSP	PVDF	108	0.22	80
	FGLP	PVDF/PE[a]	130	0.20	70
Gore	Gore-Tex	PTFE	64	0.20	90
Sartorious	Enka	PP	100	0.1	75
Hoechst Celanese Co.	Celgard 2500	PP	28	0.05	45
	Celgard 2400	PP	25	0.02	38
3M Corporation	3ME	PP	79	0.73	85
AkzoNobel	Accurel	PP	150	0.43	70

[a]The membrane is supported by PP or PE woven.

On the contrary, PVDF can be easily dissolved in a variety of solvents at room temperature and can be produced by phase separation process. In Table 4.5, the experimental performance (vapor mass flux) of some commercial membranes, flat sheet and hollow fiber membranes is presented, measured in various operating conditions (aqueous solution concentration, hot feed and cold stream temperature, T_h, T_c, and vacuum pressure, P_v) for different MD configurations.

4.10.2 Membrane Synthesis

In recent years, the research interest has focused on the synthesis of new materials for membranes and the modification of existing membranes especially for MD process, since the commercial hydrophobic membranes have been developed for other separation processes and they do not satisfy essentially all the characteristics needed for a sufficient MD membrane. A large variety of different types of membranes has been proposed, such as hydrophobic—hydrophilic membranes, ceramic membranes, and carbon nanotubes (CNTs), which are still in the process of the wider development. The objective is to increase the membrane durability and to improve the permeation flux. In Table 4.6, the performance of some synthesized membranes in various operating conditions (aqueous solution concentration, hot feed and cold stream temperatures, T_h, T_c, and vacuum pressure, P_v) given for different MD configurations is referred.

4.10.2.1 Polymeric Membranes (Polymer Blends and Additives)

Various types of hydrophobic single layer membranes, flat sheet and hollow fiber, have been fabricated for the MD process. Among the polymeric materials used, PVDF is

Table 4.5 Reported Performance (Permeate Flux) of Commercially Available Membranes (Flat Sheet and Hollow Fiber Membranes) in Various Operating Conditions for Different MD Configurations

Reference	MD Configuration	Membrane	Operation Conditions (Feed Solution, T_h, T_c and P_v)	Flux
Andrjesdottir et al. (2013)	DCMD	PTFE (GE Energy)	15 ppt. NaCl, 60°C, 20°C	30 kg m^{-2} h^{-1}
Hwang et al. (2011)	DCMD	PTFE (GE Osmonics)	6 wt% NaCl, 60°C, 20°C	21.5 L m^{-2} h^{-1}
Cath et al. (2004)	DCMD	PTFE (GE Osmonics)	0.6 g L^{-1} NaCl, 40°C, 20°C	24.7 kg m^{-2} h^{-1}
Hsu et al. (2002)	DCMD	PTFE (Millipore)	5 wt% NaCl, 45°C, 20°C	39 kg m^{-2} h^{-1}
Schofield et al. (1990)	DCMD	PVDF (Millipore)	30 wt% NaCl, 80°C, 21°C	56 kg m^{-2} h^{-1}
Cath et al. (2004)	DCMD	PP (GE Osmonics)	0.6 g L^{-1} NaCl, 40°C, 20°C	11 kg m^{-2} h^{-1}
Song et al. (2008)	DCMD	HF-PP (Membrana)	City water, 90°C, 20°C	50 kg m^{-2} h^{-1}
Criscuoli et al. (2008)	DCMD	HF-PP (Membrana)	Distillate water, 59°C, 14°C	25.4 kg m^{-2} h^{-1}
Al-Obaidani et al. (2008)	DCMD	HF-PP (Microdyn)	35 g L^{-1} NaCl, 75°C, 15°C	13 kg m^{-2} h^{-1}
Teoh et al. (2008)	DCMD	HF-PP (Hyflux)	3.5 wt% NaCl, 75°C, 20°C	8.2 kg m^{-2} h^{-1}
Cipollina et al. (2012)	AGMD	PTFE (Gore)	35 g L^{-1} NaCl, 80°C, 20°C	10.5 L m^{-2} h^{-1}
Liu et al. (1998)	AGMD	PTFE (Millipore)	3 wt% NaCl, 75°C, 20°C	17 kg m^{-2} h^{-1}
Hsu et al. (2002)	AGMD	PTFE (Millipore)	5 wt% NaCl, 45°C, 20°C	5.1 kg m^{-2} h^{-1}
Banat and Simandl (1998)	AGMD	PVDF (Millipore)	3 wt% NaCl, 80°C, 20°C	7 kg m^{-2} h^{-1}
Khayet et al. (2000)	SGMD	PTFE (Gelman)	Distillate water, 65°C, 20°C	18.7 kg m^{-2} h^{-1}
Khayet et al. (2002)	SGMD	PTFE (Gelman)	Distillate water, 70°C, 20°C	16.2 kg m^{-2} h^{-1}
Mericq et al. (2009)	VMD	PTFE (Millipore)	300 g L^{-1} NaCl, 65°C, 100 Pa	18 L m^{-2} h^{-1}
Mericq et al. (2009)	VMD	PTFE (Desal. Inc.)	300 g L^{-1} NaCl, 65°C, 100 Pa	68 L m^{-2} h^{-1}
Criscuoli et al. (2008)	VMD	HF-PP (Membrana)	Distillate water, 70°C, 10 Pa	56.2 kg m^{-2} h^{-1}

Table 4.6 Reported Performance (Permeate Flux) of Synthesized Membranes (Flat Sheet and Hollow Fiber Membranes) in Various Operating Conditions for Different MD Configurations

Reference	MD Configuration	Membrane	Operation Conditions (Feed Solution, T_h, T_c, and P_v)	Flux
Tomaszewska (1996)	DCMD	PVDF/Li/DMA	1–2% NaCl, 60°C, 20°C	9.7 kg m^{-2} h^{-1}
Wang et al. (2008)	DCMD	PVDF/NMP/EG	3.5 wt% NaCl, 80°C, 17.5°C	41.5 kg m^{-2} h^{-1}
Feng et al. (2004)	DCMD	PTFE-co-TFE	0.3 M NaCl, 55°C, 25°C	7.3 kg m^{-2} h^{-1}
Feng et al. (2006)	DCMD	PTFE-co-HFP	0.3 M NaCl, 55°C, 25°C	14.5 kg m^{-2} h^{-1}
Khayet et al. (2005)	DCMD	PTFE/SMMs/PEI	0.5 M NaCl, 45°C, 35°C	12.6 kg m^{-2} h^{-1}
Qtaishat et al. (2009a)	DCMD	PTFE/SMMs/PEI	0.5 M NaCl, 65°C, 15°C	20.9 kg m^{-2} h^{-1}
Qtaishat et al. (2009b)	DCMD	PTFE/SMMs/PS	0.5 M NaCl, 50°C, 40°C	8.3 kg m^{-2} h^{-1}
Qtaishat et al. (2009c)	DCMD	PTFE/SMMs/PES	0.5 M NaCl, 50°C, 40°C	9.4 kg m^{-2} h^{-1}
Bonyadi and Chung (2007)	DCMD	(HF) PVDF/PVDF	3.5 wt% NaCl, 90°C, 16.5°C	55 kg m^{-2} h^{-1}
Garcia-Payo et al. (2009)	DCMD	(HF) PVDF-co-HFP	Distillate water, 45°C, 18°C	1.4 kg m^{-2} h^{-1}
Li et al. (2003)	DCMD	(HF) PE	35 g L^{-1} NaCl, 60°C, –	0.8 L m^{-2} h^{-1}
Li et al. (2003)	DCMD	(HF) PP	35 g L^{-1} NaCl, 60°C, –	0.3 L m^{-2} h^{-1}
Li and Sirkar (2004)	DCMD	(HF) PP	1 wt% NaCl, 90°C, 15–17°C	80 kg m^{-2} h^{-1}
Larbot et al. (2004)	DCMD	(HF) Ceramic Al_2O_3	1 M NaCl, 95°C, 5°C	5.4 kg m^{-2} h^{-1}
Larbot et al. (2004)	DCMD	(HF) Ceramic ZrO_2	1 M NaCl, 95°C, 5°C	6.9 kg m^{-2} h^{-1}
Cerneaux et al. (2009)	DCMD	Ceramic TiO_2	0.5 M NaCl, 95°C, 5°C	20 L m^{-2} d^{-1}
Cerneaux et al. (2009)	DCMD	Ceramic ZrO_2	0.5 M NaCl, 95°C, 5°C	95 L m^{-2} d^{-1}
Cerneaux et al. (2009)	AGMD	Ceramic TiO_2	0.5 M NaCl, 95°C, 5°C	20 L m^{-2} d^{-1}
Cerneaux et al. (2009)	AGMD	Ceramic ZrO_2	0.5 M NaCl, 95°C, 5°C	113 L m^{-2} d^{-1}
Li et al. (2003)	VMD	(HF) PE	35 g L^{-1} NaCl, 60°C, 8 Pa	4 L m^{-2} h^{-1}
Li et al. (2003)	VMD	(HF) PP	35 g L^{-1} NaCl, 60°C, 8 Pa	2.9 L m^{-2} h^{-1}
Li and Sirkar (2005)	VMD	(HF) PP	1 wt% NaCl, 85°C, 60–66 cm Hg	71 kg m^{-2} h^{-1}
Jin et al. (2008)	VMD	(HF) PPESK/TFA	0.5 wt% NaCl, 40°C, 0.078 MPa	3.7 L m^{-2} h^{-1}
Cerneaux et al. (2009)	VMD	Ceramic TiO_2	0.5 M NaCl, 40°C, 300 Pa	146 L m^{-2} d^{-1}
Cerneaux et al. (2009)	VMD	Ceramic ZrO_2	0.5 M NaCl, 40°C, 300 Pa	180 L m^{-2} d^{-1}

considered the most known due to its excellent chemical and thermal resistance and that it can be dissolved in common organic solvents at low temperatures.

Ortiz de Zárate et al. (1995) prepared asymmetric flat sheet PVDF membranes from binary solutions of PVDF/dimethyl acetamide (DMAC) or PVDF/dimethyl formamide (DMF), by using the phase inversion technique. They observed that increasing the PVDF concentration in the solution both the pore size and the porosity decreased, while no improvement in permeate flux was detected. Tomaszewska (1996) studied the effect of lithium chloride (LiCl) concentration in the PVDF/DMAC and PVDF/DMF solutions. The increase of LiCl concentration from 0 to 3 wt% led to the enhancement of both the pore size and the porosity, as well as of the permeate flux in a DCMD configuration. However, the LEP_w and the mechanical resistance were decreased.

In parallel, hollow fiber membranes were prepared using different techniques. Fujii et al. (1992a) fabricated porous hollow fiber PVDF membranes by the dry/ wet spinning method using the dimethyl sulfoxide (DMSO) as a solvent. These membranes exhibited pore sizes smaller than those of microfiltration membranes. The melt-extruded/cold-stretching technique was used by Li et al. (2003) to prepare PE and PP hollow fiber membranes for water desalination by DCMD and VMD. PE membranes presented higher permeate flux values compared to PP membranes in both configurations. Moreover, Wang et al. (2008) used NMP (N-methyl-2-pyrrolidone) as solvent and ethylene glycol as nonsolvent additive to synthesize a PVDF hollow fiber membrane by the dry/wet spinning method. The fabricated membrane exhibited a very narrow pore size distribution and an external thin layer over the porous substrate. This membrane structure resulted in a decrease of the mass transfer resistance and therefore to an enhancement of the permeate flux.

In addition to homopolymers of PTFE, PVDF, and PP, membranes for the MD process can be prepared from their copolymers with enhanced hydrophobicity and durability. Feng et al. (2004) prepared asymmetric flat sheet membranes from the PVDF-co-tetrafluoroethylene (PVDF-co-TFE) by the phase inversion method. The membranes presented higher permeate flux values for the DCMD configuration compared to those of the PVDF membranes (Fig. 4.21A). The same research group fabricated membranes from copolymer PVDF-co-hexafluoropropylene (PVDF-co-HFP) with the same method (Feng et al., 2006). These membranes are more hydrophobic, have higher solubility, and present higher mass flux of vapor compared to those of PVDF membranes. Also, Garcia-Payo et al. (2009, 2010) prepared hollow fiber PVDF-co-HFP membranes by dry/wet spinning method using DMAC as a solvent and polyethylene glycol (PEG) as a nonsolvent additive (Fig. 4.21B).

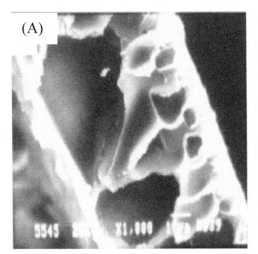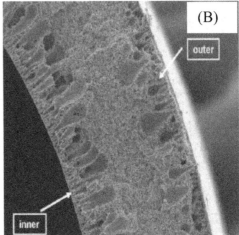

Figure 4.21 SEM images of the cross-section of hollow fiber membranes: (A) PVDF-*co*-TFE membrane (Feng et al., 2004) and (B) PVDF-*co*-HFP membrane (Garcia-Payo et al., 2009).

4.10.2.2 Composite Membranes (Multilayer, Hydrophilic/Hydrophobic)

The use of composite membranes in MD process was first reported by Cheng and Wiersma (1982) in a series of patents. The membranes that prepared consisted of a hydrophobic layer of PTFE or PVDF and a hydrophilic layer of cellulose acetate, polysulfone, or cellulose nitrate. Kong et al. (1992) modified a hydrophilic microporous cellulose nitrate membrane surface via plasma polymerization using two monomer systems: the OFCB and VTMS/CF$_4$. A trilayer membrane, with a hydrophilic layer sandwiched between two hydrophobic layers, was prepared and tested in a DCMD device. This membrane exhibited similar DCMD behavior to the typical DCMD behavior that frequently observed when it is used as a hydrophobic single layer membrane.

Khayet et al. (2005) and Qtaishat et al. (2009a) were the first that presented the concept of hydrophobic/hydrophilic membrane. A composite flat sheet membrane was prepared by phase inversion method, wherein a hydrophobic surface modifying macromolecule (SMM) was blended with a hydrophilic base polymer, polyetherimide (PEI). As a result, the membrane top layer becomes hydrophobic, while the bottom layer becomes hydrophilic. It was shown that this type of membrane satisfies all the requirements of high permeability MD membranes, since the permeate flux in DCMD was greater than that of commercial PTFE membranes (Fig. 4.22A). Beside PEI membranes (Qtaishat et al., 2009a), polysulfone (PS) and polyethersulfone (PES) flat sheet membranes were modified using different types of SMMs, solvents and additives in order to optimize the performance of the composite hydrophobic/hydrophilic membrane (Qtaishat et al., 2009b,c).

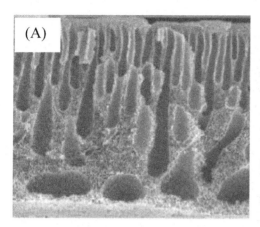 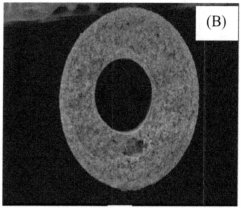

Figure 4.22 SEM images: (A) cross-section of hydrophobic/hydrophilic PTFE/SMMs/PEI membrane (Khayet et al., 2005) and (B) cross-section of dual layer hydrophobic/hydrophilic PVDF/PAN/NMP membrane (Bonyadi and Chung, 2007).

The effect of the coating materials and the heat treatment in hollow fiber membranes was investigated by Fujii et al. (1992a,b). Silicone rubber, polyketone, and poly (1-trimethylsilyl-1-propyne) (PMSP) were used as coating materials in the bores of PVDF membranes. It was observed that the water permeability decreased after the coating. On the contrary, the thermal treatment resulted in the increase of the membrane hydrophobicity from 94°C to 102°C. Li and Sirkar (2004 and 2005) developed composite hollow fiber membranes for use in DCMD and VMD devices. The membranes were commercial hollow fiber PP membranes coated on their external surface with a variety of ultrathin fluorosilicone layers via the plasma polymerization technique. The DCMD and VMD experiments, carried out in the temperature range 60−90°C, showed high values of permeate flux and temperature polarization coefficient (93−99%), as well as absence of pore wetting.

Novel hydrophobic/hydrophilic hollow fiber PPESK membranes were prepared by coating the internal surface with silicone rubber and sol−gel polytrifluoropropyl siloxane (Jin et al., 2008). The composite silicone rubber membrane exhibited high permeate flux value, $3.5 \, \text{L m}^{-2} \, \text{h}^{-1}$, for VMD and 99% salt rejection. Moreover, the coextrusion dry/jet wet spinning method was applied for the first time by Bonyadi and Chung (2007) to prepare a dual layer hydrophobic/hydrophilic membrane. For the outer hydrophobic layer a PVDF/NMP solution was used, while for the inner hydrophilic layer a PVDF/PAN/NMP solution (Fig. 4.22B). A permeate flux of $55.2 \, \text{kg m}^{-2} \, \text{h}^{-1}$ and a salt separation factor of 99.8% were obtained at 90°C for 3.5 wt% NaCl aqueous solution. It was claimed that the proposed membrane exhibited higher DCMD performance compared to the previous reported membranes.

4.10.2.3 Ceramic Membranes

The ceramic membranes are inorganic membranes known for their mechanical strength, thermal and chemical stability. Their application ensures the long operational life and robustness of the process. In the ceramic membranes mainly oxides of aluminum, silicon, titanium, and zirconium are involved with the hydroxyl group at the surface to render the membrane as hydrophilic. Due to their hydrophilic nature, these membranes have not gained much popularity for MD applications. However, some efforts have been attempted to modify their hydrophilic surface.

Larbot et al. (2004) used ceramic hollow fiber Al_2O_3 and ZrO_2 membranes (pore size: 200 and 500 nm), grafted by fluoroalkylsilanes for water desalination by DCMD configuration. After grafting, the measured water contact angle was higher, which is the indication of membrane hydrophobicity. The grafted Al_2O_3 membranes with pore size of 200 nm exhibited permeate flux of $163.2 \, L \, m^{-2} \, d^{-1}$ and salt rejection factor up to 99%. On the contrary, the modified ZrO_2 membranes, pore size of 500 nm, showed higher permeate flux of $202 \, L \, m^{-2} \, d^{-1}$ and 99.5% salt rejection.

Recently, a comparison was carried out between hydrophobic ZrO_2 (pore size 50 nm) and TiO_2 (pore size 5 nm) tubular ceramic membranes, tested in various configurations of MD process (DCMD, AGMD, and VMD) (Cerneaux et al., 2009). The internal surface of membranes was modified by grafting C_8F_{17}. The highest permeate fluxes were obtained by the modified ZrO_2 membrane, 180, 95, and $113 \, L \, m^{-2} \, d^{-1}$, for the three process configurations. In all cases, the salt rejection was higher than 99%.

4.10.2.4 Carbon Nanotubes

CNTs present a growing interest in their application at MD process, due to their unique mechanical, chemical, and physical properties. The fast transport of water vapor molecules inside the nanotubes, their potential to change the water—membrane interface to stop the penetration of the liquid phase, as well as the preferential transport of vapors through the pores have encouraged the incorporation of CNTs into the membrane matrix. Dumee et al. (2010, 2011) reported on a series of CNTs, mounted on a paper-like structure, Bucky Paper (BP). This is self-supported, very thin membranes held together with Van der Waals forces. Even if their performance is similar to that of PTFE membranes (permeability $3.3 \times 10^{-12} \, kg \, m^{-2} \, s^{-1} \, Pa^{-1}$ and NaCl rejection 95%), they have a relevant short lifetime, due to the cracks that are locally formed, reducing the membrane permeability (Fig. 4.23A). The application of CNT-based membranes has caused a considerable increase in performance for aqueous NaCl solutions (Dumee et al., 2011; Roy et al., 2014). In Table 4.7, the properties

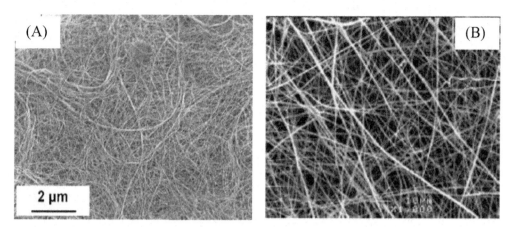

Figure 4.23 SEM images: (A) surface of carbon nanotube membrane, BP (Dumee et al., 2010) and (B) surface of PVDF nanofiber membrane (Feng et al., 2008).

Table 4.7 Characteristic Properties and Reported Performance of Carbon Nanotube Membranes (Mar-Camacho et al., 2013)

CNT Membrane Type	Thickness, δ (μm)	Pore Size, d_p (nm)	Porosity, ε (%)	Flux (kg m^{-2} h^{-1})	NaCl Rejection (%)
BP	55	25	90	12	94
Sandwiched BP	140	25	90	15	95.5
PTFE/BP	105	25	88	7.75	99
Alkoxy-silane BP	62	23	90	9.5	98.3

and the performance of CNT membranes for water desalination are presented (Mar-Camacho et al., 2013).

4.10.2.5 Electrospun Membranes

A relatively new category of membranes are nanofiber membranes formed by the electrospinning method. In this technique, fibers are spun under the pressure and electric field and form a nonwoven mat. That mat shows very high porosity, excellent hydrophobicity, and very good interconnectivity, characteristics that make these membranes ideal candidates for desalination. Feng et al. (2008) investigated the fabrication of PVDF nanofiber membranes via electrospun method (Fig. 4.23B). In AGMD conducted experiments, permeate flux equal to 11.5 kg m^{-2} h^{-1} and NaCl rejection factor equal to 98.5% were obtained. It was claimed that MP is comparable to that of commercial microfiltration membranes.

REFERENCES

Alkhudhiri, A., Darwish, N., Hilal, N., 2012. Membrane distillation: a comprehensive review. Desalination. 287, 2–18.

Alklaibi, A.M., 2008. The potential of membrane distillation as a stand-alone desalination process. Desalination. 223, 375–385.

Alklaibi, A.M., Lior, N., 2004. Membrane distillation desalination: status and potential. Desalination. 171, 111–131.

Al-Obaidani, S., Curcio, E., Macedonio, F., di Profio, G., Al-Hinai, H., Drioli, E., 2008. Potential of membrane distillation in seawater desalination: thermal efficiency, sensitivity study and cost estimation. J. Membr. Sci. 323, 85–98.

Andrjesdottir, O., Ong, C.L., Nabavi, M., Parades, S., Khalil, A.S.G., Michel, B., et al., 2013. An experimentally optimized model for heat and mass transfer in direct contact membrane distillation. Int. J. Heat Mass Transfer. 66, 855–867.

Bahmanyar, A., Asghari, M., Khoobi, N., 2012. Numerical simulation and theoretical study in simultaneously effects of operating parameters in direct contact membrane distillation. Chem. Eng. Proc. 61, 42–50.

Banat, F.A., Simandl, J., 1993. Membrane distillation for dilute ethanol separation from aqueous streams. J. Membr. Sci. 163, 333–348.

Banat, F.A., Simandl, J., 1996. Removal of benzene traces from contaminated water by vacuum membrane distillation. Chem. Eng. Sci. 51, 1257–1265.

Banat, F.A., Simandl, J., 1998. Desalination by membrane distillation: a parametric study. Sep. Sci. Technol. 33, 201–226.

Banat, F., Jumah, R., Garaibeh, M., 2002. Exploitation of solar energy collected by solar stills for desalination for membrane distillation. Renewable Energy. 25, 293–305.

Banat, F., Jwaied, N., Rommel, M., Koschikowski, J., Wieghaus, M., 2007a. Desalination by a "compact SMADES" autonomous solar powered membrane distillation unit. Desalination. 117, 29–37.

Banat, F., Jwaied, N., Rommel, M., Koschikowski, J., Wieghaus, M., 2007b. Performance evaluation of a "large SMADES" autonomous desalination solar-driven membrane distillation plant in Aqaba, Jordan. Desalination. 117, 17–28.

Bandini, S., Saavedra, A., Sarti, G.C., 1997. Vacuum membrane distillation: experiments and modeling. AIChE J. 43, 398–408.

Basini, L., D'Angelo, G., Gobbi, M., Sarti, G.C., Gostoli, C., 1987. A desalination process through sweeping gas membrane distillation. Desalination. 64, 245–257.

Bergman, T.L., Lavine, A.S., Incropera, F.P., Witt, D.P., 2011. Fundamentals of Heat and Mass Transfer. 7th Edition J. Wiley & Sons, New York.

Bier, C., Plantikow, U., 1995, Solar powered desalination by membrane distillation (MD). In: Proc. IDA World Congress on Desalination and Water Sciences, Abu-Dhabi, November 18–24, 1995, Vol. V (1985), pp. 397–410.

Bird, R.B., Stewart, W.E., Lightfoot, E.N., 2002. Transport Phenomena. 2nd ed. John Wiley & Sons Inc, USA.

Boddeker, K.W., 1990. Terminology in pervaporation. J. Membr. Sci. 51, 259–272.

Bodell, B.R., 1968. Distillation of saline water using silicon rubber membrane. US Patent no. 3,361,645.

Bonyadi, S., Chung, T.S., 2007. Flux enhancement in membrane distillation by fabrication of dual layer hydrophobic–hydrophilic hollow fiber membranes. J. Membr. Sci. 306, 134–146.

Calabro, V., Drioli, E., Matera, F., 1991. Membrane distillation in the textile wastewater treatment. Desalination. 83, 209–224.

Cath, T.Y., Adams, V.D., Childress, A.E., 2004. Experimental study using direct contact membrane distillation: a new approach to flux enhancement. J. Membr. Sci. 228, 5–16.

Cerneaux, S., Struzynska, I., Kujawski, W.M., Persin, M., Larbot, A., 2009. Comparison of various membrane distillation methods for desalination using hydrophobic ceramic membranes. J. Membr. Sci. 337, 55–60.

Chafid, A., Al-Zahrani, S., Al-Otaibi, M.N., Hoong, C.F., Lai, T.F., Prabu, M., 2014. Portable and integrated solar-driven desalination system using membrane distillation for arid remote areas in Saudi Arabia. Desalination. 345, 36–49.

Chang, H., Chang, C.-L., Ho, C.D., Li, C.-C., Wang, H.P., 2011. Experimental and simulation study of an air gap membrane distillation module with solar absorption function for desalination. Desalin. Water Treat. 25, 251−258.

Chang, H., Lyu, S.G., Tsai, C.-M, Chen, Y-H, Cheng, T.W., Chou, Y.-H., 2012. Experimental and simulation study of a solar thermal driven distillation desalination process. Desalination. 286, 400−411.

Chiam, C.K., Sarbatly, R., 2013. Vacuum membrane distillation processes for aqueous solution treatment. Chem. Eng. Proc. 74, 27−54.

Cheng, D.Y., Wiersma, S.J., 1982. Composite membrane for a membrane distillation system. US Patent Serial No. 4,316,772.

Cipollina, A., Koschikowski, J., Gross, F., Pfeifle, D., Rolletschek, M., Schwantes, R., 2011. Membrane distillation: solar and waste heat driven demonstration plants for desalination. In: Proceedings of the International Workshop on Membrane Distillation and Related Technologies, October 9−12, Ravello (Italy).

Cipollina, A., Di Sparti, M.G., Tamburini, A., Micale, G., 2012. Development of a membrane distillation module for solar energy seawater desalination. Chem. Eng. Res. Des. 90, 2101−2121.

Criscuoli, A., Carnavale, M.C., Drioli, E., 2008. Evaluation of energy in membrane distillation. Chem. Eng. Proc. 47, 1098−1105.

Curcio, E., Drioli, E., 2005. Membrane distillation and related operations—a review. Sep. Pur. Rev. 34, 35−86.

Ding, Z., Liu, L., El-Bourawi, M.S., Ma, R., 2005. Analysis of a solar-powered membrane distillation system. Desalination. 172, 27−40.

Drioli, E., Wu, Y., Calabro, V., 1987. Membrane distillation in the treatment of aqueous solutions. J. Membr. Sci. 33, 277−284.

Drioli, E., Criscuoli, A., Curcio, E., 2002. Integrated membrane operations for seawater desalination. Desalination. 147, 77−81.

Duan, S.H., Ito, A., Ohkawa, A., 2001. Removal of trichloroethylene from water by aeration, pervaporation and vacuum membrane distillation. J. Chem. Eng. Jpn. 34, 1069−1073.

Dumee, L.F., Sears, K., Schutz, J., Finn, N., Huynh, C., Hawkins, S., et al., 2010. Characterization and evaluation of carbon nanotube bucky-paper membranes for direct contact membrane distillation. J. Membr. Sci. 351, 36−43.

Dumee, L.F., Campbell, J.L., Sears, K., Schutz, J., Finn, N., Duke, M., et al., 2011. The impact of hydrophobic coating on the performance of carbon nanotube bucky-paper membranes in membrane distillation. Desalination. 283, 64−67.

El-Bourawi, M.S., Ding, Z., Ma, R., Khayet, M., 2006. A framework for better understanding membrane distillation separation process. J. Membr. Sci. 285, 4−29.

Fane, A.G., Schofield, R.W., Fell, C.J.D., 1990. The efficient use of energy in membrane distillation. Desalination. 64, 231−243.

Fath, H.E.S., Elsherbiny, S.M., Alaa, A.H., Rommel, M., Wieghau, M., Koschikowski, J., et al., 2008. PV and thermally small-scale stand-alone solar desalination system with very low maintenance needs. Desalination. 225, 58−69.

Feng, C., Shi, B., Li, G., Wu, Y., 2004. Preparation and properties of microporous membrane from poly (vinylidene fluoride-co tetrafluoroethylene) for membrane distillation. J. Membr. Sci. 237, 15−24.

Feng, C., Wang, R., Shi, B., Li, G., Wu, Y., 2006. Factors affecting pore structure and performance of poly(vinylidene fluoride-co-hexafluoro propylene) asymmetric porous membranes. J. Membr. Sci. 277, 55−64.

Feng, C., Khulbe, K.C., Matsuura, T., Copal, R., Kaur, S., Ramakrishna, S., 2008. Production of drinking water from saline water by air-gap membrane distillation using polyvinylidene fluoride nanofiber membrane. J. Membr. Sci. 311, 1−6.

Findley, M.E., 1967. Vaporization through porous membranes, I & EC. Process Design Dev. 6 (2), 226−230.

Findley, M.E., Tanna, V.V., Rao, Y.B., Yeh, C.L., 1969. Mass and heat transport in evaporation through porous membranes. AIChE J. 15, 483−489.

Franken, A.C., Nolten, J.A.M., Mulder, M.H.V., Bargeman, D., Smolders, C.A., 1987. Wetting criteria for the applicability of membrane distillation. J. Membr. Sci. 33.

Franken, A.C.M., Rippenger, S., 1988. Terminology for membrane distillation. Eur. Soc. Membr. Sci. Technol. 11.

Frikha, N., Matlaya, R., Chouachi, B., Gabsi, C., 2014. Simulation of an autonomous solar vacuum membrane distillation for seawater desalination. Desalin. Water Treat. 52 (7–9), 1725–1734.

Fujii, Y., Kigoshi, S., Iwatani, H., Aoyama, M., 1992a. Selectivity and characteristics of direct contact membrane distillation type experiment. I. Permeability and selectivity through dried hydrophobic fine porous membranes. J. Membr. Sci. 72, 53–72.

Fujii, Y., Kigoshi, S., Iwatani, H., Aoyama, M., Fusaoka, Y., 1992b. Selectivity and characteristics of direct contact membrane distillation type experiment. II. Membrane treatment and selectivity increase. J. Membr. Sci. 72, 73–89.

Garcia-Payo, M.C., Essalhi, M., Khayet, M., 2009. Preparation and characterization of PVDF-HFP copolymer hollow fiber membranes for membrane distillation. Desalination. 245, 469–473.

Garcia-Payo, M.C., Essalhi, M., Khayet, M., 2010. Effects of PVDF-HFP concentration on membrane distillation performance and structural morphology of hollow fiber membranes. J. Membr. Sci. 347, 209–219.

Gekas, V., 1988. Terminology for pressure-driven membrane operations. Desalination. 68, 77–92.

Gryta, M., Tomaszewska, M., Grzechulska, J., Morawski, A.W., 2001. Membrane distillation of NaCl solution containing natural organic matter. J. Membr. Sci. 181, 279–287.

Hanemaaijer, H., Van Heuven, L.W., Method for the purification of a liquid by a membrane distillation, in particular for the production of desalinated water from seawater or brackish water or process water. US Patent 6,716,355 B1 2004.

Hanemaaijer, J.H., Van Medevoort, J., Jansen, A.E., 2006. Memstill membrane distillation. A future desalination technology. Desalination. 199.

Hanemaaijer, J.H., Jansen, A., Van Medevoort, J., De Jong H., Van Sonsbeek E., Koele, et al., Membrane distillation method for the purification of a liquid. US Patent 20100072135 A1, 2010.

Harryson, A.C., Jonsson, C., Wimmerstedt, B., 1986. Membrane distillation—A theoretical study of evaporation through micro porous membranes. Lund University Report.

Hinai, H.A., Drioli, E., 2008. Potential of membrane distillation in seawater desalination: thermal efficiency, sensitivity study and cost estimation. J. Membr. Sci. 323, 85–98.

Hogan, P.A., Sudjito, Fane, A.G., Morrison, G.L., 1991. Desalination by solar heated membrane distillation. In: Proc. 12th Int. Symposium on Desalination and Water Reuse, Malta, April 15–18, 1991, Vol. 1, pp. 81–90.

Hsu, S.T., Cheng, K.T., Chiou, J.S., 2002. Seawater desalination by direct membrane distillation. Desalination. 143, 279–287.

Hwang, H.J., He, K., Gray, S., Zhang, J., Shik Moon, Il, 2011. Direct Contact Membrane Distillation (DCMD): experimental study on the commercial PTFE membrane and modelling. J. Membr. Sci. 371, 90–98.

Jin, Z., Yang, D.L., Jian, S.H.X.G., 2008. Hydrophobic modification of poly (phtalazinone ether sulfone ketone) hollow fiber membrane for vacuum membrane distillation. J. Membr. Sci. 310, 20–27.

Jonsson, A.S., Wimmerstedt, R., Harrysson, A.C., 1985. Membrane distillation—A theoretical study of evaporation through microporous membranes. Desalination. 56, 237–249.

Khayet, M., Godino, M.P., Mengual, J.I., 2000. Nature of flow in sweeping gas membrane distillation. J. Membr. Sci. 170, 243–255.

Khayet, M., Godino, M.P., Mengual, J.I., 2002. Thermal boundary layers in sweeping gas membrane distillation process. Pro. Syst. Eng. 48, 1488–1497.

Khayet, M., 2011. Membranes and theoretical modeling of membrane distillation. Adv. Colloid Inter. Sci. 184, 56–88.

Khayet, M., Velazquez, A., Mengual, J.I., 2004. Direct contact membrane distillation of humid acid solutions. J. Membr. Sci. 240, 123–128.

Khayet, M., Mengual, J.I., Matsuura, T., 2005. Porous hydrophobic/hydrophilic composite membranes. Application in desalination using direct contact membrane distillation. J. Membr. Sci. 252, 101–113.

Kimura, S., Nakao, S.I., Shimatani, S.I., 1987. Transport phenomena in membrane distillation. J. Membr. Sci. 33, 285–298.

Kong, Y, Lin, X., Wu, Y., Cheng, J., Xu, J., 1992. Plasma polymerization of octafluorocyclo-butane and hydrophobic microporous composite membranes for membrane distillation. J. Appl. Polym. Sci. 46, 191–199.

Koo, J., Han, J., Sohn, J., Lee, S., Hwang, T.-M., 2013. Experimental comparison of direct contact membrane distillation with vacuum. Desalin. Water Treat. 51 (31–33), 6299–6309.

Koros, W.J., Ma, Y.H., Shimidzu, T., 1996. Terminology for membranes and membrane processes. J. Membr. Sci. 120, 149–159.

Koschikowski, J., Wieghaus, M., Rommel, M., 2003. Solar thermal-driven desalination plants based on membrane distillation. Desalination. 156, 295–304.

Koschikowski, J., Wieghaus, M., Rommel, M., 2009. Membrane distillation. In: Cipollina, A., Micale, G., Rizzuti, L. (Eds.), Seawater Desalination. Springer–Verlag, Berlin & Heidelberg.

Kubota, S., Ohta, K., Hayano, I., Hirai, M., Kikuchi, K., Murayama, Y., 1988. Experiments on seawater desalination by membrane distillation. Desalination. 69, 19–26.

Kullab A., Liu C., Martin A.R., 2005. Solar desalination using membrane distillation-Technical evaluation case study. In: Proc. ISES Meeting, Orlando, FL, August6–12, 2005, Paper No. 1395.

Larbot, A., Gazagnes, L., Krajewski, S., Bukowska, M., Kujawski, W., 2004. Water desalination using ceramic membrane distillation. Desalination. 168, 367–372.

Lawson, K.W., Lloyd, D.R., 1996. Membrane distillation. I: Module design and performance evaluation using vacuum membrane distillation. J. Membr. Sci. 120, 111–121.

Lawson, K.W., Lloyd, D.R., 1997. Review: membrane distillation. J. Membr. Sci. 124, 1–25.

Li, B., Sirkar, K.K., 2004. Novel membrane and device for direct contact membrane distillation – based desalination process. Ind. Eng. Chem. Res. 43, 5300–5309.

Li, B., Sirkar, K.K., 2005. Novel membrane and device for vacuum membrane distillation-based desalination process. J. Membr. Sci. 257, 60–75.

Li, J.M., Xu, Z.K., Yuan, W.F., Xiang, H., Wang, S.Y., Xu, Y.Y., 2003. Microporous polypropylene and polyethylene hollow fiber membranes. Part 3. Experimental studies on membrane distillation for desalination. Desalination. 155, 153–156.

Liu, G.L., Zhu, C., Cheung, C.S., Leung, C.W., 1998. Theoretical and experimental studies on air gap membrane distillation. Heat Mass Transfer. 34, 329–335.

MacGregor, R.K., Emery, A.P., 1969. Free convection through vertical plane layers: moderate and high Prandtl number fluids. J. Heat Transfer. 91, 391–401.

Mar-Camacho, L., Dumee, L., Zhang, J., de-Li, J., Duke, M., Gomez, J., 2013. Advances in membrane distillation of water desalination and purification application. Water. 103. Available in <www.mdpi.com/journal/water>.

Martinez-Diez, L., Vazquez-Gonzaalez, M.I., 1999. Temperature and concentration polarization in membrane distillation of aqueous salt solutions. J. Membr. Sci. 156, 265–273.

Martinez-Diaz, L., Florido-Diez, F.J., 2001. Desalination of brines by membrane distillation. Desalination. 137, 267–273.

Mason, E.A., Malinauskas, A.P., 1983. Gas Transport in Porous Media: The Dusty-Gas Model. Elsevier, Amsterdam, the Netherlands.

Meindersma, G.W., Guijt, C.M., de Haan, A.B., 2006. Desalination and water recycling by air gap membrane distillation. Desalination. 187, 291–301.

Mericq, J.P., Laborie, S., Cabassud, C., 2009. Vacuum membrane distillation for an integrated seawater desalination process. Des Water Treat. 9, 287–296.

Mericq, J.P., Laborie, S., Cabassud, C., 2010. Vacuum membrane distillation of seawater reverse osmosis brines. Water Res. 44, 5260–5273.

Nene, S., Kaur, S., Sumod, K., Joshi, B., Raghavarao, K.S.M.S., 2009. Membrane distillation of raw cane-sugar syrup and membrane clarified sugarcane juice. Desalination. 147, 71–77.

Ortiz de Zárate, J.M., Pena, L., Mengual, J.I., 1995. Characterization of membrane distillation membranes prepared by phase inversion. Desalination. 100, 139–148.

Phattaranawik, J., Jiraratananon, R., 2001. Direct contact membrane distillation: effect of mass transfer on heat transfer. J. Membr. Sci. 188, 137–143.

Phattaranawik, J., Jiraratananon, R., Fane, A.G., 2003. Heat transport and membrane distillation coefficients in direct contact membrane distillation. J. Membr. Sci. 212, 177–193.

Present, R.D., 1958. Kinetic Theory of Gases. McGraw–Hill, New York.

Qin, Y., Wu, Y., Liu, L., Cui, D., Zhang, Y., Liu, et al., 2010. Multi-effect membrane distillation process for desalination and concentration of aqueous solutions of non-volatile or semi-volatile solutes. In: AIChe Annual Meeting, November 10.

Qtaishat, M., Rana, D., Khayet, M., Matsuura, T., 2009a. Preparation and characterization of novel hydrophobic/hydrophilic polyetherimide composite membranes for desalination by direct contact membrane distillation. J. Membr. Sci. 327, 264–273.

Qtaishat, M., Khayet, M., Matsuura, T., 2009b. Novel porous composite hydrophobic/hydrophilic polysulfone membranes for desalination by direct contact membrane distillation. J. Membr. Sci. 341, 139–148.

Qtaishat, M., Matsuura, T., Khayet, M., Khulbe, K.C., 2009c. Comparing the desalination performance of SMM blended polyethersulfone to SMM blended polyetherimide membranes by direct contact membrane distillation. Desalin. Water Treat. 5, 91–98.

Roy, S., Bhadra, M., Mitra, S., 2014. Enhanced desalination via functionalized carbon nanotube immobilized membrane in direct contact membrane distillation. Sep. Purif. Technol. 136, 58–65.

Saffarini, R.B., Summers, E.K., Arafat, H.A., Lienhard, V.J.H., 2012. Technical evaluation of standalone solar powered membrane distillation systems. Desalination. 286, 332–341.

Sakai, K., Muroi, T., Ozawa, K., Takesawa, S., Tamura, M., Nakaue, T., 1986. Extraction of solute-free water from blood by membrane distillation". Trans. Am. Soc. Artif. Intern. Organs. 32, 397–400.

Sarti, G.C., Gostoli, C., Bandini, S., 1993. Extraction of organic components from aqueous streams by vacuum membrane distillation. J. Membr. Sci. 80, 21–33.

Schofield, R.W., Fane, A.G., Fell, C.J.D., 1987. Heat and mass transfer in membrane distillation. J. Membr. Sci. 33, 299–313.

Schofield, R.W., Fane, A.G., Fell, C.J.D., 1990a. Gas and vapor transport through microporous membranes. I. Knudsen–Poiseuille transition. J. Membr. Sci. 53, 159–171.

Schofield, R.W., Fane, A.G., Fell, C.J.D., 1990b. Gas and vapor transport through microporous membranes. II. Membrane distillation. J. Membr. Sci. 53, 173–185.

Schofield, R.W., Fane, A.G., Fell, C.J.D., Magoun, R., 1990. Factors affecting flux in membrane distillation. Desalination. 77, 279–294.

Shneider, K., Holz, W., Wollbreck, R., Ripperger, S., 1988. Membranes and modules for transmembrane distillation. J. Membr. Sci. 39, 25–42.

Smolders, K., Franken, A.C.M., 1989. Terminology on membrane distillation. Desalination. 72, 249–262.

Song, L., Ma, Z., Liao, X., Kosaraju, P.B., Irish, J.R., Sirkar, K.K., 2008. Pilot plant studies of novel membranes and devices for direct contact membrane distillation-based desalination. J. Membr. Sci. 323, 257–270.

Summers, E.K., Lienhard V, J.H., 2013. A novel solar-driven air-gap membrane distillation system. Desalin. Water Treat. 51, 1344–1351.

Teoh, M.M., Bonyadi, S., Chung, T.S., 2008. Investigation of different hollow fiber module designs for flux enhancement in the membrane distillation process. J. Membr. Sci. 311, 371–379.

Tomaszewska, M., Gryta, M., Morawski, A.W., 1995. Study on the concentration of acids by membrane distillation. J. Membr. Sci. 102, 113–122.

Tomaszewska, M., 1996. Preparation and properties of flat-sheet membranes from polyvinylidene fluoride for membrane distillation. Desalination. 104, 1–11.

Udriot, H., Araque, A., Von Stokar, U., 1994. Azeotropic mixtures may be broken by membrane distillation. Chem. Eng. J. 54, 87–93.

Van-Haut, A, Henderyckx, Y, 1967. The permeability of membranes to water vapor. Desalination. 3 (2), 169–173.

Wang, K.Y., Chung, T.S., Gryta, M., 2008. Hydrophobic PVDF hollow fiber membranes with narrow pore size distribution and ultra-skin for the fresh water production through membrane distillation. Chem. Eng. Sci. 63, 2587–2594.

Weyl, P.K., 1967. Recovery demineralised water from saline water. US Patent no. 3,340,186.

Wieghaus, M., Koschikowski, J., Rommel, J., 2008. Solar membrane distillation ideal for remote areas. Desalin. Water Reuse. 18, 37–40.

Zakrzewska-Trznadel, G., Harasimowicz, M., Chmielewski, A.G., 1999. Concentration of radioactive components in liquid low-level radioactive waste by membrane distillation. J. Membr. Sci. 163, 257–264.

Zhang, J., Dow, N., Duke, M., Ostarcevic, E., Li, J.D., Gray, S., 2010. Identification of material and physical features of membrane distillation membranes for high performance desalination. J. Membr. Sci. 349, 295–303.

Zolotarev, P.P., Ugrosov, V.V., Volkina, I.B., Nikulin, V.N., 1994. Treatment of waste water for removing heavy-metals by membrane distillation. J. Hazard Mater. 37, 7–82.

Zrelli, A., Chaouachi, B., Gabsi, S., 2014. Simulation of a solar thermal membrane distillation: comparison between linear and helical fibre. Desalin. Water Treat. 52 (7–9), 1683–1692.

CHAPTER FIVE

Humidification—Dehumidification

5.1 INTRODUCTION

Humidification—Dehumidification (H/D) is a procedure where a carrier gas, usually air, is loaded with water vapors, until saturation and then by cooling the humid air is dehumidified producing fresh water. Solar distillation is in fact an H/D procedure (not a method) where the warm air is humidified, moving upwards, inside the still, by density differences, is cooled by coming in contact with the cooler transparent cover and produces a clear concentrate, the fresh water. In a solar still, the whole procedure takes place in the same space, the distillation chamber, single or multiple effect (for more details, see chapter: Solar Distillation—Solar Stills).

The H/D method operates usually with forced fluid circulation and the procedure of humidification and dehumidification takes place in two different chambers (called also towers, columns, stacks, canals, etc.): the air humidification (evaporation) and air dehumidification (condensation) chambers. This separation permits the recovering of condensation heat which returns back to the system. The whole system of H/D operates at temperatures below the boiling point of water (80—85°C) and at ambient pressure. It can be operated in most of the cases by conventional fuels, waste heat, or solar energy. The principle has been developed trying to solve the major problems of conventional solar stills that is the considerable energy loss (sensible heat and condensation enthalpy) and their low efficiencies. The method results in an overall increase of efficiency in comparison to conventional solar distillation unit of the same exactly size. H/D method is considered a promising technique for small and medium capacities. By the time being the method is still under investigation not yet commercialized totally.

From the historical point of view, the first prototype pilot unit was installed in Puerto Peñasco, New Mexico, USA, as a Project study of University of Arizona and the Georgia Institute of Technology (Hodges et al., 1966). The system was called *"Humidification Cycle Distillation."* Grune (1970) presented a multiple effect H/D procedure called the *"Multiple Effect Humid Cycle"* and later on Larson et al. (1989) gave details of the conventional H/D method. He refers "This new technology, referred to herein as *The Carrier—Gas Process*, is basically unique and innovative distillation process operating at temperatures below boiling and ambient pressures." The same method, where H/D processes are simultaneously performed in one continuous contact chamber was called also *"Dewvaporating"* (Hamieh et al., 2000) a name which comes from merging *Dew*

Thermal Solar Desalination

and *evaporating*. Independently, of all these titles, or differences in technical arrangements today the name *Humidification—Dehumidification* is acceptable.

Many other investigators published works on this method. The first patent on this method was granted to Chafik (1996).

5.2 DEFINITIONS

Two new terms, used in conventional distillation methods, must be referred here. They are used seldom in solar fired distillation:

Gained output ratio (GOR) is the amount of desalinated water (m_d), in kg, produced per kg of heating steam (m_{st}), ie, m_d/m_{st}. It is mainly used in MSF, MED, and TVC distillation systems, where usually is in the range of 0.8—0.9 of the effect number used.

$$\text{GOR} = \dot{m}_d/\dot{m}_{st} = (1 - c_{fw}/c_{br})\frac{\dot{m}_{fw}}{m_d} \tag{5.1}$$

$$\text{GOR} = \dot{m}_d h_{ev}/\dot{m}_{fw} c_{pw}(T_{h-in} - T_{h-out}) \tag{5.2}$$

where c_{fw} and c_{br} are concentration of feed water and brine, respectively; $\dot{m}_{fw}$ is mass flow rate of feed water; $\dot{m}_d$ is mass flow rate of distillate; and T_{h-in} and T_{h-out} are the inlet and outlet temperatures in the humidifier, respectively.

Performance ratio R is another term used in conventional distillation procedures. It is expressed as $R = lb$ of desalinated water per *1000 BTU* or in metric system the amount of desalinated water produced per 2326 kJ of input heat.

5.3 GENERAL OPERATION PRINCIPLES

5.3.1 The Setup of Single Effect Conventional H/D System

A single effect conventional H/D system may consist of the following components.

1. The two chambers (or towers):
   a. The air humidification (water evaporation) chamber, where heated air flows upwards in a counter-flow mode to the trickling downwards warm seawater. For large surface contact area, air—seawater, the humidification chamber is filled with increasing interface contact material. Any type of gas—liquid contact equipment can be used for this purpose such as tray towers, packed towers, and pad humidifiers. Water is evaporated and air is loaded with water vapor up to saturation conditions
   b. The dehumidification (vapor condensation) chamber where air is cooled by the incoming seawater feed in a heat exchanger condensing water vapor

forming distilled water. Heat exchangers incorporate fins presenting large surface contact area.

2. Heating of the circulating air or of the feed seawater, depending from the mode of fluid circulation to the desired temperature. This can be achieved by any conventional heat source or by a solar collector field. Operation temperature (70–80°C) is in favor of solar energy system

3. From a packed material inside the humidification chamber to create large contact area of seawater–air interface. Packed material may be Rusching rings, saddles, plastic honeycomb material, or wooden material (Fig. 5.1). For very small capacity systems where evaporation and condensation take place in one chamber but separately, tissues, fleeces, etc. are used

4. A heat exchanger or coils, depending on the systems size, inside the dehumidification chamber to preheat the incoming seawater by the condensing water vapor

5. For smooth and continuous operation of solar fired systems, a storage tank may be necessary

6. Piping for seawater and desalinated water circulation, pumps and the auxiliary components, as valves, fittings, etc.

7. Control units, measurement devices, etc.

There exist various combinations of chamber arrangement:
- Two totally separated towers as are presented schematically in Fig. 5.2
- Two separate chambers, for evaporation and condensation in one device, as is shown in Fig. 5.3. This arrangement is suitable for small capacity solar powered systems
- The dewevaporation system with two separate but in continuous contact towers
- Various other arrangements with adapted parts, eg, compressors, etc., as are described later.

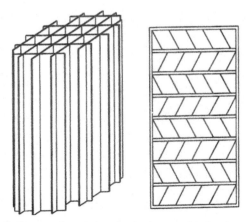

Figure 5.1 Two different forms of wooden packed material for humidification towers (right, Nawayseh et al., 1999 and left, Al-Hallaj, 1994).

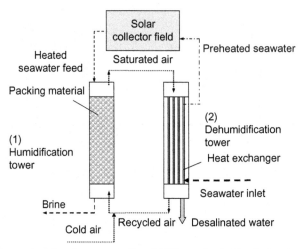

Figure 5.2 Operation principle of a conventional H/D method. Circulation is of closed air—open water cycle.

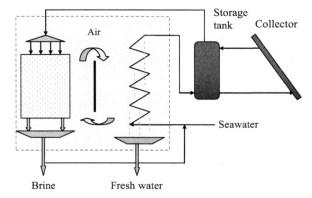

Figure 5.3 A small capacity H/D installation using condensation evaporation procedure in one space and as packing material rough tissue or fleece.

5.3.2 Basic Operation Principles

The basic operation principle of the conventional method is the evaporation of seawater or any other aqueous solution, into the evaporation chamber (air humidification).

This is achieved by air mass flow upwards, countercurrently to the downwards trickling hot seawater. In Fig. 5.2, a schematic of the tower configuration is presented. According to this concept, seawater enters the condensation stack (2) where it is preheated in a heat exchanger or coils, by the condensing vapor and then is forwarded to the collector field to increase its temperature up to 50–80°C, according to the intensity of solar radiation. Younis et al. (1993) refer that by using flat plate collectors or solar ponds the maximum achieved temperature is 70–80°C. Heated seawater enters the top of the evaporation column (1) and trickles downwards the packed material inside the evaporation chamber. The cooled air from the condensing column enters the evaporation column at the bottom and by flowing upwards is heated and humidified up to saturation. This is a procedure of closed air—open seawater cycle. Independently of the operation mode inside the towers pressure drop must be maintained as minimum as possible.

Various modes of operation exist usually differing according to the type of the circulation loop, the arrangement of the evaporation/condensation units and the packing material. The main circulation arrangement may be of:

- **Closed water—open air cycle (CWOAC)** where brine recirculates up to a certain concentration that will not permit scale formation. Air circulates through the hot seawater up to saturation and after unloading its humidity by cooling in the dehumidification chamber is rejected to the environment. This operation mode can incorporate thermal storage for seawater

- **Closed air—open water cycle (CAOWC)** where seawater circulation is once through and then rejected. Air is circulated continuously by natural or forced movement depending on the systems configuration. Occasionally air is added to the air circulation loop due to its mass loss (Figs. 5.2 and 5.3).

Heating for both cycles may be for water or for air. Various parameters affect these two circulation cycles. Narayan et al. (2010) studied the cycles and report that the performance depends on the mass flow rate ratio, ie, the ratio of mass flow rate of water at the inlet of the humidifier to the mass flow rate of dry air through the humidifier.

Various submodes of fluid circulation or of different techniques improving the performance were studied. They include (Narayan et al., 2010):

- High efficiency air-heated cycle
- Multiextraction air-heated cycle
- Subatmospheric pressure air-heated cycle
- Vacuum pressure cycle
- Thermal vapor compressor H/D.

In general, the air heating procedure is preferable as it is more simple, more flexible, and economical but also less efficient than water heating.

The tower concept may be applied for larger capacities. Packing material may be Rusching rings, saddles, or any other shape or materials similar to those used in vapor—liquid transfer operations. Small capacity installations combine the evaporation/condensation devices in one chamber separated by an intermediate plate as is a general layout of a unit shown in Fig. 5.3. This configuration uses normally as packing material fleeces, stems, and/or tissues. More recent applications use pad humidifiers which are compact devices. The photographs of Figs. 5.4A and 5.4B are both of a multiple effect humidification (MEH) pilot plant installed in Tunisia. Pad humidifiers are constructed of sheets of rigid corrugated cellulose material (Chafik, 2004;

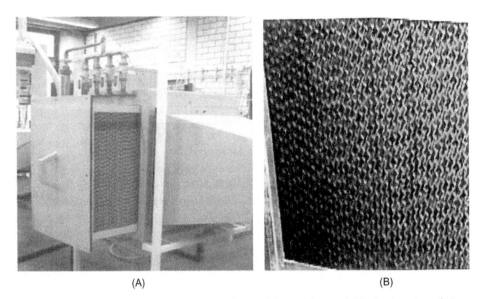

Figure 5.4 New concepts of humidifiers. (A) The pad humidifier and (B) the interior of the pad humidifier which consists of tree cassettes made of corrugated cellulose material presenting the increased wetted surface (Chafik, 2004; Ben Amara et al., 2004a).

Ben Amara et al., 2004a). Multiple effect humidification/dehumidification (MEH) concept is used to increase productivity. It consists of a series of evaporation towers as is described later.

The process of H/D presents several attractive features such as (Müller-Holst, 2001):

- Separation at low temperatures, usually 65−80°C
- Simple design and in addition small capacity units can be manufactured and installed locally due to modest level of technology employed
- Has capacity flexibility
- Operates at ambient pressure
- The separation of heating surface from the evaporating zone protects the heating surface from scale deposits and corrosion
- In comparison to conventional solar distillation units of the same capacity, the H/D units are compact without a need of large installation surface areas.

5.3.3 Humidifying Step—$h-x$ Diagram

The humidification technique depends on the property of the air that can be mixed with large amounts of vapour (1.168 m^3 at 75°C and 1000 m^3 at 75°C of air) can carry about 0.5 kg of water vapor. This means that per kg of air, as mean amount, about 0.5 kg of vapor are mixed. The various air/vapor mixtures as well as air at saturation point can be identified from enthalpy−humidity ($h-x$) diagrams (Mollier diagrams used in drying methods) or any psychrometric chart. Water vapor in these diagrams (ie, humidity of the air) is expressed as g of water vapor per kg of dry air.

Let us follow the paths in an $h-x$ diagram (Fig. 5.5). Air having initial temperature $T_{in} = 20°C$ and initial absolute humidity $x = 5$ g kg^{-1} is heated up to $80°C$. During heating the enthalpy of the air is increasing but humidity remains unchanged. This is presented in the diagram by the line ab (Fig. 5.5). The heated air is then humidified, adiabatically, to saturation conditions. This is pointed out by line bc which moves about parallel to the enthalpy line of the diagram and intercepts the saturation curve of the diagram $(\varphi = 1.0)$ at c. Humidity at point of saturation c is $x = \sim 25.5$ g kg^{-1}. Thus the gained humidity is $25.5 - 5.0 = 20.5$ g kg^{-1}. The procedure is endothermic and is followed by a temperature decrease. Temperature of the humid air is at this point $30°C$. If the available solar radiation can heat the air only to $60°C$ (point e in the diagram), then the air will be humidified up to saturation condition, point f at the saturation curve.

The air is loaded at saturation with 19.5 g kg^{-1} humidity and its temperature is $24°C$. Gained humidity is 14.5 g kg^{-1}. The humidifying steps in a humidifier by using air as heat carrier at temperatures less than $100°C$ were presented initially by Chafik (1999, 2003, 2004). By this procedure, the air is heated stepwise in a collector up to temperatures between $50°C$ and $80°C$ and the preheated air is humidified by adding seawater into the humidifier.

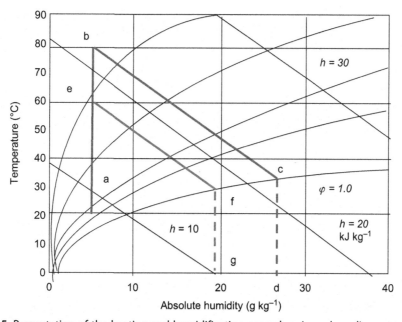

Figure 5.5 Presentation of the heating and humidification procedure in an $h-x$ diagram.

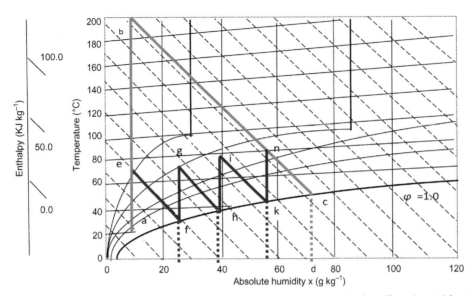

Figure 5.6 Presentation in humidity–enthalpy diagram of the multiple effect humidification operation.

This reheating procedure can be repeated many times and is in fact the principle of multiple effect humidification process (MEH). Air can be heated to higher temperatures say 200°C and then humidified. This is shown in the $h-x$ diagram presented in Fig. 5.6 by the lines *ab* and *bc*. Flat plate collectors cannot heat air higher than 80–85°C but the same result may be achieved by stepwise heating and humidifying the air, as is indicated in Fig. 5.6 by the lines *efg–ghi* and *ikn*. Air having absolute humidity of 10 g kg^{-1} is heated up to 70°C at first step and humidified to saturation state. Its temperature is ~30°C and humidity content ~28. Further it is reheated up to 75°C ($x=\sim40$ g kg^{-1}, $T=\sim40$°C), humidified and again heated to 79°C ($x=\sim55$ g kg^{-1}, $T=\sim43$°C) and humidified. At the end it is reheated to 82°C and humidified achieving the same results of humidity content of ~72 g kg^{-1} and $T=48$°C. The gained humidity is $72-10=62$ g kg^{-1}.

5.4 MATHEMATICAL MODELS

5.4.1 Conventional Single Effect H/D Process

There exist many mathematical models and equations concerning H/D method, many of which are addressed to a special device configuration or mode of operation.

Ettouney (2005) presents the energy balance of the humidifier (evaporator) as a function of pressure drop inside the column. For the following model, a steady state operation and negligible losses to the environment are assumed and that the air stream leaving the humidification chamber is saturated:

$$\dot{m}_{air}(h_{air2} - h_{air1}) = \dot{m}_{sw}c_{psw}(T_{sw2} - T_{sw1}) \tag{5.3}$$

where $\dot{m}_{air}$ and $\dot{m}_{sw}$ are mass flow rate of carrier air and seawater feed (kg s^{-1} Pa^{-1}). Air enthalpy h_{air} is a function of its temperature T_{air} and humidity x_{air}:

$$h_{air} = (c_{pair} + c_{pvap}x_{air})T_{air} + h_{ev}x_{air} \tag{5.4}$$

where h_{ev} (h_{fg}) is evaporation enthalpy (latent heat of evaporation). Absolute humidity of air is expressed by the relative air humidity ϕ as:

$$x_{air} = 0.62198 \frac{\phi p_s}{(p_{atm} - p\varphi)} \tag{5.5}$$

where p_s is pressure at the dry bulb temperature (saturation pressure) and p_{atm} the ambient pressure.

In the condenser (dehumidification chamber), the energy balance is (Ettouney, 2005):

$$\dot{m}_{sw}c_{p-sw}(T_{sw1} - T_{sw2}) = m_{air}(h_{air1} - h_{air2}) \tag{5.6}$$

$$\dot{m}_{sw}c_{psw}(T_{sw-out} - T_{sw-in}) = U_cA_cT_{c-ln} \tag{5.7}$$

where U_c (W m^{-2} K^{-1}) is overall heat transfer coefficient at the condenser, A_c is the condenser area, and T_{c-ln} is the mean logarithmic temperature in the condenser expressed as:

$$T_{c-ln} = \frac{(T_{air} - T_{sw-out}) - (T_{air} - T_{sw-in})}{\ln[(T_{air-out} - T_{sw-out})/(\overline{T}_{ait} - T_{sw-in})]} \tag{5.8}$$

Overall heat transfer coefficient in the condenser is expressed by the tube radius r (outer and inner diameters, respectively) of the heat exchanger as:

$$\frac{1}{U_c} = \frac{1}{h_{in}} \frac{r_{out}}{r_{in}} + R_f + \frac{r_{out}\ln(r_{out}/r_{in})}{\lambda_c} + \frac{1}{h_{out}} \tag{5.9}$$

where R_f is a parameter of the fouling resistance (m^2 kW^{-1}) and λ (W m^{-1} K^{-1}) is thermal conductivity. The distillate production based on humidity balance is:

$$\dot{m}_d = \dot{m}_{air}(x_{out} - \overline{x}_{con}) \tag{5.10}$$

For an open seawater–closed air system, inside the humidifier, seawater trickles downwards through the packing material, in a flow rate of m_{sw} (kg m^{-2} s^{-1}), countercurrently to the air stream which has a flow rate of m_{air} $(\mathrm{kg\,m^{-2}\,s^{-1}})$. The temperature of bulk seawater is T_{sw} and that of the bulk air T_{air}.

5.4.1.1 For the Humidifier Tower

Seawater and air streams flow are presented in Fig. 5.6. In each point of the column, seawater—air interface is in equilibrium at temperature T_{int}. For an infinite volume in the column of height dz and cross-sectional area A (Fig. 5.7), heat and mass balance can be expressed by the following equations. For this system, Hodges et al. (1966) and Younis et al. (1993) present a mathematical model based on seawater—air interface.

Bulk seawater transfers both mass and heat to the air stream decreasing its enthalpy and mass. Thus at the seawater—air interface, heat transfer from seawater to the air is:

$$-d\left(m_{sw}h_{ev-sw}\right) = h_{h-\mathrm{inf}}(T_{sw} - T_{\mathrm{inf}}) = \overline{\dot{m}}_{sw}C_{p-sw}dT_{sw} \tag{5.11}$$

For the air stream correspondingly:

$$\dot{m}_{air}dh_{h-air} = h_{h-ai}(T_{\mathrm{inf}} - T_{air}) = k_{hum}(h_{\mathrm{inf}} - h_{air}) \tag{5.12}$$

and for the humidity at the interfacial, from seawater to air:

$$-d\left(\dot{m}_{hum}h_{hum}\right) = h_{h-hum}(T_{\mathrm{inf}} - T_{air}) \tag{5.13}$$

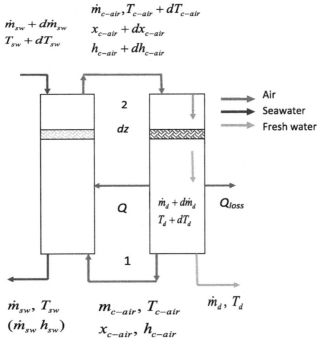

Figure 5.7 Mass and energy balance in the columns.

For the total height Z of the column between points 1 and 2, the balance is (Fig. 5.7), assuming that heat and mass coefficient remain constant along the column:

$$\dot{m}_{air}(h_{air2} - h_{air1}) = \dot{m}_{sw}c_{psw}(T_{sw2} - T_{sw1}) \tag{5.14}$$

The necessary height of the column and the corresponding surface area can be calculated by the following two expressions, respectively:

$$Z_{ev,tower} = \int_{1}^{2} dz = \int_{ait1}^{air2} \frac{\dot{m}_{air}dh_{air}}{k_{hum}(h_{int} - h_{air})} dz \tag{5.15}$$

$$A_{ev-tower} = \frac{\dot{m}_{air}}{k_{hum}} \int_{hair1}^{air2} \left(\frac{1}{h_{int} - h_{air}}\right) dh_{air} \tag{5.16}$$

The relation between air enthalpy at the main stream and at the interface air/seawater is given as:

$$(h_{air} - h_{int})/(T_{air} - T_{int}) = h_{h-hum}/k_{hum} \tag{5.17}$$

Eq. (5.17) can be solved graphically, in an enthalpy–temperature diagram, by using mean values of mass airflow and humidity mass transfer coefficient. Eq. (5.17) when plotted in an enthalpy–temperature diagram gives a straight line for air called "*evaporator operating line*" (line *acb* in Fig. 5.8) which for the humidification column

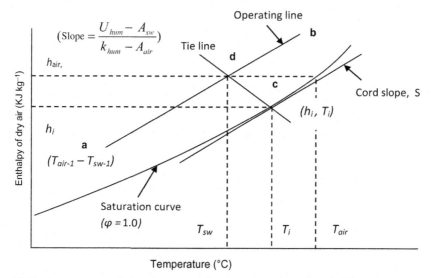

Figure 5.8 Temperature–enthalpy diagram with operating and tie line for the evaporation column (Hodges et al., 1966; Younis et al., 1993).

is situated on the right of the saturation air—vapor curve. Any point on this line is related to a point on the saturation curve ($\varphi = 1.0$) by a joining line, cd, the "*tie line*." The slope of air versus T_{sw}, the operation line, is given by the following equation:

$$-h_{h-\text{int}}/k_{hum} = (h_{\text{int}} - h_{air})/(T_{\text{int}} - T_{sw}) \qquad (5.18)$$

where

$c_{p\text{-}sw}$	Seawater specific heat	$\text{J kg}^{-1}\text{K}^{-1}$
$h_{h\text{-}air}$	Heat transfer coefficient of the air	$\text{W m}^{-2}\text{K}^{-1}$
$h_{h\text{-}inf}$	Heat transfer coefficient at the interface	$\text{W m}^{-2}\text{K}^{-1}$
$h_{ev}\ (h_{fg})$	Enthalpy of evaporation (latent heat of evaporation)	J kg^{-1}
h_{air}	Enthalpy of air at dry basis	J kg^{-1}
h_{sw}	Enthalpy of seawater	J kg^{-1}
h_{hum}	Enthalpy of humidity in air	J kg^{-1}
$h_{h\text{-}hum}$	Heat transfer coefficient of humidity in humid air	$\text{W m}^{-2}\text{K}^{-1}$
k_{air}	Mass transfer coefficient of air	$\text{kg m}^{-2}\text{s}^{-1}$
k_{hum}	Mass transfer coefficient with humidity as driving force	$\text{kg m}^{-2}\text{s}^{-1}$
$\dot{m}_{air}$	Air mass flow rate per unit of cross-section area	$\text{kg m}^{-2}\text{s}^{-1}$
$\dot{m}_{sw}$	Seawater mass flow rate per unit of cross-section area	$\text{kg m}^{-2}\text{s}^{-1}$
$\dot{q}_{sen}, \dot{q}_{ev}$	Sensitive and evaporation heat flow rate, respectively	W m^{-2}
T_{air}	Temperature of the air	K
T_{inf}	Temperature at the interface	K
x	Absolute humidity	g kg^{-1}
φ	Relative humidity of the humid air	—

5.4.1.2 For the Dehumidifier

The condenser stack may be packed with fined heat exchanger tubes. The seawater stream flows upwards inside the tubes. Saturated hot air circulates downwards at the outer fined tube surface where the water vapor is condensing flowing also downwards with the air (Fig. 5.9). The heat of the humid air is transferred to the metallic surface of the tubes as sensible or latten heat of condensation and then through the tube walls by conduction and to the flowing seawater by convection. The effective heat transfer rate to the metallic surface, neglecting the formed condensate film, is:

$$q = \dot{m}_{air}h_{air} = k_{aw-air}(h_{air} - h_{inf})sA_{air} \qquad (5.19)$$

and the heat transfer rate of the flowing seawater inside the tubes is:

$$\dot{m}'_{sw}dT_{sw} = U_h(T_{inf} - T_{sw})dA_{air} \qquad (5.20)$$

where A_{sw} and A_{air} are the seawater and the air side surface area including fins and bare tube surface, respectively. U is the overall heat transfer coefficient from air to seawater through the tube walls. $\dot{m}_{sw}$ and $\dot{m}_{air}$ are mass flow rate of seawater and

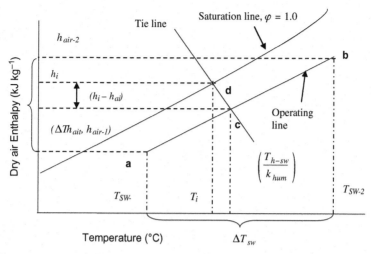

Figure 5.9 Temperature—enthalpy diagram for the condensation stack with operating and tie lines (Hodges et al., 1966; Younis et al., 1993).

air, respectively, in $kg\ s^{-1}$. Overall heat transfer coefficient can be expressed by the resistance to heat transfer as:

$$U_h = \frac{1}{(R_{wl} + R_f + R_{sc} + 1/h_{h-sw})} \qquad (5.21)$$

where R is the corresponding resistance to heat transfer and the subscripts wl, f, and sc denote tube wall, fins and scale or fouling, respectively. h_{h-sw} is heat transfer coefficient, inside tube, from wall to flowing seawater.

A graphical presentation, of the condensation column, similar to Fig. 5.8 is presented in Fig. 5.9. The operating line of the condenser stack lies left of the saturation curve and is expressed by the balance equation:

$$\dot{m}_{sw}c_{p-sw}(T_{sw1} - T_{sw2}) = m_{air}(h_{air1} - h_{air2}) \qquad (5.22)$$

where (1) is the entering point into the condensing column of the saturated air and (2) refers to the outlet point of the humid air. The slope of the tie line is expressed as:

$$(h_{air} - h_{int})/(T_{sw} - T_{int}) = U_h A_{sw}/k_{hum} A_{air} \qquad (5.23)$$

where the ratio of seawater surface to the air surface A_{sw}/A_{air} in the condenser column must be known (ie, A_{sw}/A_{air}, in the condenser column, must be known).

The height of the condensation stack is calculated as:

$$Z_{cond,t} = \frac{\dot{m}_{air}}{k_{sw-air}} \int_{h_{air1}}^{h_{air2}} \frac{dh_{air}}{(h_{air} - h_{sw})} dz \qquad (5.24)$$

where m_{air} is the air mass (kg m^{-2} s^{-1}) and $K_{sw\text{-}air}$ (kg m^{-2} s^{-1}) is overall mass transfer coefficient. It is calculated as:

$$k_{sw-air} = \frac{1}{1/k_{hum} + sA_{air}/A_{sw}U_{hum}}$$ (5.25)

The term s is the slope of the saturation curve, and is calculated as:

$$s = \frac{h_{air} - h_{int}}{(T_{air} - T_{sw}) - (T_{int} - T_{sw})} = \frac{h_{air} - h_{int}}{T_{air} - T_{int}}$$ (5.26)

Similar to Eq. (5.15), the surface area for the air is:

$$A_{cond,air} \frac{\dot{m}_{air}}{k_{sw-air}} \int_{h_{air1}}^{h_{air2}} \left(\frac{1}{h_{air} - h_{sw}}\right) dh_{air}$$ (5.27)

The distillate production rate (kg h^{-1}) is calculated from the following expression:

$$\dot{m}_d = \sum_{j=1}^{n} m_j \Delta x_{airj}$$ (5.28)

where Δx presents the change of absolute humidity of air between inlet and outlet in the condensation column.

5.4.2 The Dewvaporation Process

This configuration of H/D method was presented by Beckman and Hamieh (1999) and Hamieh et al. (2000, 2001). The dewvaporation technique consists of two continuous contacting towers (Fig. 5.10) which focus on an innovative heat driven procedure. The dew formation side of the tower is slightly hotter than the evaporation site. Air circulating from the top to the bottom is cooled producing fresh water allowing the condensation heat to be transferred through the heat transfer wall to the evaporation site. The air in the evaporation site, entering at the ambient temperature, circulates from the bottom of the tower to the top rising its temperature, evaporating water from the falling film of seawater and wetting the heat transfer wall.

Xiong et al. (2005, 2006) observed that conventional H/D process performed in two separate columns presents a limited humidification effect on the carrier gas and considerable loss of heat of condensation. H/D in two towers in contact as the dew evaporating technique in Fig. 5.10 has advantages. The chambers are called the tube and shell components of the tower and present the mass and energy balance of the system.

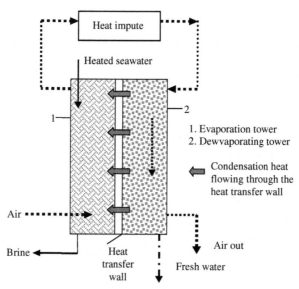

Figure 5.10 The dewvaporation principle of operation (Hamieh et al., 2000).

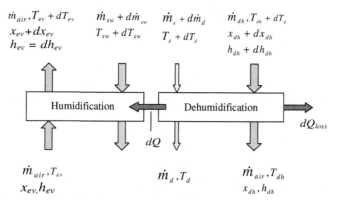

Figure 5.11 Presentation of circulation streams in a differential height of a continuous contacting tower.

For a differential height of the tower dz, heat and mass transfer for the tube and shell side of the column are presented schematically in Fig. 5.11. The corresponding equations are:

$$d(\dot{m}_{air}h_{air})_{hum} = d(\dot{m}_{sw}c_{psw}T_{sw}) + d\dot{Q} \qquad (5.29)$$

$$d(\dot{m}_{air}h_{air})_{dh} = d(\dot{m}_{d}c_{pd}T_{d}) + d\dot{Q} + dQ_{loss} \qquad (5.30)$$

where the subscripts *hu* and *dh* denote the humidification and dehumidification side of the tower.

From the mass balance of seawater in the tube and shell side of the towers result in the following equations:

$$dm_{sw} = (m_{air}dx_{air})_{hum}$$ (5.31)

$$dm_d = (m_{air}dx_{air})_{dh}$$ (5.32)

Eq. (5.31) can be expressed by the humidity difference between the water surface of the falling film and the bulk air as the driving force:

$$(m_{air}x_{air})_{hum} = k_m(x_{sw} - x_{hum})dA_M$$ (5.33)

where k_m is mass transfer coefficient (kg m^{-2} s^{-2}) and dA_m the corresponding surface area where vaporization takes place. Mass transfer coefficient k_m as a mean value of the whole column is presented by the mean logarithmic humidity difference $\Delta x_{\ln-m}$:

$$k_m = \frac{\dot{m}_{air}(x(T_w) - x(T_{in}))_{hum}}{A_m \Delta x_{\ln-m}}$$ (5.34)

In this process, the mass transfer coefficient was correlated with the specific air mass velocity w_{air} (kg m^{-2} s^{-1}) and the film flow rate m_f (kg m^{-1} s^{-1}). The multiple regression analysis yielded the following relationship:

$$k_m = 0.0021 w_{air}^{1.27} \dot{m}_f^{-0.29}$$ (5.35)

5.5 MULTIPLE EFFECT HUMIDIFICATION–DEHUMIDIFICATION (MEH)

This procedure refers to the use of two or more humidification units. By this mode of operation, air mass after each humidification step is lead back to the collector field to be reheated as was presented by Chafik (2004) and analyzed in Section 5.3.1.

In multiple effect H/D, air is loading stepwise with water vapor up to saturation conditions. As it is obvious, the highest the temperature of air or seawater the highest is the vapor loads of air per unit air mass. In solar installations, air and/or seawater cannot be heated at higher temperatures (about 80°C) and the multiple humidification offers the possibility to air to be loaded, by reheating, with as much as possible vapor. This is the advantage of MEH. Chafik (2002, 2003, 2004) describes an MEH plant based on the principle of H/D using a heat recovering system of sensible heat, in heat exchangers, which reduces air temperature almost to its initial temperature. A state of the art on H/D is presented by Bourouni et al. (2001) and Mathioulakis et al. (2007).

One of the recent multiple effect H/D systems described by Ben Amara et al. (2004a) and Houcine et al. (2006) is a pilot plant installed in the "Institute National de Research Scientific et Technique—INRST," in Tunisia (Fig. 5.12) with the use of a

(A)

(B)

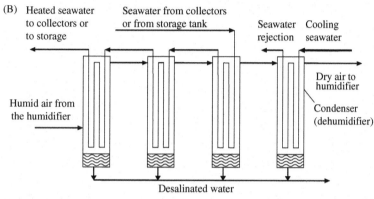

Figure 5.12 (A) The H/D pilot plant built in INRST, Tunisia. (B) Flow diagram of the heat exchanger humidifier section with heat recovery system (Houcine et al., 2006).

new prototype solar collector having high efficiency. In the humidifier, finned tube heat exchangers which simultaneously act as heat recovery system are used (Fig. 5.13). The absolute humidity at saturated conditions is given by the following expression:

$$x_{sat} = \cfrac{0.622}{\cfrac{p}{9 \times 10^{-6}T^4 - 0.0003T^3 + 0.0406T^2 - 0.1178T + 10.738} - 1} \tag{5.36}$$

where p is ambient pressure (atm) and T is the temperature of saturated air (°C). The pad humidifier shown in Fig. 5.4 has an efficiency based on humidity content of the air:

$$\eta_{hum} = (x_{out} - x_{in})/(x_{our-s} - x_{in}) \tag{5.37}$$

where x_{out} is the humidity of saturated air leaving the last humidifier and x_{in} is the humidity of air entering the first humidifier. For the solar prototype collector the efficiency is expressed as:

$$\eta_c = -0.4985u_o \frac{T_{c-ab} - T_a}{G} + 0.637 = -4.985 \frac{T - T_{am}}{G} - 0.637 \tag{5.38}$$

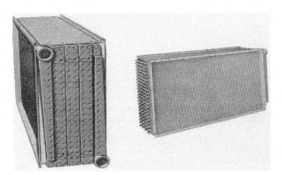

Figure 5.13 The heat exchanger used in the MEH system as humidifier and as heat recovering device (Houcine et al., 2006).

where the above equation is the characteristic equation of the new type solar collector especially designed for this system. The term u_o is the system's coefficient having a value of $10 \, m^2 \, W^{-1} \, K^{-1}$ and T_{c-ab} is collector's absorber temperature. The term $u_o(T_{c-ab} - T_a)$ is the reduced collector temperature and T_m is collectors air mean logarithmic temperature.

The solar prototype collector was designed and constructed especially for the solar MEH desalination system. The collector consists of a polycarbonate plate with insert dark aluminum strips (Chafik, 1999, 2002). The developed new collector has the following selection criteria (Chafik, 2003; Ben Amara et al., 2004b):

- A minimum ratio of price to efficiency
- A maximum life time. A 10-year operation minimum life time
- Resistance against sea climate. There exist no corroding polymer or steel parts
- Hermetic tightness
- Simple design easy to fabricate in local workshops.

The efficiency of this prototype collector is reported by Chafik (2004) who gives an equation similar to Eq. (5.38):

$$\eta = 10\left[(T_m - T_a)/G\right] + 0.6 \tag{5.39}$$

An approximate expression for the calculation of the number of effects needed is presented by Grune (1970). The expression is derived from the experimental results in a pilot H/D installation.

$$\text{No. of effects} = \frac{\Delta T_{1-2}}{\Delta T_{3-4}} = \frac{T_{air2} - T_{air1}}{T_{c-out} - T_{sw}} \tag{5.40}$$

In this system, seawater introduced to the dehumidification heat exchanger chambers at ambient temperature 25°C is preheated up to 60°C and introduced to collector field increasing its temperature to 66°C, then flows from the top of the humidifier downwards to its bottom. The saturated air stream enters the humidification chamber at 28°C flowing upwards, countercurrently to seawater, from the bottom to the top of the humidifier. Its temperature, at any point of the humidifier, is near the corresponding seawater temperature. Saturated air leaves the humidifier top at 64°C. The approximate number of effects needed is: (64−28)/(66−60) = 36/6 = 6 effects.

Most of the H/D processes operate with air as gas carrier. Abu-Arabi and Reedy (2003) studied other gas carriers such as hydrogen, helium, neon, nitrogen, oxygen argon, and carbon dioxide. They concluded that lower molecular weight gases, such as hydrogen and helium, are preferable due to higher heat transfer rates. Better mass transfer coefficients were achieved with carbon dioxide. Carbon dioxide, under the same conditions, produced also more desalinated water.

5.5.1 H/D System Using Two Collector Fields

The previously described H/D procedure is based on the air recirculation mode, where seawater may be heated by a solar collector field or any other heat source of low temperature, such as solar ponds, waste heat, can be used. Fig. 5.14 presents an H/D system where both air and seawater feed are heated in separate collector fields. The evaporator is a rectangular horizontal chamber. As packing material it uses parallel wooden plates covered with cotton tissue. The unit operates both in closed or open air cycle and for the evaporator the efficiency is defined as:

$$\eta = \frac{x_{in} - x_{out}}{x_{in} - x_{s-in}} \tag{5.41}$$

where x_{in} is the humidity of air at the inlet, x_{out} at the outlet of the evaporator, and x_{st-in} is the saturation humidity that corresponds to the actual humid process (Orfi et al., 2004).

5.6 OTHER CONCEPTS OF THE H/D METHOD

5.6.1 The Desiccant Absorption/Desorption H/D Process

Ettouney (2005) presents mathematical equations for two new concepts, ie, the *"desiccant absorption/desorption procedure"* (HDD) and the *"Humidification membrane drying"* methods. The schematic flow diagrams are presented in Figs. 5.15 and 5.17.

Desiccant absorption/desorption is a method which uses a liquid desiccant, eg, a solution of lithium bromide or a solid one, eg, zeolite. The system (Fig. 5.15) consists

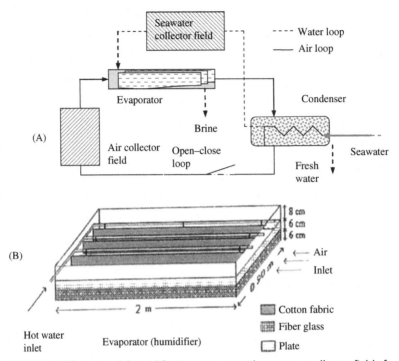

Figure 5.14 (A) Humidification—dehumidification system with separate collector fields for seawater and air heating. The evaporation chamber is horizontal. (B) Sketch of the horizontal humidifier with wooden supports of the cotton tissue (Orfi et al., 2004).

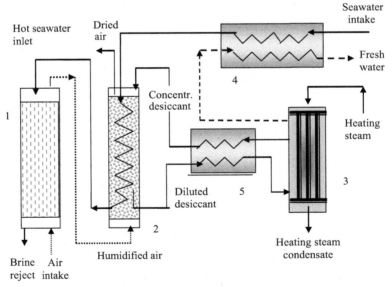

Figure 5.15 The new concept on H/D method using desiccant absorption/desorption technique: (1) humidifier, (2) absorber, (3) desorber, (4) condenser, and (5) heat exchanger (Ettouney, 2005).

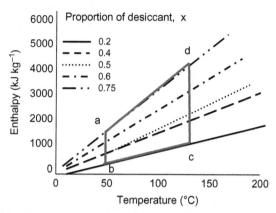

Figure 5.16 The variation of LiBr enthalpy as a function of the boiling temperature and concentration (Ettouney, 2005).

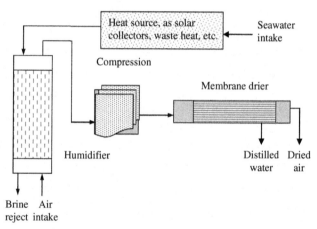

Figure 5.17 The new concept on H/D using selective membrane for air dehumidification (Ettouney, 2005).

of a large number of units: an air humidification chamber (1), similar to conventional humidification process, absorber (2) and desorber (3) beds, a water vapor condenser chamber (4), and a desiccant heat exchanger (5). Due to the unsteady state of this procedure, two beds are used, one for absorption and the other for desorption. The procedure produces large amounts of heat which can be used for feed water preheating. The desiccant may be regenerated by any external heat source, eg, steam, etc. A packed or tray absorption unit may increase absorption performance. Absorption/desorption units operate under isothermal conditions, ie, desiccant inlet and outlet temperatures in each unit are equal. Fig. 5.16 presents the fluctuation of LiBr as a function of temperature and concentration. Points a–d represent inlet and outlet from the absorber and inlet and outlet from the desorber units and x is the mass fraction of desiccant. This H/D method has in general high efficiency.

Water mass balance in the desiccant absorption unit is:

$$\dot{m}_{ds}(S_{ds-o} - S_{ds-i}) = m_{air}(x_o - x_h) \tag{5.42}$$

where S_{ds-out}, S_{ds-i} are the outlet and inlet salinities of seawater in the desiccant, $\dot{m}_{ds}$ (kg s^{-1} Pa^{-1}) is the desiccant mass flow rate, and x_o and x_h are humidity at the outlet of the unit and at the humidifier.

The energy balance of the heat exchanger for diluted/concentrated desiccant is:

$$\dot{m}_{ds}(h_{d-o} - h_{air-i}) = (\dot{m}_{ds} - \dot{m}_d)(h_{d-i} - h_{air-o}) \tag{5.43}$$

where h_{d-o} indicates enthalpy of distillate output, h_{d-i} and h_{air-o} are enthalpies of desiccant inlet and air outlet, respectively.

The heat transfer area A_{st} of the heating steam tubes in the disrober (3 in Fig. 5.15) is calculated from the steam mass flow rate (kg s^{-1}) and the overall heat transfer coefficient U_{st} (Ettouney, 2005):

$$A_{st} = m_{st}h_{st}/U_{st}(T_{st} - T_{ds-0}) \tag{5.44}$$

where U is overall heat transfer coefficient in each system and subscripts st and ds are referred to heating steam and to desiccant.

5.6.2 Membrane Drying Humidification

Membrane drying humidification is another new approach for H/D procedure. Humidification procedure is similar to the conventional H/D method. The humidified air steam leaving the humidifier is compressed and passes through the membrane drying unit where humidity permeates selectively to the permeate side of the membrane living fresh water condensate. Special drying membranes are used available commercially. This procedure is especially suitable for small capacity installations for arid and remote regions, as solar energy can be used for seawater heating and PV for driving the compression unit (Fig. 5.17).

5.6.3 H/D Under Varied Humidification Cycle

To get the optimum performance in an H/D closed air cycle, air leaving the dehumidification unit and entering the humidification chamber should contain the minimum possible humidity. It should be noted that humidity ratios are higher at pressures lower than the atmospheric pressure. This led to a new combined cycle where humidification operates under subpressure conditions and dehumidification at higher pressure. This mode of operation is presented schematically in Fig. 5.18. Air leaving humidification chamber unit is compressed increasing its energy load.

The air enters the condensation chamber at compressed pressure and at the exit of the condenser is expanded to environmental pressure. This mode of cycle minimizes

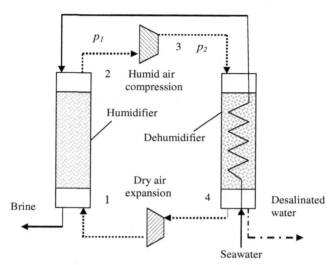

Figure 5.18 Schematic diagram of humidification/dehumidification with variety pressure cycle.

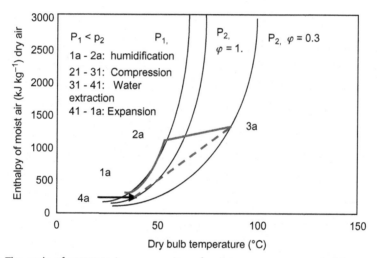

Figure 5.19 The cycle of compression—expansion of variety pressure presented in a psychrometric chart of the operation mode (Narayan et al., 2010).

air humidity at the exit of condenser and maximizes air humidity load at the humidifier exit (Narayan et al., 2010).

The varied pressure cycle is presented at the psychrometric diagram in Fig. 5.19. GOR of the cycle is given as:

$$GOR = \dot{m}_d h_{gh} / W_{in} \tag{5.45}$$

where W_{in} is the compressor work, m_s mass flow rate of distillate, and h_{gh} enthalpy of humid air.

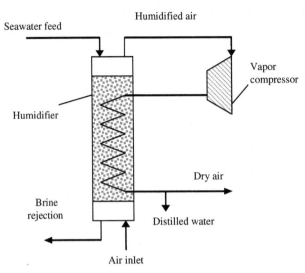

Figure 5.20 Schematic of an H/D method with thermocompression.

5.6.4 H/D Systems with Vapor Compression

This system was proposed by Vlachogiannis et al. (1999). The thermocompression cycle is presented in Fig. 5.20 schematically. This setup includes a conventional humidifier, a thermocompressor, and a heat exchanger. The humidified air is heated during compression and flowing downwards through the heat exchanger tubes heats up the incoming seawater stream (Fig. 5.20). Instead of a heat exchanger this may be achieved by the use of a separate chamber, as by conventional systems, where air is compressed in a thermocompressor and heated in an air heater before entering the humidifier. GOR of this modified H/D method is (Narayan et al., 2010):

$$\text{GOR} = \dot{m}_d h_{gh}/(\dot{Q}_{in} + \dot{Q}_{st}) \tag{5.46}$$

where $\dot{Q}_{in}$ and $\dot{Q}_{st}$ are the heat of the air heater and the input heat by the steam, respectively.

5.7 SOLAR H/D SYSTEMS WITH STORAGE TANKS

Solar desalination systems can achieve higher productivities if operated continuously. By using storage tanks, a 24-h operation is feasible for most solar desalination systems. Ben-Bacha et al. (2007) present mathematical models and a simulation procedure that predict the behavior and energy balance in a combined H/D-storage tank system. Three types of storage tanks were studied:

1. Storage tank with two internal coil heat exchangers connected to solar collector and to evaporation tower.
2. Storage tank is connected directly, in contact, to another storage tank. A coiled heat exchanger is connected to the evaporation tower.
3. The storage tank is built up as a divided tool with internal coil heat exchanger connected to the evaporation tower.

The divided hot water storage tank showed higher operating efficiency and better productivity of distilled water.

5.8 THE ECONOMICS OF THE H/D METHOD

The cost effectiveness of an H/D technique system, based on the present value cost (PVC) of the plant, is presented by Al-Hallaj et al. (2006):

$$PVC = I_{inv} + \frac{c + \overline{MR}}{PVF} \tag{5.47}$$

where PVF is the present value factor:

$$PVF = \frac{[(1+r)^n - 1]}{[r(r+1)^n]} \tag{5.48}$$

The product cost C_d of the system is:

$$C = PVC / \sum_{i=1}^{n} m_d \tag{5.49}$$

where

c	Cost for the employs
C	Cost of water unit product
I_{inv}	Systems investment cost
MR	Maintenance and parts replacement cost
m_d	Desalinated water, $m^3\,y^{-1}$
r	Rate of yearly interest
n	Number of useful years of operation

5.9 COUPLING SOLAR STILLS OR H/D SYSTEMS TO GREENHOUSES

In arid regions lacking fresh water, a combination of a solar still or an H/D system with a greenhouse, as a greenhouse environment has to be humid, may provide both fresh water and some crops cultivation. In the paper by Kabeel and Almagar (2013), an interesting description is given for the combination of desalination and greenhouse including economics.

These systems are rather new developments. Among the first papers on this idea of combining a solar still with a greenhouse system were presented by Trombe and Foëx (1961) and Selçuk and Tran (1975). Chaibi (2000) presented a detailed analysis of a seawater distillation-greenhouse system. He pointed out that "*Apart from economical and sociological problems associated with the operating of solar desalination systems a number of technical problems can cause operational difficulties.*"

The last decade studies that were performed on pilot plants and/or commercial size installations proved that this technique creates cool humid optimum environmental conditions for crop cultivation and simultaneously water production for irrigation and domestic consumption.

Fig. 5.21 presents the sketch one of the first combined H/D-greenhouse installations (Paton and Davies, 1996; Goosen et al., 2003; Davies and Paton, 2005; Paton, 2001), suitable for small remote communities in arid zones. The thermodynamic simulation on the influence of greenhouse-related parameters of this combined system is presented by Sablani et al. (2003) and Goosen et al. (2003). They presented the thermodynamic and economic analyses of the system for various climatic scenarios. Fig. 5.22 shows a sketch of a solar H/D system combined with a greenhouse (Goosen et al., 2003).

A detail description of an H/D-greenhouse system is presented in Fig. 5.23 (Tahri et al., 2005). Seawater or well water is pumped and stored, after suitable filtration,

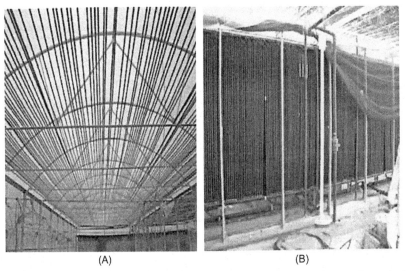

(A)   (B)

Figure 5.21 (A) The roof tubes for heating of seawater directly by solar radiation through the transparent greenhouse roof. (B) The condenser (Tahri et al., 2005).

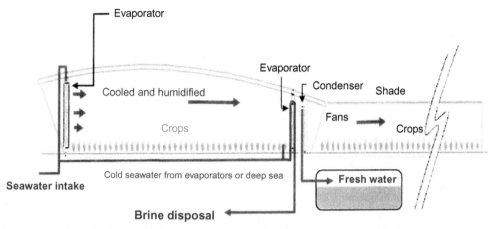

Figure 5.22 Sketch of a solar humidification/dehumidification system combined with a greenhouse (Goosen et al., 2003).

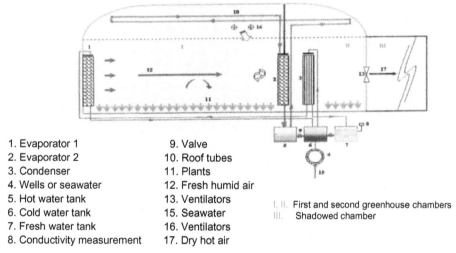

1. Evaporator 1	9. Valve
2. Evaporator 2	10. Roof tubes
3. Condenser	11. Plants
4. Wells or seawater	12. Fresh humid air
5. Hot water tank	13. Ventilators
6. Cold water tank	15. Seawater
7. Fresh water tank	16. Ventilators
8. Conductivity measurement	17. Dry hot air

I, II. First and second greenhouse chambers
III. Shadowed chamber

Figure 5.23 Schematic diagram of solar H/D-greenhouse system (Tahri et al., 2005).

into the deep water storage tank (6). Stored feed water is pumped to the condenser (3) heat exchanger tubes where is preheated and then to the first evaporator (1). The brine leaving the evaporator 1 returns to brine storage tank (5). The evaporator consists of a lattice of honeycomb cardboard and consist the front wall of the greenhouse. Seawater flowing downwards the evaporator 1 cools and humidifies the air stream entering the greenhouse. Humidified air loads chamber I environment with the necessary humidity for the plants cleaning simultaneously the atmosphere from the released carbon dioxide. A second evaporator (2) moistens the air up to saturation point. Feed seawater for the second evaporator is heated directly by circulation into the solar heating pipes (10) placed along of the transparent top of the chamber. Produced fresh water from the second evaporator is stored directly to the fresh water storage tank (8) and is used for the plant irrigation and/or for domestic purposes. Cold air leaving the condenser enters section III and either is rejected to the environment or is used to a second greenhouse system.

REFERENCES

Abu-Arabi, M.K., Reedy, K.V., 2003. Performance of a desalination process based on humidification—dehumidification cycle with different carrier gas. Desalination. 156, 281—293.

Al-Hallaj, S., 1994. Solar desalination with humidification—dehumidification cycle: heat and mass transfer studies. MSc Thesis. Jordan University of Science and Technology, Jordan.

Al-Hallaj, S., Parekh, M.M., Farid, J.R., Selman, J.R., 2006. Solar desalination with humidification—dehumidification cycle: Review and economies. Desalination. 195, 169—186.

Beckman, J., Hamieh, B., 1999. Carrier gas desalination analysis using H/D cycle. Chem. Eng. Commun. 177, 183—193.

Ben Amara, M, Houcine, I, Guizani, A, Maâlej, M, 2004a. Theoretical and experimental study of a humidifier used in a seawater desalination process. Desalination. 168, 1—12.

Ben Amara, M, Houcine, I, Guizani, A, Maâlej, M, 2004b. Comparison of indoor and outdoor experiments on a newly designed air solar plate collector used with the operating condition of a solar desalination process. Desalination. 168, 81—88.

Ben Bacha, H., Dammak, T., Ben Abdalah, A.A., Maalej, A.Y., Ben Dhia, H., 2007. Desalination with solar collectors and a storage tank: modeling and simulation. Desalination. 206, 341—352.

Bourouni, K., Chiabi, M.T., Tandrist, L., 2001. Water desalination by humidification and dehumidification of air: state of the art. Desalination. 137 (1—3), 167—176.

Chaibi, M.T., 2000. An overview of solar desalination for domestic and agriculture water needs in remote arid areas. Desalination. 127, 119—133.

Chafik E., 1996. Eine Verfahren zur Entzalzung von Meerwasser mit Hilfe von Solarenenergie. German Patent No. DE 196 20 214A1.

Chafik E., 1999. Ein Solar Kollektor zur Erwärmungvon Luft oderWasserströme. German Patent No. 299 16087 4, 1999.

Chafik, E., 2002. A new seawater desalination process using solar energy. Desalination. 156, 333—348.

Chafik, E., 2003. A new type of seawater desalination plants using solar energy. Desalination. 156, 333—348.

Chafik, E., 2004. Design of plants for solar desalination using multi-stage heating/humidifying technique. Desalination. 168, 55—717.

Davies, P.A., Paton, C., 2005. The seawater greenhouse in the United Arab Emirates: thermal modelling and evaluation of design options. Desalination. 173 (2), 103—111.

Ettouney, H, 2005. Design and analysis of humidification dehumidification desalination process. Desalination. 183, 341—352.

Goosen, M.F.A., Sablami, S.S., Paton, C., Perret, J., Al-Nuaimi, A., Haffar, I., et al., 2003. Solar energy desalination for arid coastal regions: development of a humidification—dehumidification seawater greenhouse. Solar Energy. 75, 413—419.

Grune W.N., 1970. Multiple effect humidity process. In: Proc. 3rd Int. Symposium on Fresh Water from the Sea, Vol. 1, pp. 655—668.

Hamieh, B.H., Beckman, J.R., Ybarra, M.D., 2000. The dewvaporation tower: an experimental and theoretical study with economic analysis. Int. Desalin. Water Reuse Quarterly. 10 (2), 30—43.

Hamieh, B.H., Beckman, J.R., Ybarra, M.D., 2001. Brackish water desalination using a 20 ft^2 dewvaporation tower. Desalination. 140, 217—226.

Hodges C.N., Tompson T.L., Groh J.E., Frieling D.H., 1966. Solar distillation utilizing multiple effect humidification, University of Arizona, Energy Lab. of Institute of Atmospheric Physics, Final Report, 160 pp.

Houcine, I., Ben Amara, M., Guizani, A., Maâlej, M., 2006. Pilot plant testing of a new solar desalination process by a multiple-effect humidification technique. Desalination. 196, 105—124.

Kabeel A.E., Almagar A.M., Seawater greenhouse in desalination and economics. In: 17th International and Water Technology Conference, IWTC17, Istanbul, November 5—7, 2013.

Larson R., Albers W., Beckman J., Freeman S., 1989. The carrier—gas process—A new desalination and concentration technology. In: Proc. 4th World Congress on Desalination and Water Reuse, Kuwait, November 4—9, 1989, Vol. 1, pp. 119—138.

Mathioulakis, E., Delyannis, E., Belessiotis, V., 2007. Desalination by alternative energy: review and state-of-the-art. Desalination. 204, 346—365.

Müller-Holst H., 2001. Multi-effect humidification seawater desalination using solar energy or waste heat. Presented at the SODESA Workshop, Agricultural University of Athens.

Narayan, G.P., Sharqawy, M.H., Lienhard, J.H., Zubair, V.S.M., 2010. Thermodynamic analysis of humidification—dehumidification desalination cycles. Desalin. Water Treat. 16, 339—353.

Nawayseh, N.K., Farid, M.M., Al-Hallaj, S., Al-Tamimi, A.R., 1999. Solar desalination based on humidification process. I: Evaluating the heat and mass transfer coefficient. Energy Conv. Manage. 40, 1423—1439.

Orfi, J., Laplante, H., Marmouch, H., Galanis, N., Bamhamou, B., Ben-Nastrallah, S., et al., 2004. Experimental and theoretical study of a humidification—dehumidification water desalination system using solar energy. Desalination. 168, 151—159.

Paton C., Davies P.A., 1996. The seawater greenhouse for arid lands. Presented at Mediterranean Conf. on Renewable Energies, Santorini, Greece.

Paton A.C., 2001. Seawater greenhouse development for Oman. Thermodynamic modelling and economic analysis. MEDRC Project-97-005b.

Sablani, S.S., Goosen, M.F.A., Paton, C., Sharyya, W.H.A., Al- Hinay, H., 2003. Simulation of freshwater production using a H/D seawater greenhouse. Desalination. 159, 283—288.

Selçuk, M.K., Tran, V.V., 1975. Solar stills for agriculture purposes. Solar Energy. 17, 103—109.

Tahri T., Abdul-Wahab S.A., Bettahar A., Douani M., Al-Hinai H., Al-Mulla Y., 2005. Desalination of seawater using a H/D seawater greenhouse. Presented at ISES Congress, 2005.

Trombe F., Foëx M., 1964. Utilization of solar energy for simultaneous distillation of brackish water and air-conditioning of hot-houses in arid regions. In: UN Proc. the Conference on New Sources of Energy, Rome, August 1961, Vol. 6, pp. 104—109.

Vlachogiannis, M., Bobtozoglou, V., Geogalas, C., Litinas, G., 1999. Desalination by mechanical vapor compression. Desalination. 122, 35—42.

Xiong, R., Wang, S., Wang, Z., 2006. A mathematical model for a thermally coupled humidification-dehumidification desalination process. Desalination. 196, 177—187.

Xiong, R.H., Wang, S.C., Xi, L.X., Wang, Z., Li, P.L., 2005. Experimental investigation of a baffled shell and tube desalination column using the humidification-dehumidification process. Desalination. 180, 253—261.

Younis, M.A., Darwish, M.A., Juwayhel, F., 1993. Experimental and theoretical study of a humidification—dehumidification desalting system. Desalination. 94, 11—24.

CHAPTER SIX

Indirect Solar Desalination (MSF, MED, MVC, TVC)

6.1 INTRODUCTION

The direct solar or solar-powered desalination (also called solar-driven or indirect solar desalination) is a dual-type desalination—solar energy conversion, where two interconnected systems, the desalination system and the solar energy collection system, have to provide the optimum operational efficiency and the optimum economic conditions of operation and maintenance.

The conventional desalination systems are technologically and operationally matured, commercialized, usually of large capacity and are relatively compact. Their operation is achieved either with the utilization of heat energy for the satisfaction of the thermal loads (distillation methods) or with electric current (membrane methods). The solar systems producing electricity and/or heat are also matured and technologically well-documented systems.

In this chapter, the solar thermal systems of the conventional distillation methods such as multistage flash (MSF), multiple effects distillation (MED), mechanical vapor compression (MVC), and thermal vapor compression (TVC) will be presented.

Using solar energy in solar-powered desalination systems, the same constrains which apply to the conventional systems operated by fossil fuels need to be considered, especially for the large-scale installations, as outlined in chapter "Humidification—Dehumidification (H/D)." In solar-powered desalination systems, flat-plate collectors or concentrating collectors can be used such as parabolic troughs, parabolic dishes, and Fresnel mirrors, which can provide the best operation conditions. These are generally dual systems of medium capacity, for example, for a desalination system of small ($<10 \text{ m}^3 \text{ day}^{-1}$) up to medium ($<10-1000 \text{ m}^3 \text{ day}^{-1}$) capacity.

Solar-powered desalination is usually suitable for application in remote areas, islands, and deserts which are either in close proximity to the sea or have sources of brackish water while there is no available electrical grid or possibility of using thermal energy of suitable temperature level.

Thermal Solar Desalination

6.1.1 The Available Solar Energy

Solar irradiation is the most important source of permanent and inexhaustible energy supply. Large quantities of solar energy are produced at the surface of the sun and diffused to all directions in space. From this quantity only a small fraction reaches the surface of the earth, with different intensities on different places of the planet. The energy that reaches the outer surface of the atmosphere is ~ 1370 W m^{-2}, while on the surface of the earth its intensity is only $100-1000$ W m^{-2}, with the largest values recorded in areas around the equator. Due to the fact that the intensity of solar radiation falling on the surface of the earth is of low density, relatively large areas are required for its collection. Irrespective of the low density, the total energy falling on the whole surface of the earth is much bigger than the energy used worldwide. The quantity which passes through the atmosphere of the earth is equal to 170 trillion kWh, a quantity that is 5000 times larger than the sum of all types of energies on the earth. All this energy is not utilizable at the level of the earth. A large percentage, $\sim 47\%$, is converted to low-temperature heat and is reemitted to the space. About 23% is spent for the water cycle, that is, the evaporation from the water surfaces of the earth, while $\sim 30\%$ is reflected back to space. A small fraction $<0.5\%$ is the indirect solar energy which appears as wind energy, wave energy, and stored as photosynthetic energy of vegetation.

To design, install and operate a solar energy system, knowledge of the availability of the incident solar energy is important. The solar radiation is limited by its nature, is dispersed on all the surface of the earth, and generally it has a relatively low and variable intensity. Moreover, it is available only during daytime and is influenced by the prevailing weather conditions of each location. Obviously these characteristics have a direct effect on the smooth operation of solar systems. This has as a consequence, that although it is available free of charge, its conversion to useful heat and/or electric energy to constitute a relatively expensive process. This is due to the fact that the corresponding technology requires expensive equipment while its storage is an additional cost for the energy transformation system.

Therefore, for those working with solar energy, the knowledge of the mean weather conditions, the geographical characteristics and the mean intensity of solar radiation during the various seasons of a year, for a particular area, is an important factor for the potential optimum design and optimum operation of a solar system.

Connecting a thermal desalination system to a solar energy installation, the following must be taken in consideration:

- Solar energy can be collected only during the daytime. This renders the operation of the desalination system as intermitted, unless a proper arrangement of heat storage is used
- The collected solar energy is variable and only a part of it, $50-68\%$ $(0.7-0.98$ kW m$^{-2})$ can be considered that is collected as useful energy.

These properties of solar energy constitute a disadvantage for the production of freshwater with conventional desalination methods, because in desalination systems which operate with conventional fuels or slightly altered fuels used in connection with solar systems, the thermal performance and the production of desalinated water is optimum only when the system is in continuous and noninterrupted operation. Therefore, the use of a proper storage medium of heat utilizes the optimum performance of the system.

Detailed description of collection of solar energy is beyond the scope of this book. Proper references and details for the collection of solar energy and the design of solar systems are given by Duffie and Beckmann (2006, 2013) and Kalogirou (2004, 2009, 2013).

6.1.2 Solar Collectors

Solar collectors are devises that collect the solar radiation converting it to thermal energy of a fluid which can be used directly or stored for use during cloudy periods or nighttime.

There are various types of solar collectors the use of which is determined from the required temperature of the desalination system and their financial availability. Higher temperature collectors are used for the production of high-pressure steam used for the production of electrical energy. The produced subcooled steam, which is formed at the expansion end of the turbine, is suitable for the thermal supply of desalination units. Depending on the supplied temperature, these are distinguished into low-, medium- and high-temperature solar collector units. The following types of collectors have been practically examined to be connected with desalination units.

6.1.2.1 Solar Ponds

Solar ponds (or salt gradient solar ponds). These are artificial lakes and constitute the simplest type of solar collector. They consist of layers of water at different concentration of salts. Generally there is no large application of solar ponds in desalination systems. The greatest temperature of the heat supplied is $\sim 90°C$.

6.1.2.2 Flat-Plate Collectors

Flat-plate collectors consist of a black absorbing plate, at the bottom of which pipes or channels are installed. This is mounted in a hermetically sealed frame with transparent cover which is penetrable to the solar radiation at the top and good thermal insulation at the bottom. The solar energy is absorbed from the black surface and supply heat to a fluid which is circulated in the pipes or the channels. The heat is transferred to a process connected to the collector or it is stored. Flat-plate collectors supply heat at generally low temperatures up to about a maximum temperature of $90°C$, and typically at a mean temperature of $70°C$. Theoretically are connected with solar stills, with humidification/dehumidification (H/D) units, membrane distillation (MD) units, and low-temperature multiple effect distillation (LT-MED) units, which are modified for low temperatures. The MED unit in this case does not comprise more than 10 effects.

6.1.2.3 Evacuated Tube Collectors
Evacuated tube collectors are flat devises which consist of cylindrical absorbing surfaces or tubes with internal fins installed in an evacuated tube to reduce the convection losses. They produce temperatures of 80°C up to 120°C (Fig. 6.1).

6.1.2.4 Concentrating Solar Collectors
Concentrating collectors or concentrating reflectors are systems that reflect or refract the incident solar radiation from one reflective "aperture" of surface A_a (m^2) to an "absorber" (or receiver) of area A_r (m^2), where $A_a \gg A_r$ applies. The optical concentration is $C_o = G_r / G_a$ while the geometric concentration is $C_c = A_r / A_a$, where G_r

Figure 6.1 Evacuated tube collectors field.

$(\mathrm{W\ m}^{-2})$ and G_a $(\mathrm{W\ m}^{-2})$ are the intensities of incident solar radiation to the areas of "aperture" and "absorber," respectively. There are various geometric forms of concentrating collectors. The main types which are used for desalination, are:

Parabolic trough collectors (PTC). These are concentrating systems of linear focus (Fig. 6.2) producing steam or hot water typically up to $375°C$ while more recent installations reach temperatures up to $560°C$. They have a concentration factor of solar energy of $C_c = 20-100$. They can be used for desalination, cooling, and production of electricity. Generally their performance is higher at temperatures of $150-190°C$. They usually use one-axis tracking of the sun. Darwish (2011) reports that from all concentrating reflectors which convert solar energy to thermal energy, the parabolic troughs are the only concentrating systems which have reached commercial maturity with well-documented references concerning availability and reliability.

Figure 6.2 (A) Parabolic trough with solar energy evacuated tube absorber. (B) Aerial photograph solar parabolic trough field. The system is installed in 2007 at Boulder, Nevada. Its capacity is 64 MW with annual production of 130 GWh. The surface area is equal to 1.2 km^2 and consists of 700 concentrating troughs.

Figure 6.3 Linear receiver of linear Fresnel type mirrors.

Figure 6.4 A field of point focus collector.

Fresnel concentrators/refractors. These consist of smooth optical segmented surfaces. Depending on the location and type of reflector, refract or reflect the incident solar radiation. These are also linear type focusing systems (Fig. 6.3).

Parabolic dish collectors. These are concentrating systems of point focusing. They produce either hot water under pressure or high-temperature steam. The temperatures produced are between $\sim 375°$C and $2000°$C (Fig. 6.4).

Figure 6.5 Photograph of the first installed tower system with 1818 heliostats, capacity of 10 MW. The installation is in Barstow, California, USA (Skinrood, 1982).

Figure 6.6 Central tower system with heliostat mirrors. Pictorial view of the first commercial European and international installation (2008). The system is located at Seville, Spain, and produces 11 MW electrical power (Anonymous, 2008).

Central receiver system. This consists of a field of heliostats. Heliostats are large flat or slightly curved mirrors which are mounted on metallic structures. The mirrors are installed in a field so as all to focus on a boiler located at the top of a tower. They produce steam of high temperature, $\sim 1000°C$, used for the production of electricity. Theoretically, the produced electricity can be used in desalination for reverse osmosis (RO), ED and MVC, or if bleed steam of the right temperature is produced, for distillation systems (Figs. 6.5 and 6.6).

6.2 SHORT HISTORICAL REVIEW

Solar energy is the oldest form of energy used as heat from the prehistoric times by humanity. It is used for the heating of dwellings and mainly for the drying of ceramics and food, a practice applied even up today in some areas of the earth.

The conversion of solar energy to heat, to mechanical energy and then to electricity with the use of various devices began from the end of 1700 (Belessiotis and Delyannis, 2006; Delyannis, 2003). Though there are references for the potential use of solar energy from antiquity, while a little after the middle of the 20th century, following intensive studies, a lot of installations of solar energy conversion to heat, as steam, and then to electric energy were erected. Today universally there are a lot of installations producing steam for use in the production of electricity. During the last few years, mainly after 2008 up today, a large number of solar electricity productions have been installed mainly in the United States and Spain, while a large number of solar electricity systems are under design and/or construction, but not at the moment in applications combined with desalination systems. A few examples of concentrating collectors are shown in Figs. 6.2–6.6.

6.3 DEFINITIONS AND NOMENCLATURE

Some definitions, like the gain output ratio (GOR—see section 5.2), are already being given in the previous chapters. For the units which operate with conventional fuels, the main definitions are given here while some definitions are given specifically for the solar systems.

Aperture, A_a, is the area of a collector which receives the utilizable incident solar radiation or the reflective surface or a mirror which receives the incident solar radiation and reflects it to the area of the receiver of the radiation (absorber).

Receiver, A_r, is an apparatus, that is, a tube, a heat exchanger, or a boiler, depending on the type of the concentrating mirror, on the surface of which solar radiation reflected from the aperture of the collector falls and from which the heat is collected from a suitable fluid (Figs. 6.2–6.6).

Absorber (absorbing surface) refers to a surface, usually of black color or selectively optically processed, which absorbs the incident solar radiation either coming from sun directly or from a reflection at the mirror of a concentrating solar collector.

Concentration ratio (CR) is the degree of salinity to which the rejected brine is condensed c_{br}, to the original salinity of seawater input c_{sw}, $CR = c_{br}/c_{sw}$. Usually it is kept to the limits of $c_{br}/c_{sw} = 2.5$ to avoid scale deposits.

Output is the produced desalinated water per hour or per day ($kg\,h^{-1}$, $kg\,day^{-1}$ or $l\,h^{-1}$, $l\,day^{-1}$).

Effect refers to a unit of a system of apparatuses. They consist of evaporation—condensation units which are set out in series forming a system of multiple distillation effects (MED).

Stage: In the MSF distillation method, they are chambers where the seawater as it flows is flashed with the production of brine and steam.

Specific heat, c_p: In the distillation methods, the specific heat of freshwater and seawater at operating conditions, that is, in usual temperatures, pressures, and salt concentration c_s of the mixture, taken at the mean temperature which for routine operations the following values are considered safe:

For seawater: $3.995 \times 10^3 \, \text{J kg}^{-1} \, \text{K}^{-1}$
For desalinated water: $4.184 \times 10^3 \, \text{J kg}^{-1} \, \text{K}^{-1}$

Enthalpy of evaporation, h_v, is the amount of heat required to evaporate a unit of mass of a fluid. It is estimated at the temperature of the balance point of steam with the water mixture boiled.

Heliostat is a flat or slightly curved concave mirror reflecting the incident solar radiation on an absorber located at the top of a tower (Figs. 6.5 and 6.6).

Central receivers or the tower concept is a system which comprises a heliostat field. The arrangement and orientation of heliostats allows the reflection of the incident solar radiation to focus on a point, to the heat exchanger or boiler which is located on the tower.

Performance ratio (PR). As mentioned in chapter 5 (section 5.2) PR represents the produced quantity of the distilled water per amount of input energy:

$$PR = \frac{m_d}{h_v} \approx \frac{m_d}{2330} \tag{6.1}$$

Economy ratio (ER) refers to the amount of desalinated water per kilogram of externally produced steam m_{st}, which is used for the heating of the supplied water up to the required maximum temperature T_{max}. This term is used for the determination of the heating steam economy. This economy ratio varies from ~8 up to ~15.

Fluted tubes are the longitudinal corrugated heat transfer tubes of the evaporators. They consist of grooves and peaks of different designs which enclose the external circumference of the tubes. They increase considerably the heat transfer area.

Heat recovery section is the number of uniform parts of a MSF distillation system where the stream of input seawater is heated from the condensing steam.

Heat rejection section is the part of the rejection of heat of an MSF distillation system where the heat of condensing steam is rejected from the last stage after it is cooled by the seawater.

Heat transfer coefficients, h. In distillation desalination, the following empirical coefficients can be used for routine operations, for seawater under boiling and for the condensing steam:

For seawater under boiling: $h_{bs} = 7000 \, \text{W m}^{-2} \, \text{K}^{-1}$
For the condensing steam: $h_{con} = 11,340 - 14,200 \, \text{W m}^{-2} \, \text{K}^{-1}$

Partial pressure, p_p, of the dissolved salts considered negligible because it is very small. Therefore, the partial pressure above the water solution is considered the partial pressure of the pure steam.

Recovery ratio (RR),$(\dot{m}_d/\dot{m}_f)$, is the fraction of the flow rate of the produced desalinated water to the flow rate of the feed seawater (kg s^{-1}/kg s^{-1}).

Capacity (C) is the capability of a desalination system to produce a prescribed amount of desalinated water. The capacity of a distiller is determined as the rate of steam production from the feed seawater. It is a function of the total heat transfer area, A_{ht}, temperature difference ΔT, and the total thermal resistance R_T.

$$C = \frac{A_{ht}\,\Delta T_T}{R_T h_{ev}} = \frac{h A_o\,\Delta T}{h_{ev}} = \frac{\dot{Q}}{h_{ev}} \tag{6.2}$$

where h_{ev} is the evaporation enthalpy (J kg^{-1}), h is the convection heat transfer coefficient (W m^{-2} K^{-1}), and Q is the quantity of heat passing the surface (J s^{-1}).

During distillation, typically under mean conditions (eg, for the Mediterranean water with salt level of 35‰) from 1 kg of feed seawater ~ 0.4 kg of steam are produced leaving behind as residual 0.6 kg of brine. During condensation, the steam supplies its released heat for the preheating of the feed seawater.

Desalination systems are energy intensive processes. Generally, it is estimated that for the production of 1000 m^3 desalinated water per day, 10,000 tons of oil are required per year or the respective amount in solar, thermal of electrical energy.

6.4 FACTORS INFLUENCING THE SELECTION OF THE DESALINATION SYSTEM

Theoretically speaking, it is possible to connect the two systems and the selection can be random. Practically however, the combination should offer an optimum system from the operational and financial perspectives. The experience gathered up to now has shown various applications of MSF and MED distillation methods, mainly with PTC although some pilot installations have been installed with various other combinations of reflectors desalination methods.

One way of finding the most appropriate technology is given by Liu et al. (2013) who analyzed and compared the advantages and disadvantages of the energy produced from conventional fuels and the renewable energy sources. The analysis comprises five factors such as the technological, environmental, social, financial, and the operation of the system and suggests evaluation indices which show the suitability of combination of the two different technologies.

For a conventional desalination installation which will operate with solar energy, the following have to be taken into account (García-Rodríguez, 2002):

- Thermodynamic considerations
- Special characteristics of the system
- Location and financial considerations.

The selection of the desalination system, as long as this is going to be installed in a remote area, which is the usual application, has to have the following characteristics according with (Abdul Fattah, 1986):

- Simplicity
- Easy operation
- Compatibility of the desalination system characteristics with those of the chosen location.

The disadvantages of the combination of desalination/solar fields are:

1. The two systems have high capital cost
2. They require for their optimum operation specialized personnel for the installation, operation, and maintenance for each one system of the installation separately
3. For remote or small areas near the sea, the transportation cost is high. In most of the cases, there is no suitable transportation infrastructure, for example, roads and ports.

The factors which influence the selection of the combination are given in the following sections.

6.4.1 Energy Demand

It is the most important factor because it influences the size and consequently the cost of the solar system. Typical energy values for the production of 1 m^3 desalinated water according to the desalination method are given in Table 6.1 (Ghaffur et al., 2013).

6.4.2 Water Demand

Water demand and the quality of the desalinated water is another important parameter which influences the selection of the system. It has to be clarified that distillation produces desalinated water with very low salt contents and if it is going to be used from human consumption requires mixing with appropriate quantity of water so as

Table 6.1 Energy Required to Produce 1 m^3 of Desalinated Water for Various Desalination Technologies (Ghaffur et al., 2013)

Method	Thermal Energy (kW$_{th}$ m^{-3})	Electrical Energy (kWh$_e$ m^{-3})	Total Energy (kWh$_T$ m^{-3})
MED	4–7	1.5–2	5.5–9
MSF	7–12	2.5–4	10–16
SWRO	–	3–4[a]	3–4
BWRO	–	0.25–2.5	0.5–2.5

[a]Including energy recovery system.

the product to contain, according to the WHO (World Health Organization), at least <100 ppm total dissolved solids (TDS). In RO, the contents of desalinated water in salts are adjusted (for potable water 300–500 ppm TDS) from the suitable flow of feed water. The feed water has a minimum influence on the selection of the method for the various desalination methods.

6.5 FACTORS INFLUENCING THE SELECTION OF THE SOLAR SYSTEM

For a solar system, the area of the installation of the solar system is of great importance. Factors that influence the selection of the installation areas include (El-Nashar, 2001a):

- The morphology of the area because most solar installations require comparatively large flat surfaces for the installation of the collectors. Therefore, the land area must have a relatively low cost while at the same time it should not be far away from the inhabited area
- The intensity of solar radiation influences directly the economy of the system; for this reason the areas of the "sun belt" which usually suffer from lack of water are the best for the installation of solar desalination systems. Suitable areas are those with large solar potential of the order of 1880–2000 $kWh\,m^{-2}\,year^{-1}$. It is also important to know the percentage of diffuse solar radiation which usually varies from 10% to 20% during the sunshine days and somewhat bigger for the overcast days
- The two systems have to be installed if possible one after the other so as to avoid heat losses and to minimize the piping system and other connecting sections.

6.5.1 Suitability of Solar Collectors for Desalination

Theoretically, all concentrating solar technologies producing electricity and heat are considered suitable for connection with desalination systems. Practically, the performance, the cost of combination systems with thermal desalination or hybrid methods, and the technology which is employed for the production of power have to be taken into account. Various possibilities are shown in Fig. 6.7.

From the various types of solar energy collectors for desalination, the most suitable are considered the flat-plate collectors and the evacuated tube collectors (Fig. 6.1), and when the desalination system operates at temperatures >>90°C the line focus concentrating collectors, like the PTC and the Fresnel system, are considered suitable.

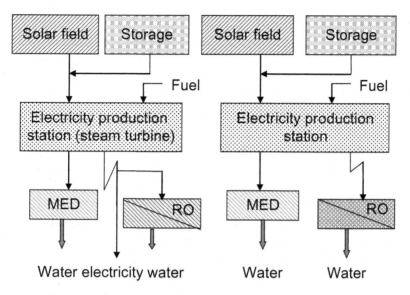

Figure 6.7 Different configuration combinations for the production of solar electricity and desalination (Trieb and Müller-Steinhagen, 2008).

The parabolic trough system is considered, at the moment, as the most suitable for the operation of thermal desalination methods of the distillation types of MSF, MED, TVC, and MVC and/or other distillation methods of special designs (García-Rodríguez et al., 1999, 2002; Darwish, 2011). For the selection of the most suitable collector system, the following have to be taken into account:
- Required operation temperature
- Ratio between the direct and total solar radiation
- Environmental temperature
- Radiation permanence
- Techno-economic factors.

6.6 CONVENTIONAL DESALINATION SYSTEMS—DISTILLATION METHODS

In this section the basic facts of distillation desalination methods (MED, MSF, TVC, and MVC) which operate with conventional fuels will be analyzed as well as their potential to be combined and operated with solar energy. Basically, distillation desalination operations operate with input steam supplied from a solar collectors system, either as steam or bleed steam. Therefore, there is no difference in the operation of distillation systems except from the type of combustible matter used for the production of steam.

6.6.1 Multiple Effect Distillation

This distillation technique is considered as the most suitable to be connected to a solar concentrating collector installation. Moreover it is found that the most effective combination is with PTC (Darwish, eolss). Generally the MED installations are of lower cost than the MSF ones because they demand much smaller heat transfer areas (Fig. 6.8). Therefore, they are more compact for the same productivity. They operate for a long time period (6–24 months) without a need for cleaning and produce good-quality desalinated water. In comparison with the MSF installations, they have better thermal performance and low consumption of electricity for the operation of the pumps, etc. (1.5–2 kW per ton of water). From the thermodynamic point of view is a more attractive system for the following cases (Darwish, 1987, eolss):

- The use of corrugated tubes, especially in the vertical arrangement, provides greater heat transfer coefficients, by minimizing their areas
- The pumping power is smaller than the one required by the MSF systems because the feed flow rate (almost double the rate of production) is smaller than the recirculation flow rate in the MSF units (with ∼7 times the product water)
- Single pass feed seawater allows higher temperatures at the first effect because it has a lower salt concentration than the respective first stage of the MSF system.

The whole system consists of a series of chambers, the evaporation/condensation effects, where each consecutive effect is at a slightly lower pressure than the previous one. For the input of heat from waste heat or from a solar energy system, the MED distillers have suitably configured so as to operate with low-pressure steam and low temperatures of the order of 70°C (LT-MED). In this case, the only restriction that exists is the number of effects which typically can be up to 10.

Each evaporation/condensation effect contains tube bundles in which the circulating seawater is preheated from the sensible heat of steam which are condensed at the external surface of the tubes. Different tube configurations have been developed, for example, long tube evaporators, submersed tube evaporators and horizontal tubes. Today, most MED distillers are made with horizontal tubes (Fig. 6.9).

For conventional MED distillers operating with conventional fuels for the horizontal tubes of the first effect, heating steam is supplied from an external source, usually at a temperature of 120°C. The preheated seawater is fed from the top part of the first chamber (or effect) flows externally of the tubes and is heated up to the temperature T_{max} and is partially evaporated. The steam vapors thus formed are fed to the tubes of the second effect where they are used as the heating medium for the heating of the down-flow seawater which flows externally on the tubes. This flow of the various fluids continues up to the last effect, the nth, of the system from where the steam is transferred to the condenser, the effect $n + 1$, where it is condensed to produce desalinated water. The heating steam in effects 2 up to n is partly

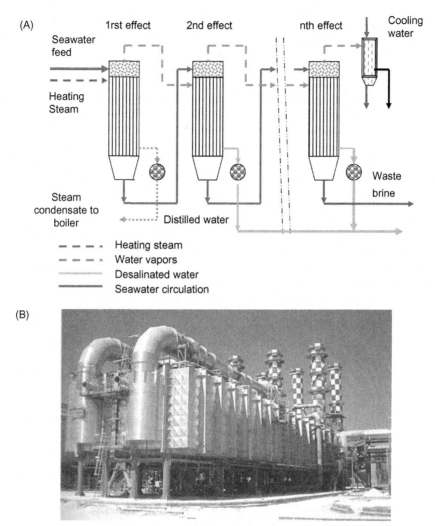

Figure 6.8 (A) Principle of operation of MED technique with vertical tubes. (B) Industrial conventional installation of MED technique. It produces 272,520 m³ day⁻¹ desalinated water from seawater with salinity of 42–46‰ (installation June 2008, Bahrain, UAE) (IDA Desalination Year Book, 2008–09).

condensed inside the tubes to form desalinated water and fed through the various effects to the condenser (Fig. 6.9) which is cooled down from a separately fed seawater which is then rejected. The brine is also transferred from effect to effect

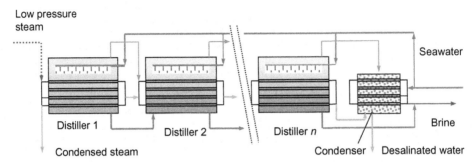

Figure 6.9 Schematic diagram of the conventional multiple effect distillation (MED).

condensed little by little and is finally rejected or partly recirculated after it is mixed with the feed seawater.

Following are the main arrangements of seawater and brine: forward feed, the backward feed, parallel without heaters, and the parallel with heaters.

6.6.1.1 The Temperature Distribution

The effective operation of conventional MED distillation units depends on small temperature differences from effect to effect taking always into consideration the brine boiling temperature rise, ΔT_{el} in each effect. In Fig. 6.10, the temperature distribution of an MED system with n effects is given as well as of the rejection part (condenser). In Fig. 6.11, the shape of a compact MED configuration with vertical arrangement of the effects is given together with the respective temperatures profile.

6.6.1.2 Brief Mathematical Analysis

The performance ratio (PR) varies from 0.85 to 0.95 and can be estimated for n number of effects and mean thermal performance per effect η_{th} as:

$$\text{PR} = \frac{\eta_{th}(1-\eta_{th}^n)}{(1-\eta_{th})} \tag{6.3}$$

The total required heat is equal to:

$$\sum \dot{Q} = hA(T_s - T_T)\frac{n}{n+1} \tag{6.4}$$

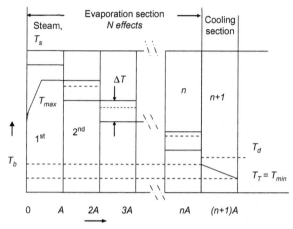

Figure 6.10 Temperature distribution in an *n* effect MED system.

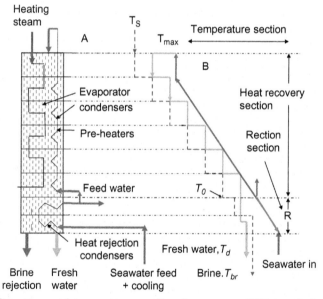

Figure 6.11 Configuration and temperature profile of a compact MED installation with effects in vertical arrangement. *Courtesy: Sasakura Co. Ltd.*

The total heat transfer coefficient is given from:

$$U_T = R_s + R_{br} + R_w + R_{sc} = \frac{1}{\dfrac{1}{h_s} + \dfrac{1}{h_{br}} + \dfrac{s_w}{\lambda_w} + \dfrac{s_{sc}}{\lambda_{sc}}} \qquad (6.5)$$

While the surface economy as a function of the number of effects n, the flow rate of the desalinated water $\dot{m}_d$, and the total area A_o are related from:

$$\frac{\dot{m}_d}{A_o} = \left[\left(\frac{h}{h_{ev}} \right) \left(\frac{n}{n+1} \right) \right] (\Delta T_o - n T_{el}) \qquad (6.6)$$

where

A, A_o	Effect area, total area of effects, respectively (it is considered that the effects have equal surface)	m^2
ΔT_o	Maximum temperature difference of the system	K
h	Heat transfer coefficient	$W\,m^{-2}\,K^{-1}$
h_{br}, h_s	Film heat transfer coefficient on the brine side and the steam side	$W\,m^{-2}\,K^{-1}$
λ_{sc}, λ_w	Conductivity coefficient for the scale deposits and the tube wall, respectively	$W\,m^{-1}\,K^{-1}$
n	Number of effects	—
R_{br}, R_s	Film thermal resistance from the brine side and the film on the steam side, respectively	$m^2\,kW^{-1}$
R_{sc}, R_w	Thermal resistance of the scale deposits and the tube wall, respectively	$m^2\,kW^{-1}$
s_w, s_{sc}	Thickness of the tube wall and the scale deposits, respectively	M
T_s, T_T, T_{el}	Temperature of the heating steam, feed seawater at the uptake point, and the brine boiling point elevation, respectively	K

The economy of the MED system and consequently the cost of the desalinated water are influenced from the number of effects. Theoretically, the optimum number of effects has to allow the flow of heat in each effect with a minimum temperature difference ΔT_{min} of the produced steam in one effect and condensed in the next effect at a lower temperature. This temperature difference has to be at least equal to the brine boiling point elevation ΔT_{el}. The brine boiling point elevation typically varies from $0°C$ to $0.7°C$. Because the temperature difference ΔT_{min} practically offers bigger heat flux, this temperature is kept at the level of $\leq 5\Delta T_{min}$.

In an MED system, the number of effects is directly related to the performance ratio (PR) while for any surface of each effect, this ratio never approaches the number n. The optimum number of effects is obtained from the balance between capital cost and the operation cost.

6.6.1.3 Low-Temperature MED Systems

As was mentioned earlier, MED systems have been developed which operate with low-temperature steam, at a temperature of about $\sim 70°C$. They are suitable for operation with solar heat and are connected with flat-plate solar collectors, evacuated tube collectors, and parabolic troughs. The advantage they have by operating at low temperature is that scale deposits and corrosion are minimized. Kronenberg (1995) presented the advantages of LT-MED as steady, flexible, and reliable systems compared to other distillation methods. The combination of LT-MED with a diesel or gas turbine is proposed and he reported that "experience has shown that these systems have superior technological features compared to other desalination systems."

6.6.2 Multistage Flash

When brine water is heated up, under a certain pressure, at a temperature which is a little lower than its boiling point and subsequently introduced in a chamber which is at a suitable under-pressure, a sudden explosive (or vigorous) boiling occurs. Bubbles are formed in the whole mass of the water and part of the water boils instantaneously whereas boiling continues until equilibrium with the produced steam is achieved. Evaporation reduces the temperature of the brine in the chamber, called stage in this case, because evaporation is a process which absorbs energy. The remaining salty water is supplied to the next effect, which is at even lower pressure, where it is released again for producing steam. This process is repeated in the following stages, each one of which is at a little lower pressure from the previous one, while the drop of temperature is proportional. Fig. 6.12 shows schematically the principle of operation of a distillation system based on the MSF method, whereas a real system is shown in Fig. 6.13.

There are two types of flow arrangements in MSF distillation systems:

1. Single pass, where brine is disposed from the last stage of the rejection section. This system has been applied in a few MSF installations because it presents a number of disadvantages, for example, it requires a greater number of flash chambers

2. Recirculating, where part of the brine recirculates after it is mixed with the feed seawater input. This operation found a bigger practical application.

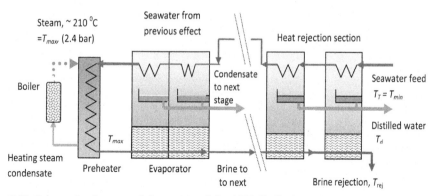

Figure 6.12 Schematic diagram of the stages of the MSF distillation system.

Figure 6.13 Industrial installation of an MSF system. It comprises four units with a capacity of 68,130 m³ day⁻¹ each. (Installation June 2008 in Qatar, UAE) (IDA Desalination Year Book, 2008−09).

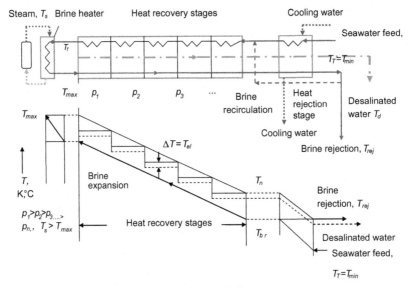

Figure 6.14 Schematic diagram of various water stream flows.

Fig. 6.14 shows the schematic flow diagram of the seawater feed, recirculation of part of the brine, and of the desalinated water. The sloped continuous lines show the temperature drop of the brine stream while the discontinuous (broken) lines the change of desalinated water temperature in the heat recovery and heat rejection sections. The cold seawater is pumped at the inlet of the condensation pipes of the heat rejection section and is used for the cooling of the steam of the last stage. The feed seawater, after it is treated with chemicals for the avoidance of scale deposits and after deaeration, for the removal of dissolved oxygen and carbon dioxide to avoid corrosion, it is introduced to the heat recovery section tubes where it is heated from the heat emitted from the condensing steam. Finally, coming out from the heat recovery section, it is heated up to the required maximum temperature T_{max}, from steam produced by a nearby boiler. Before entering the first flash stage, it passes through a flow control valve which helps to keep the right pressure so as not to have seawater boiling. Subsequently the seawater feed enters the first stage where it is flashed forming steam. As the water is flashed, the brine is cooled down and passing through a pressure control system and enters the next stage which is at a lower pressure than the previous one. This process is repeated sequentially up to the final stage, that is, the rejection stage, which is the colder stage. The water quantity which is flashed at each stage is proportional to the enthalpy of the saline mixture or to the temperature difference between the brine and the distillate.

The steam formed condenses to the external surface of the tubes of the heat exchangers from the feed seawater which is heated little by little as it passes from stage to stage. Steam requires some residence time in each stage so as to achieve equilibrium conditions. For a given flow rate, the residence time depends on the length of the stage (chamber), the interface geometry of the two phases, and the degree of the turbulence, that is, generally the residence time depends on the size of the flash chamber.

Each evaporation stage consists of a housing shell which contains the brine, the produced steam and tubes bundle where the steam is condensed while heating the feed seawater circulating inside the tubes. The arrangement of the tubes bundle can be cross-flow or parallel to the brine flow as it is shown in Fig. 6.15.

The number of stages in the MSF distillation system is limited by the pressure difference between the stages.

6.6.2.1 Thermal Analysis of an MSF Distillation System

The desalination systems of MSF require detailed analysis of heat and mass balance, whereas for the design the financial factor has to be counted as well. The balance includes the temperatures, pressure, and the heat profile of each stage of the system. Generally, for the MSF distillation method, regarding the handling of the heat transported, it can be considered that it consists of three sections:

1. The heat input section for the feed of external steam, from a boiler, usually saturated for the corresponding temperature
2. The heat recovery section where the evaporation heat is recovered in the condensers of the various stages
3. The heat rejection section which retains the thermodynamic process, by reducing the temperature and the pressure. It is located at the last effect of the system.

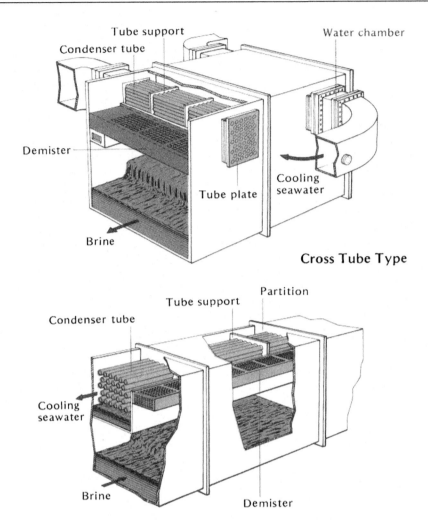

Figure 6.15 MSF stage system: (A) cross-flow tube bundles and (B) long tube bundles, parallel to flow (Mitsubishi Heavy Industries pamphlet).

The transfer of heat is estimated from the well-known heat transferred equation taking into account the mean logarithmic temperature difference:

$$\dot{Q} = hA\,\Delta T_{\mathrm{m}} = hA\frac{T_{\mathrm{f}} - T_{\mathrm{br}}}{\ln[(T_{\mathrm{cv}} - T_{\mathrm{br}})/(T_{\mathrm{cv}} - T_{\mathrm{f}})]} \tag{6.7}$$

where T_{f} is the feed water temperature of the brine which recirculates at its exit from the condenser from stage 1, T_{br} is the brine temperature which recirculates and enters the condenser of the heat recovery section, T_{cv} is the temperature of the condensing steam, and h is the convection heat transfer coefficient.

The overall heat transfer coefficient can be determined by summing up the thermal resistances between the condensing steam and the cooled brine stream. The temperature differences between flashed brine and the condensing steam are difficult to be determined because of the boiling point elevation, and for this reason at each stage, a reduction of pressure equivalent to the raise of the boiling point elevation becomes accepted.

The minimum temperature is related with the maximum temperature T_{max}, and the lower temperature $T_T = T_{min}$, of the system, where T_T is the temperature of the seawater feed. The following relation theoretically applies (Khan, 1986):

$$\dot{Q}_{min} = W_{min}\left(\frac{T_{max}}{T_{max} - T_T}\right) \qquad (6.8)$$

where W_{min} is the minimum required theoretical work for the desalination process. Practically is the preferable to utilize the brine rejection temperature as it is higher than the temperature of the seawater feed $T_{rej} > T_T$, because with the brine rejection there is a small reduction of the brine sensible heat.

The minimum work required for the separation depends on the salt content of the water and is given by:

$$-W_{min} = \int_{c_a}^{c_b} RT \ln(p/p_o) dV \qquad (6.9)$$

where p is the partial pressure of steam of the brine water and p_o is the partial pressure of the clean water at the temperature of the mixture. dV is the change of volume in the system whereas c_a, c_b are the respective concentrations of the salts.

The performance ratio (PR) is related to the number of stages as the number of heat recovery and heat rejection stages is divided by the number of heat rejection stages which are represented by the corresponding stage temperatures (Fig. 6.16):

$$PR = \frac{(T_{max} - T_T)}{(T_n - T_T)} \qquad (6.10)$$

where T_n is the exit temperature from the heat recovery section as it enters the heat rejection section and T_T is the temperature of the cold seawater feed as it is taken from the sea.

Recirculation can be applied as long as the concentrations of salts at the rejection section which form scale deposits have not reached a critical point. In the operation mode with recirculation, the concentration of the hotter stage of the system is much

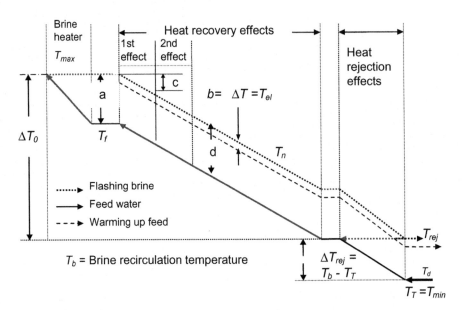

Figure 6.16 Temperature profile of a typical MSF system with heat recovery stages and brine recirculation.

greater than the concentration of the seawater. The recirculation mass flow rate of the brine $\dot{m}_{rec}$ is found from the difference of the maximum temperature and that of the desalinated water T_d.

$$\dot{m}_{rec} = \frac{h_d}{c_p(T_{max} - T_d)} \dot{m}_d \qquad (6.11)$$

where c_p is the specific heat of the recirculated brine, h_d is the enthalpy of the desalinated water, and $\dot{m}_d$ is the flow rate of the produced desalinated water.

6.6.3 Mass and Energy Balance

In a typical MSF system with brine recirculation, the thermal losses are assumed to be negligible, as they are not more than 1–2% of the input thermal energy and for routine calculations they can be omitted. For the estimation of the total heat loss from all stages, the following equation can be used (Darwish, eolss):

$$\dot{Q}_{Tloss} = \dot{m}_{rec}c_p \sum_{i=1}^{n}(1 - n_i)\Delta T_i = \dot{m}_{rec}c_p(T_{max} - T_T)(1 - \eta_m) \qquad (6.12)$$

where η_m is the mean performance of all stages and $\dot{m}_{rec}$ (kg s^{-1}) is the recirculated brine flow rate.

6.6.3.1 Heat and Mass Balance at the Water Cooling Section

The flow rate of the required cooling water $\dot{m}_{col}$, is determined from the fraction of the rejected mass of seawater and the quantity of brine which recirculates, determined from the concentration range in salts of the rejected brine:

$$\dot{m}_{col} = \frac{\dot{Q} - \dot{m}_{rej}c_p(T_{rej} - T_f) - \dot{m}_d c_p(T_{rej} - T_T)}{c_p(T_b - T_T)} \tag{6.13}$$

where $\dot{Q}$ is the energy entering the system from the heating steam and $\dot{m}_{rej}$ is the flow rate of the rejected cooling water.

The temperatures in the figure are:

a   Temperature difference between the entry and exit of the brine heater, $\Delta T_h = T_{max} - T_f$
b ΔT = Temperature difference between the flashed brine and desalinated water $\approx T_{el}$
c ΔT_{stg} = Temperature difference between the entry in a stage and the entry to the next stage
d ΔT_k = Temperature difference between flashed brine—seawater

The quantity of the recirculated brine is estimated from:

$$\dot{m}_{rec} + \dot{m}_{sw} + \dot{m}_d = \dot{m}_{rec} + \dot{m}_{rej} \tag{6.14}$$

The thermal balance in the various sections is the following.
Thermal balance in the recirculation section:

$$Q_{col} = (\dot{m}_f + \dot{m}_{rej})c_p(T_{max} - T_d) \tag{6.15}$$

Thermal balance in the heat recovery section:

$$Q_R = \dot{m}_d h_d = (\dot{m}_f + \dot{m}_{rej})c_p(T_{max} - T_d) \tag{6.16}$$

Thermal balance in the brine heater section:

$$Q_h = \dot{m}_{st}h_{ev-st}(\dot{m}_f + \dot{m}_{rej})c_p(T_{max} - T_d) \tag{6.17}$$

Whole system thermal balance:

$$\dot{m}_{st}h_{ev-st} = \dot{m}_f(T_{max} - T_f) = \dot{m}_f c_p(T_n - T_f) \tag{6.18}$$

where m_{st} is the heating steam mass flow rate and h_{ev-st} is the heating steam evaporation enthalpy.

6.6.3.2 Heat Transfer Coefficients

In bibliography there are many equation types for the heat transfer coefficients, which usually are used for the estimation of the heat transfer areas. Henning and Wagnik (1995) compared the data from bibliography, for the MSF systems, both for the water side and for the steam side at the condensers. This search was performed mainly for large capacity MSF systems. The calculations showed that the bibliography data give significant differences for the heat transfer area and thus differences in costs. Additionally, they proposed original equations for the estimation of the heat transfer in the condenser pipes of the MSF distillers.

The heat transfer coefficient for the water circulating inside the tubes in the tube bundle for turbulent flow is given by:

$$h_{sw} = 0.0267 \frac{\lambda_{sw}}{d_{in}} Re^{0.8} Pr^{0.4} \qquad (6.19)$$

The Prandl number is determined from the mean physical properties of the seawater. At the outer surface of the tubes wall, the coefficient for the condensing steam is estimated from the Nusselt number for horizontal pipes as:

$$h_{con} = 0.843k \left[\frac{\lambda_{con}^3 \rho_{con}^2 h_{ev}}{d_{out} \eta_{con} \Delta T x} \right]^{0.25} \qquad (6.20)$$

where

d_{in}, d_{out}	Internal and external tube diameter	m
ΔT	Temperature difference between the seawater and the condensing steam	K
h_{ev}	Evaporation enthalpy = condensation enthalpy	$J\,kg^{-1}$
η_{con}	Condensate dynamic viscosity	Pas
$\lambda_{sw}, \lambda_{con}$	Conduction heat transfer coefficient of water and condensate, respectively	$W\,m^{-1}\,K^{-1}$
ρ_{con}	Condensate density	$kg\,m^{-3}$
x	Number of pipes in the bundle	—

Symbol k in Eq. (6.20) is the "turbulence factor," which is a correction factor that determines the deviation from the turbulent flow conditions. For the laminar flow, which happens usually during condensation where a condensate film is formed, factor k has a smallest value. For condensation with the formation of drops (dropwise condensation), the heat transfer coefficient has a large value, while for a pure dropwise condensation factor $k = 1$. In practice the usual case is a combined formation of film/dropwise condensation. The factor k is a function of the number of tubes and the arrangement of the tube bundle. It is estimated from the following empirical equation:

$$k = 1.238 + 0.3538x + 0.001570x^2 \qquad (6.21)$$

Eq. (6.21) is applied for steam which condenses and is free from dissolved gases, as gases generally significantly reduce the heat transfer coefficient. Eqs. (6.20) and (6.21) are accurate and give the optimum heat transfer coefficients for bundles from 10 to 16 tubes.

6.6.3.3 Temperature Ranges

The temperature difference between the maximum T_{max} of the feed water entering the first stage and the temperature of the seawater feed entering the brine heater is called "terminal temperature difference" and is a function of the minimum temperature difference and the number of stages:

$$(T_{max} - T_f) = [(T_{stg} - T_{el}) + (T_{max} - T_f)]/n \qquad (6.22)$$

In the tube bundle, the temperature differences are very small and for the estimations of the heat transfer of this section, the mean temperatures are used:

$$T_m = (T_{max} - T_T - T_{el}) - \left(\frac{2n + 1}{2n}\right)\left(\frac{\dot{m}_d}{\dot{m}_f + \dot{m}_{rec}}\right)\frac{h_f}{c_p} \qquad (6.23)$$

In the brine heater, the temperature differences are large and the mean logarithmic values are used:

$$\Delta T_m = \frac{T_{max} - T_f}{\ln[(T_{rej} - T_f)/(T_{rej} - T_{max})]} \qquad (6.24)$$

Or can be given as a function of the seawater flow rate as:

$$\Delta T_m = \frac{(T_{max} - T_T) - [\dot{m}_d/(\dot{m}_f + \dot{m}_{rec})](h_f/c_p)}{\ln\{(T_{st} - T_T) - [\dot{m}_d/(\dot{m}_f + \dot{m}_{rec})](h_f/c_p)\}/(T_{st} - T_{max})} \qquad (6.25)$$

The necessary heater and condenser area of the first stage is estimated from the following equations, respectively (Darwish, eolss):

$$A_h = \frac{\dot{m}_{rst}h_{ev}}{U_{h-br}(T_{sr} - T_{max})}\ln\frac{T_{st} - T_f}{T_{st} - T_{max}} \qquad (6.26)$$

$$A_1 = \frac{\dot{m}_f c_p}{U_c}\ln\frac{T_{v1} - T_2}{T_{v2} - T_1} \qquad (6.27)$$

where T_1, T_2 are the initial and final temperatures (temperature increase $\Delta T = T_2 - T_1$) of the water feed to the condenser tubes because of the supplied heat from the condensing steam on the tube outside area of the condenser of stage 1.

The condensation factors in the brine heater c_{h-bw} and in the brine rejection section c_{bd} are, respectively:

$$c_{h-bw} = \frac{(\dot{m}_{rec}/\dot{m}_d) - 1}{\dot{m}_{ev}/\dot{m}_d}c_{bd} \qquad (6.28)$$

$$c_{bd} = \frac{\dot{m}_f/\dot{m}_d}{(\dot{m}_f/\dot{m}_d) - 1}c_{h-bw} \qquad (6.29)$$

where

$\dot{m}_{bd}$	Rejected brine mass flow rate	kg s^{-1}
$\dot{m}_d$	Desalinated water mass flow rate	kg s^{-1}
$\dot{m}_f$	Seawater feed mass flow rate	kg s^{-1}
$\dot{m}_{rec}$	Recirculate mass flow rate	kg s^{-1}
$c_{h\text{-}br}$	Condensation factor of seawater in the heater	kg s^{-1}
c_{bd}	Condensation factor of brine in the brine rejection	kg s^{-1}
U_{av},	Mean overall heat transfer coefficient and the brine overall heat	W m^{-2} K^{-1}
$U_{h\text{-}br}$	transfer coefficient, respectively	

Darwish (eolss) gives a detailed analysis of the MSF systems, for one, two, and many stages of systems as well as for one pass or with brine recirculation systems. For the recirculation conditions, he gave the following performance equations:

$$\text{PR} = \frac{\dot{m}_d}{\dot{Q}/2330} = \frac{2330\dot{m}_d}{\dot{m}_{rec}c_p(T_{max} - T_f)/\eta_{st}} \tag{6.30}$$

$$\text{GOR} = \frac{\dot{m}_s}{\dot{m}_{st}} = \frac{\dot{m}_d}{\dot{Q}/h_{st}} = \frac{h_{st}\dot{m}_d}{\dot{m}_{rec}c_p(T_{max} - T_f)/\eta_{st}} \tag{6.31}$$

$$\frac{\dot{Q}}{\dot{m}_d} = \frac{\dot{m}_{rec}c_p(T_{max} - T_f)/\eta_{st}}{\dot{m}_d} \tag{6.32}$$

where η_{st} is the efficiency of the heating steam. For the brine recirculation conditions, he gave the ratios of the system flow rates as:

$$\frac{\dot{m}_f}{\dot{m}_d} = \frac{(\dot{m}_{rec}/\dot{m}_d) - 1}{(\dot{m}_{rec}/\dot{m}_d)(1 - c_f/c_{br}) - 1} \tag{6.33}$$

$$\frac{\dot{m}_{rec}}{\dot{m}_d} = 0.5 + \frac{h_{ev}}{c_p(T_{max} - T_n)} \tag{6.34}$$

$$\frac{\dot{m}_d}{\dot{m}_{st}} = \frac{\dot{m}_d}{\dot{Q}_{st}/h_{ev\text{-}st}} = \frac{\dot{m}_d h_{ev\text{-}st}}{\dot{m}_{rec}c_p(T_{max} - T_f)/\eta_{st}} \tag{6.35}$$

The mean efficiencies of the heat recover−rejection sections, respectively, are:

$$\eta_{rec} = \frac{1}{n-j}\sum_{i=1}^{n-j} n_i \tag{6.36}$$

$$\eta_j = \frac{1}{j}\sum_{n-j+1}^{n} n_i \tag{6.37}$$

where

h_{ev-st}	Condensate evaporation enthalpy	$J\,kg^{-1}$
i	Stage number of the recovery section	—
$\dot{Q}_{st}$	Heat supplied from the heating steam	$J\,s^{-1}$
j	Number of rejection stages	—
η_{st}	Heating steam efficiency	—
c_f, c_b	Concentration of salts in feed seawater and the maximum brine concentration, respectively	‰

6.6.4 Distillation with Vapor (Re)Compression (MVC−TVC)

In the systems with vapor recompression, the salty feed water is heated from the noncondensed steam which is compressed mechanically (MVC) or thermally (TVC). The resulting vapor during the evaporation of a salty mixture is superheated because of the brine boiling point elevation and has a lower pressure from the clean water saturation pressure. If this vapor is compressed at a higher pressure, its temperature is increased because of the supplementary energy input. By increasing its pressure and temperature at the desired level, it can be used as a source of heat for the evaporation of the salty mixture (brine water or seawater).

Brine, for example, coming from the Mediterranean seawater having double salt content (70‰) boils at 101.05°C at atmospheric pressure and condenses at 100°C. For the heating steam in order to be condensed at 101.05°C has to be compressed up to 104.35 which is at pressure 1.03 kPa higher than the atmospheric.

In a practical application, there are two methods of vapor compression, MVC and the TVC which differ only in the way the vapor is compressed. Distillation with vapor recompression is a technique which demands only auxiliary supply of heat from an external source for its operation. Distillation systems are constructed with horizontal tubes as in this form the maximum overall effective heat transfer coefficient U_{eff} is taken because of the simultaneous boiling and condensation at the heat transfer areas. An MVC system is shown in Fig. 6.17.

6.6.4.1 Mechanical Vapor Compression

In this system of a series of evaporation stages, steam from the last stage is directed to a compressor and compressed mechanically (from an engine, steam turbine, etc.) increasing its pressure from p_1 to p_2 and consequently its temperature. Subsequently, it is fed in the first stage of the evaporator where it is used to heat and evaporate part of the seawater feed while at the same time it loses its superheating and condense. The condensate is mixed with the stream of the condensate coming from the evaporator.

As it is also shown in Fig. 6.18, part of the seawater feed is preheated in a heat exchanger from the brine and the desalinated water from temperature $T_T = T_{min}$ at a

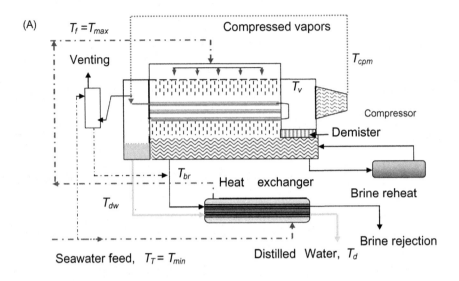

Figure 6.17 (A) Schematic representation of the circulating cycle in a one-stage MVC system. (B) MVC distillation system (IDE Technologies Ltd, Israel) (Lokiec and Ophir, 2007).

temperature $T_f = T_{max}$, the evaporation inlet temperature which is slightly lower than the brine boiling point elevation. The brine is directed to the heat exchanger at a temperature T_{br}, and it is cooled and rejected at a temperature $T_{rej} \approx T_d$.

Steam produced in the evaporator during the cascading of the seawater in the last stage is fed at the saturation temperature $T_{sat} = T_1 - T_{el}$ to the compressor where it is compressed at a temperature T_{com}, increasing its enthalpy. Subsequently it is expanded in the condenser tubes up to the temperature T_{com}, heating at the same time the down-flow seawater externally of the tubes.

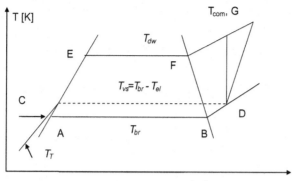

Figure 6.18 T–S for the mechanical compressor of the MVC method.

MVC Mathematical Analysis

The work produced by the compressor is estimated from the following simple expressions which are based on the T–S diagram (Fig. 6.18).

$$W_c = p_1 V_1 \frac{(p_2/p_1)^2 - 1}{2(p_2/p_1)} = R T_{vs} \frac{(p_2/p_1)^2 - 1}{2(p_2/p_1)} \qquad (6.38)$$

where p_1, p_2 is the pressure before and after the compression, T_{vs} is the steam saturation temperature, and V_1 is the volume of the steam before compression. The previous equation can be expressed in terms of the temperature of each cycle as:

$$W_c = c_p T_{vs} \exp[h_v(T_{dw} - T_{vs} + T_{el})]/c_p T_{vs}^2 \qquad (6.39)$$

For typical operating conditions, that is $(T_{dw} - T_{vs}) < 5°C$ and $T_{vs} < 110°C$, the previous equation is simplified as:

$$W_c = h_v(T_{dw} - T_{vs} + T_{el})/T_{vs} \qquad (6.40)$$

where

c_p	Steam specific heat	J kg^{-1} K^{-1}
h_v	Steam enthalpy	J kg^{-1}
T_{dw}	Steam condensing temperature after compression $\approx$ desalinated water temperature	K
T_{vs}	Superheated steam temperature in the evaporator	K

The minimum work required W_{min} is:

$$W_{min} = \dot{m}_d(h_{dw} - h_{vs}) = \dot{m}_d c_p(T_{dw} - T_{vs}) = m_d c_p T_{el} \qquad (6.41)$$

Eq. (6.41) can be applied in typical MVC systems where $(T_{dw} - T_{vs} = T_{el})$ applies and the feed temperature of the water T_f is equal to the brine temperature T_{br} in the evaporator. The parameters h_{dw} and h_{vs} are the enthalpies of steam after the compressor and the superheated steam in the evaporator, respectively.

The heat that the compressed steam must have in order to heat at the appropriate temperature from the feed seawater is given by:

$$\dot{Q}_f = \dot{m}_d(h_{com} - h_{dw}) = \dot{m}_d c_p[(T_{com} - T_{dw}) + h_v] \qquad (6.42)$$

Whereas the heating required for the increase of the feed water temperature in the heat exchanger is:

$$\dot{Q}_{he} = \dot{m}_T c_p(T_f - T_T) = \dot{m}_{br} c_p(T_{br} - T_d) + \dot{m}_d c_p(T_{dw} - T_d) \qquad (6.43)$$

The isentropic process is shown in the (T$-$S) diagram, of the Fig. 6.18. Line AB presents the evaporation of the seawater at a constant temperature. Line CD shows the steam produced in the evaporation chamber at constant temperature. The temperature of the steam is $T_{vs} = T_{br} - T_{el}$. At point D the steam is in equilibrium with the heated brine. The steam at temperature T_{vs} is compressed until its temperature is increased at T_{com} while increases at the same time its enthalpy up to h_{com}. The temperature T_{com} has to increase so as to give a temperature difference in the evaporation chamber $\Delta T = T_{dw} - T_{vs}$ which constitutes the moving power for the evaporation. Point G in the diagram represents the minimum energy required to achieve isentropic compression. Finally, line EF represents the compression of steam under constant temperature whereas the part GFE represents the specific heat required to produce 1 m^3 desalinated water ($Q_{sp}/\dot{m}_d$). The area enclosed between lines CDdFEC gives the specific isentropic work $W_{sp} = \dot{m}_d(h_{com} - h_{vs})$. The performance ratio (PR) for the MVC systems is expressed by the specific consumption of the thermal energy ($W/\dot{m}_d$) which cannot be compared directly with the term of specific consumption of thermal energy as it is used in the other distillation systems (2330 $\dot{m}_d/Q$), because the thermal energy is lower in quality from the mechanical energy and because it does not take into consideration the required auxiliary heat supplied to the system. It is expressed as:

$$PR = \frac{W}{\dot{m}_d} + \dot{Q}_{avl} \qquad (6.44)$$

where $W/\dot{m}_d$ is the total input work per unit of desalinated water and $\dot{Q}_{avl}$ is the available energy of the auxiliary source which is estimated according to the second law of thermodynamics as:

$$\dot{Q}_{avl} = Q_{aux} \frac{T_{aux} - T_a}{T_{aux}} \eta_{isn} \qquad (6.45)$$

where T_{aux} is the temperature at which the auxiliary energy is added, T_a is the environmental temperature, and η_{isn} is the isentropic efficiency of the cycle which converts the auxiliary heat Q_{aux} in equivalent work. El–Sayed (1986) gives the following performance ratio:

$$PR = \frac{h_{ev}\dot{m}_d}{W + \eta_{he}Q_{aux}}\eta_{he} \tag{6.46}$$

where η_{he} is the efficiency of the heat engine which operates the compressor. The maximum efficiency is:

$$\eta_{max} = (T_{dw} - T_T)/T_{dw} \tag{6.47}$$

where T_T is the suction temperature of the seawater which is approximately equal to that of the environment.

Darwish (eolss) gives the gain ratio (GR) as:

$$GR = \frac{1}{(\dot{m}_f/\dot{m}_d)c_p(T_b - T_f) + h_v}\,kg\,J^{-1} \tag{6.48}$$

where m_f is the feed water mass flow rate and T_b is the boiling temperature. The term GOR is not applicable in the MVC method.

6.6.4.2 Thermal Vapor Compression
A thermal TVC distillate system, also called "thermal compression or heat effusion," is similar to an MVC system. Simply the mechanical compressor is replaced by a thermal steam compressor. In the TVC method, a mixture of steam which consists of driving steam produced in a boiler and part of the steam produced in the distiller heats the feed water of the distiller. The steam condenses as saturated water with enthalpy h_{fd}. Part of the condensate $\dot{m}_{dis}$ is introduced in the boiler while the rest $\dot{m}_c$ is mixed with the desalinated water coming from the distiller $\dot{m}_{dis}$. The relation $\dot{m}_{dt} = (\dot{m}_c + \dot{m}_{dis})$ applies.

The rate of steam which is produced in the distiller at a saturation temperature T_{sat} is $\dot{m}_v$. The power steam is introduced at a high speed through an orifice and expands at a low-pressure sucking part from the vapor $\dot{m}_{suc}$ of the last stage. The streams are mixed in a chamber where the flow speed is reduced simultaneously and the appropriate pressure and temperature is obtained for the induction of the hot steam mixture in the first stage as a heating medium of the system. The rest of the steam $\dot{m}_v - \dot{m}_{suc} = \dot{m}_d$ is condensed as desalinated water. The relation $\dot{m}_v = \dot{m}_{suc} + \dot{m}_d$ applies.

The whole circuit of the circulating fluids is shown schematically in Fig. 6.19 for a distillation system of two stages.

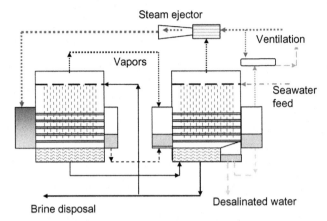

Figure 6.19 Two-stage distillation system with thermal vapor compression.

The energy balance at the steam nozzle is:

$$\dot{m}_{st}h_{st} + \dot{m}_{suc}h_{suc} = (\dot{m}_{st} + \dot{m}_{suc})h_{mix} \tag{6.49}$$

The ratio of the quantity of the suction vapor to the hot driving steam $\dot{m}_{suc}/\dot{m}_{st}$ mainly depends on the pressure of the driving steam, the suction pressure at the distiller, and the outlet pressure from the nozzle.

The steam mixture $(\dot{m}_{st} + \dot{m}_{suc})$ warms the feed seawater $\dot{m}_{sw}$ from the temperature T_{sw} at the boiling temperature T_{bo} and simultaneously evaporates part of the feed water with flow $\dot{m}_v$. The thermal balance is:

$$(\dot{m}_{st} + \dot{m}_{suc})(h_{mix} - h_{fd}) = \dot{m}_{sw}c_p(T_{bo} - T_{sw}) + \dot{m}_{suc}h_{ev} \tag{6.50}$$

The quantity of the feed water is estimated from the stream concentrations of the circulation cycle:

$$\frac{\dot{m}_{sw}}{\dot{m}_d} = \frac{x_{max}}{x_{max} - x_{sw}} \tag{6.51}$$

where x_{max} is the maximum allowable concentration of salts, which is so much so as not to form scaling.

The TVC systems have capacities in the range of $45-1500 \text{ m}^3 \text{ day}^{-1}$; they have however low performance and are usually combined with the MSF or MED distillation system, as hybrid systems.

6.7 DUAL-PURPOSE PLANTS

The dual-purpose or "cogeneration" plants are desalination MSF or MED systems or combinations of hybrid distillation systems, for example, MED/TVC or even distillation/RO, which are connected with electricity power stations. In this combination, the cost is shared in both systems which take advantage of the energy that would be rejected if the two systems were operated separately.

The principle of this combination is the use of high-temperature steam for the production of the electrical power and the low-pressure steam, which is coming from the turbine, as the heating medium for the desalination system, while desalinated water is supplied to the boiler of the electricity power plant, for the production of steam. From the thermodynamic point of view, in order for the dual-purpose plant to be effective, the ratio of the water demand to electricity demand has to be within certain acceptable limits.

To achieve the optimum performance result of this combination, it is important, as El-Nashar (2001a) mentions, that the size and the characteristics of each system to be chosen so as the demand curves to have the optimum cost. It has to be taken into account that both the water and the electricity have daily and seasonal demand variations.

The energy consumption per unit of the produced electricity and produced desalinated water is a measure of the operation fluctuating cost. In Fig. 6.20, the consumption of energy of the individual plants for the production of electricity and desalinated water and the saving in energy from the combination of their operation are shown.

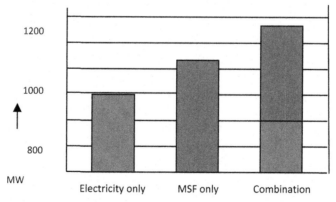

Figure 6.20 Reduction of the required energy for the operation of a combined dual-purpose plant for the production of large capacity electricity/desalination, as to the energy consumption of individual installations (Somariva, 2004).

The advantages of the dual-purpose plants are:

- The maximum efficiency is obtained when both plants operate close to their designated productivities
- There is a considerable saving in fuels
- Per unit of produced steam a bigger boiler can be used which usually has a lower cost (economy of scale)
- The produced desalinated water has a lower cost from the desalination system of the same capacity which operated independently.

The ratio of "power to water ratio = PWR" has particular importance and depends on the type of installation, and to be precise, from the type of turbine used in the power production system as well as from the performance ratio (PR) of the desalination system. The bigger the distillation system performance ratio, so small will be the PWR ratio and vice versa. The return ratio, power/distillate index (ie, produced power to produced distillate quantity) is (Kalamaddin et al., 1993):

$$x = MW/1000 \text{ m}^{-3} \text{ day}^{-1} \tag{6.52}$$

For typical dual-purpose plants, the ratio x varies from 0.1 to 1.0, however a mean value is taken as $x < 0.5$.

According to one methodology, the effectiveness criterion for the selection of the optimum combination of water to power demand is the comparison of the required fuel for the achievement of specific electricity loads and the driving heat of the desalination system. For the combined system and for the production of gross energy P (kW) and quantity of heat Q_p with fuel consumption $(Q_f)_{cog}$, the conserved energy is given from El-Nashar (2001b) as:

$$\Delta Q_f = \frac{Q_p}{\eta_{bl}} + \frac{P}{\eta_c} - (Q_f)_{cog} \tag{6.53}$$

The net heat rate (NHR) (MW$_{th}$/MW$_e$) is given from:

$$\text{NHR} = [(Q_f - Q_P/\eta_{bl})/P] \tag{6.54}$$

The fuel energy saving ratio (FESR) is given as:

$$\text{FESR} = \frac{\Delta Q_f}{(Q_P/\eta_{bl} + P/\eta_c)} = 1 - \frac{\eta_c/\eta_{cog}}{[1 + (\lambda_{cog}\eta_c/\eta_{bl})]} \tag{6.55}$$

where

ΔQ_f	Fuel energy saved	W
η_{bl}	Boiler efficiency	—
η_c	Efficiency of a conventional electrical power station	—
η_{cog}	Efficiency of the dual system = $P/(Q_f)_{cog}$	—
λ_{cog}	Ratio of used energy to electrical power produced	MW$_{th}$/MW$_e$

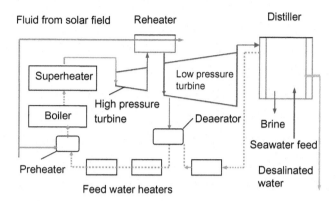

Figure 6.21 Simple schematic diagram of a solar power and desalination cogeneration system (Darwish and Darwish, 2014).

P	Net overall electricity return	kW
Q_f	Fuel heat	W
$(Q_f)_{cog}$	Fuel saved	W
Q_p	Usage heat	W

In Fig. 6.21, a schematic diagram of the flow of fluids and heat for a dual-purpose plant are shown. The plant is a desalination/power producing system which is powered from solar energy by a concentrating collector field (Darwish and Darwish, 2014).

6.8 SOLAR DESALINATION COMBINATIONS

One of the first solar systems was installed in Abu Dhabi (Fig. 6.22) which is a multiple effect stack (MES) distillation system with evacuated tube solar collector system with a designed production of $120 \, m^3 \, day^{-1}$ and typically $85 \, m^3 \, day^{-1}$ (El-Nashar, 1993, 2000a,b, 2001a) while some other combinations have been studied which concern designs of mainly pilot installations. García-Rodríguez et al. (2002) report that typical temperature for solar distillation MED is 65–75°C whereas for MSF distillation system it is 80–90°C. For these systems, flat-plate collectors are suitable, as well as evacuated tube collectors, PTC, compound parabolic collectors, and linear Fresnel mirrors with better combination being the parabolic troughs.

From the four main industrialized desalination methods, two belong to the thermal (MED and MSF) and two (MVC and RO) to the mechanical. The two thermal methods have high application reliability and are fully industrialized.

Figure 6.22 The MES installation and the evacuated tube solar collector field installed in Abu Dhabi (El-Nashar, eolss).

The two mechanical methods have medium class reliability but they are also matured and fully industrialized. The MED distillation method is more effective from the MSF distillation with respect to the primary operation energy and the consumption of electrical power has a smaller cost and lower operating temperatures. It is therefore suitable for combination with solar concentrating systems as with parabolic troughs or linear Fresnel collectors. Lately, the MED distillation system is combined either with MVC or with RO for better performance of the hybrid system of both desalination methods.

The combination of PTC with an MED distillation constitutes a system of high reliability with simple operation. The mean concentration of the parabolic trough is $C = 15-40$ and the supplied temperature is $\sim 380°C$. They rotate around a single axis and use as heat transfer fluid:

1. Synthetic oils, which however restrict the higher supplied temperature
2. Water, so as to produce steam directly in the absorbing tube of the collector, with temperatures reaching up to 400°C. García-Rodríguez and Gómez-Camacho (1999, 2001) and García-Rodríguez, 2003 proposed the use as a heat transfer fluid, water, seawater, or brine, and they performed by economic evaluation for each of the fluids.

The direct steam generation PTC has many advantages against those using thermal oil. The MED systems are more flexible than the MSF ones, can operate at part load, and have lower possibilities of scale deposits. Generally, they are considered more suitable than the MSF for relatively small capacity installations, whereas the desalination systems with TVC have lower performance from both MED and MSF systems (García-Rodríguez et al., 1999). Usually they are combined with the MED or MSF distillation systems.

The connection of the two systems is achieved with auxiliary devises whereas for the systems PTC/MED and PTC/MSF a storage tank and a conventional boiler are required. The storage tank uses oil as the operating fluid. In this case the input—output temperature difference to and from the solar collector field is $\sim 80°C$.

Darwish and Darwish (2014) report that although RO is the more efficient method and has a lower desalinated water production cost, there is still interest for the thermal methods MSF, MED, and TVC, either as single units or hybrid combinations, for connections with solar energy collection systems. The most efficient from those distillation methods is the LT-MED, which consumes approximately the same thermal energy as the MSF. The required solar energy is $250-300$ MJ m^{-3} and ~ 4 kWh m^{-3} for powering the pumps. The MED requires $1.5-2.0$ kWh m^{-3} for the circulation of the various fluids. Additionally, the operational limits presented in Table 6.2 apply.

There is a large number of publications which refer to the conventional desalination powered from solar energy and/or renewable sources of energy. A few of these papers are referred here in a random selection. Ophir and Nadav (1982) give a review of the systems used for the production of electrical power and desalinated water from solar energy. El-Nashar (1985, 2000b) described the optimization of solar desalination installations, whereas Kalogirou (1997, 2001) presented a study on the influence of the cost on the price of the desalinated water which is produced with renewable energy sources. Trieb and Müller-Steinhagen (2008) refer to the use of concentrating solar energy for the production of desalinated water in Middle East and North Africa. Ali et al. (2011) presented a techno-economic review of the indirect solar desalination for MSF and MED installations as well as for the RO and MD. Sagie et al. (2005) analyzed the commercial solar energy systems and the commercial desalination systems which can be connected with solar energy and they estimated the economics of these systems. Among the latest research papers are those of Darwish (2014) and Darwish and Darwish (2014) which refer to the combination of the solar desalination, with the use of auxiliary energy from natural gas, for an installation of power/desalinated water in Qatar (UAE).

Fig. 6.23 shows the combinations of desalination/electric power systems, according to the type of turbine in the electricity power production section.

In Fig. 6.22 one of the first installations of indirect production of desalinated water with solar energy is shown. It concerns the installation in Abu Dhabi

Table 6.2 Operational Limits for the Various Thermal Desalination Methods

Distillation Method	Heating Steam Pressure (bar)	Maximum Temperature (°C)
MSF	2–3	111–115
LT-MED	2–3	80
TVC	3–10	70

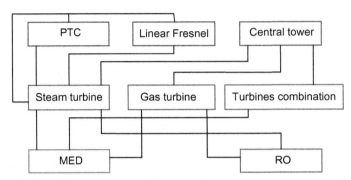

Figure 6.23 Possibility of the most suitable combinations of solar energy with desalination systems.

Figure 6.24 The hybrid desalination system "AQUASOL" with solar system, at Plataforma solar de Almeria, Spain (Alarcón et al., 2005).

(El-Nashar, 1993, 2000a,b, 2001a,b). It consists of a field of solar evacuated tube collectors and a multistage stack distillation system (MES). The evacuated tube collector field has a total surface area of 1862 m². The system is designed for 120 m³ day⁻¹ but operated at a little lower production. The seawater feed had salt concentration equal to 55‰ and evaporator inlet temperature 135°C. El-Nashar (2001a, 2003) reported that on the evacuated tubes dust and sand deposited and these reduced the efficiency of the system and that part of the produced water was used for the cleaning of the collector surface, thus reducing the available production.

Various combined types of conventional desalination/solar energy have been constructed, as the AQUASOL system shown in Fig. 6.24 (Alarcón et al., 2005). Most of these installations are of pilot type for the study of the operating conditions, whereas lately studies have been performed for installation in arid areas but they have not yet erected so as to operate. In Table 6.3 some pilot installations of conventional desalination methods are indicatively given with some operation data, whereas in Table 6.4 also indicatively some of the latest installations of H/D and MD are given which are powered indirectly with solar energy.

Table 6.3 Desalination Installations Indirectly Powered From Solar Energy (MSF, MED)

Location	Type of Desalinator	Desalinated Water (m³ day⁻¹)	Solar Collector Type
Al-Ain (UAE)[a]	20-Stage MSF	600	Flat-plate collectors
	55-Eff. MED	500	
Al-Azhar University, Gaza[a]	MSF	10	Flat-plate collectors
Area of Hzag, Tunisia[a]	Distillation	0.2	Flat-plate collectors + PV
Arabian Gulf	MED	6000	Parabolic troughs
Gran Canaria, Spain[a]	MSF	10	Low concentration
Kuwait[b]	MSF	100	Parabolic troughs
La Paz, Baja California, Mexico[c]	10-Stage MSF	10	Parabolic troughs + flat-plate collectors
La Desiré Island, French Caribbean[a]	14-Eff. MED	40	Evacuated tube collectors
PSA, Almeria, Spain[d]	MED heat pump	72	Compound parabolic collectors
Takahami Island, Japan[e]	16-Effect MES	16	Flat-plate collectors
Sulaibiya, Kuwait[f]	RO + MSF	20/25	Point focus collectors
Safat, Kuwait[a]	MSF	10	Flat-plate collectors
Um-Al-Nar, Abu Dhabi[g]	18-Eff. MES	120	Evacuated tube collectors

[a]García-Rodríguez (2007a,b)
[b]Kriesi (1984)
[c]Scholle and Schubert (1980)
[d]Alarcón et al. (2005)
[e]Delyannis (1987)
[f]Moustafa et al. (1984), Moustafa et al. (1985a), Moustafa et al. (1985b)
[g]El-Nashar (1985, 2001a)

Table 6.4 Indirect Solar-Powered Desalination Installations (H/D, MD)

Location	Type of Desalinator	Desalinated Water (m³ day⁻¹)	Solar Collector Type
Arid area, Quom, Iran[a]	2-Stage H/D	580 L day⁻¹	
Sfax, Tunisia[b]	Multiple effect humidification (MEH)	180 L day⁻¹	Original flat-plate collectors
INRST (National institute of Scientific and Technological Research), Tunisia (see Fig. 5.12)[c]	4-Stage MEH	355 kg day⁻¹	Original flat-plate collectors
Al-Hail, Muscat, Oman[d]	MEH	180 L day⁻¹	Flat-plate collectors
Red Sea, Irbid[e]	Autonomous Air gap membrane distillation (AGMD)	120 L day⁻¹	Flat-plate collectors + PV

[a]Zamen et al. (2014)
[b]Mueller-Holst and Mueller (2005)
[c]Houcine et al. (2006)
[d]Mueller-Holst and Mueller (2005)
[e]Banat et al. (2007)

REFERENCES

Abdul Fattah, A.F., 1986. Selection of a solar desalination system for supply water in arid zones. Desalination. 60, 165—189.

Alarcón, D., Blanco, J., Malato, S., Maldonato, M.I., Fernández, P., 2005. Design and set up of an hybrid solar seawater desalination system. The AQUASOL Project. ISES Meeting, Orlando, FL, Paper No. 1381.

Ali, M.T., Fath, H.S., Armstrong, P.R., 2011. A comprehensive techno-economic review of indirect solar desalination. Renew. Sust. Energ. Rev. 15 (8), 4187—4199.

Anonymous, 2008. CSP power tower. Renew. Energ. Focus. May/June, 52.

Banat, F., Jwaied, N., Rommel, M., Koschikowski, J., Wieghaus, M., 2007. Desalination by a "compact SMADES" autonomous solarpowered membrane distillation unit. Desalination. 217, 29—37.

Belessiotis, V., Delyannis, E., 2006. The history of renewable energies for water desalination. Desalination. 128, 147—159.

Darwish, A., 2011. Prospects of using energy for power and desalted water production in Kuwait. Desal. Water Treat. 36, 219—238.

Darwish, A., 2014. Thermal desalination in GCC and possible development. Desalin. Water Treat. 52, 27—47.

Darwish, M.A., 1987. On the thermodynamics of dual purpose power-desalting plants. Part 1: Using steam turbines. Desalination. 64, 151—167.

Darwish, M.A., eolss. MSF Engineering and Fundamentals of Multiple-Effect-Evaporation. Available from: http://www.desware.net/Sample-Chapters/D04/D08-048.pdf.

Darwish, M.A., Darwish, A., 2014. Solar cogeneration power-desalting plat with assisted fuel. Desal. Water Treat. 52, 9—26.

Delyannis, E., 1987. Status of solar assisted desalination. Desalination. 67, 3—19.

Delyannis, E., 2003. Historic background of desalination and renewable energies. Solar Energ. 75, 357—366.

Duffie, J.A., Beckmann, W.A., 2006, Second Edition 2013. Solar Engineering Thermal Processes. John Wiley & Sons, Hoboken, NJ, p. 908.

El-Nashar, A.M., 1985. Abu-Dhabi solar distillation plant. Desalination. 52, 217—234.

El-Nashar, A.M., 1993. An optimal design of a solar desalination plant. Desalination. 93, 597—614.

El-Nashar, A.M., 2000a. Predicting part load performance of small MED, evaporators—a simple simulation program and its experimental verification. Desalination. 130, 217—234.

El-Nashar, A.M., 2000b. Economics of a small solar-assisted multiple-effect stack distillation plant. Desalination. 130, 201—215.

El-Nashar, A.M., 2001a. Water from the sun: case study, The Abu-Dhabi solar desalination plant. Re-Focus.26—29, http://www.re-focus.net/m.

El-Nashar, A.M., 2001b. Cogeneration for power and desalination—state-of-the-art. Desalination. 134, 7—28.

El-Nashar, A.M., 2003. Effect of dust on performance of a solar desalination plant operating in Arid Desert Area. Solar Energy. 75, 421—431.

El-Nashar, A.M., (eolss). Multiple effect distillation of seawater using solar energy—the case of Abu-Dhabi solar desalination plant (eols). http://www.eolss.net/sample-chapters/c08/e6-106-30.pdf.

El-Sayed, M.M., 1986. Rational bases for designing vapor compression water desalination systems. Desalination. 58 (No 3), 213—226.

García-Rodríguez, L., 2003. Renewable energies in desalination: state-of-the-art. Solar Energ. 75 (5), 381—393.

García-Rodríguez, L., 2002. Seawater desalination driven by renewable energies: a review. Desalination. 143, 103—113.

García-Rodríguez, L., 2007a. Solar powered Rankine cycles for fresh water production. Desalination. 212, 319—327.

García-Rodríquez, L., 2007b. Assessment of most promising developments in solar desalination. In: Rizzuti, L., et al., (Eds.), Solar Desalination for 21st Century. Springer-Verlag, Berlin, pp. 355—369.

García-Rodríguez, L., Gómez-Camacho, C., 1999. Preliminary design and cost analysis of a solar distillation system. Desalination. 126, 109–114.

García-Rodriguez, L., Gómez-Vamacho, C., 2001. Perspectives of solar assisted seawater desalination. Desalination. 136, 213–218.

García-Rodríguez, L., Palmero-Marrero Ana, I., Gómez-Camacho, C., 1999. Application of direct steam generation in a solar parabolic trough collector to a multiple-effect distillation. Desalination. 125, 139–145.

García-Rodríguez, L., Palmero-Marrer, Ana I., Gómez-Camacho, C., 2002. Comparison of solar thermal technologies for application in seawater desalination. Desalination. 142, 135–142.

Ghaffur, N., Missimer, T.M., Amy, G.L., 2013. Technical review and evaluation of the economics of water desalination: current and future challenges for better water supply sustainability. Desalination. 309, 197–207.

Henning, S., Wagnik, K., 1995. Comparison of different equations for calculation of heat transfer coefficients in MSF evaporators. In: International Desalination Association (IDA) (Ed.), Proceedings of the IDA World Congress on "Desalination and Water Sciences", vol. III. Abu-Dhabi Printing and Publishing Co., Abu-Dhabi, pp. 515–524.

Houcine, I., BenAmara, M., Guizani, A., Maâlej, M., 2006. Pilot plant testing of a new solar desalination process by multiple-effect-humidification technique. Desalination. 196, 105–124.

Kalamaddin, B.A., Khan, S., Ahmed, B.M., 1993. Selection of optimally matched cogeneration plants. Desalination. 93, 311–321.

Kalogirou, S.A., 1997. Economic analysis of a solar assisted desalination system. Renew. Energ. 12 (4), 351–367.

Kalogirou, S.A., 2001. Effect of fuel cost on the price of desalination water: case for renewables. Desalination. 138, 137–144.

Kalogirou, S.A., 2004. Solar thermal collectors and application. Progr. Energ. Combust. 30, 231–295.

Kalogirou, S.A., 2009. Solar Energy Engineering: Process and Systems. first ed. Elsevier Inc, London 760 pp.; second ed., 2013, 860 pp.

Khan, A.H., 1986. Desalination Processes and Multistage Distillation Practice. Elsevier, Amsterdam, 596 pp.

Kriesi, R., 1984. Experience with solar powered $10 \, \text{m}^3 \, \text{d}^{-1}$ in Kuwait and results of up scaling experiments. Desalination. 45, 371–380.

Kronenberg, G., 1995. The LT-MED process for combined power generation and seawater desalination. Proceedings of the IDA World Congress on "Desalination and Water Sciences", vol. III, Abu-Dhabi, pp. 459–469.

Lokiec, F., Ophir, A., 2007. The Mechanical Vapor Compression: 38 Years of Experience, IDA World Congress—Maspalomas, Gran Canaria—Spain, REF: IDAWC/MP07-084.

Liu, Y., Guo, Y., Wei, Q., 2013. Analysis and evaluation of various energy technologies in seawater desalination. Desal. Water Treat. 51 (19–21), 3743–3753.

Moustafa, S., Hoeller, W., El-Mansy, H., Kamal, A., Jarrar, D., 1984. Design, specifications and application of a $100 \, \text{kW}_e$ ($700 \, \text{kW}_{th}$) cogeneration power plant. Solar Energ. 32, 263–269.

Moustafa, S.M.A., El-Mansy, H., Elimam, A., Zewen, H., 1985a. Operational strategies for Kuwait's $100 \, \text{kW}_e/0.7 \, \text{MW}_{th}$ solar power plant. Solar Energ. 34, 231–238.

Moustafa, S.M.A., Jarrar, D.I., El-Mansy, H.I., 1985b. Performance of a self regulating solar multistage flash desalination system. Solar Energ. 35, 333–340.

Mueller-Holst, H., Mueller, M., 2005. Solar thermal desalination for remote area using the multiple-effect humidification (MEH) method—from research to business. ISES Meeting, Orlando, FL, Paper No. 1149.

Ophir, A., Nadav, N., 1982. Solar energy as a source for production of power and desalinated water. Desalination. 40, 103–124.

Sagie, D., Magdenberg, E., Weinberg, J., 2005. Commercial scale solar-powered desalination, ISES Conference, Orlando, FL, Paper No. 1554.

Scholle, L.G., Schubert, K.P., 1980. Planung, bau und installation einer solaren Entsalzungsanlage. Hamburg. 11, 767–776.

Skinrood, A., 1982. Recent developments in central receivers. SUNWORLD. 6 (4), 98–101.

Somariva, C., 2004. Desalination Management and Economics. Mott MacDonald Faverham House Group, Surrey, UK, 80 pp. (booklet printed by Mott-McDonald).

Trieb, F., Müller-Steinhagen, H., 2008. Concentrating solar power for seawater desalination in the Middle East and North Africa. Desalination. 220, 165–183.

Wagnick, K., 2000. Present status of thermal seawater desalination techniques. Desal. Water Reused. 10 (1), 14–21.

Zamen, M., Soufari, S.M., Abbasian Vahdat, S., Amidpour, M., Zeinali, M.A., Izanloo, H., et al., 2014. Experimental investigation of a two-stage solar humidification–dehumidification desalination process. Desalination. 332, 1–6.

APPENDIX A

A.1 GENERAL

Table A.1 SI Base Units (Taylor and Thompson, 2008a,b)

Base Quantity	Symbol	Unit's Name	SI Unit
Length	l	meter	m
Mass	m	kilogram	kg
Time	t	second	s
Temperature, thermodynamic	T	kelvin	K
Electric current	Θ	ampere	A
Luminous intensity		candela	cd
Amount of substance	n	mole	mol

Table A.2 Derived SI Units with Special Names and Symbols (Taylor and Thompson, 2008a,b)

Physical Quantity	Special Symbols	Name	Unit Symbol	Related SI Unit
Capacitance	F	farad	$C\,V^{-1}$	$A^2\,s^4\,kg^{-1}\,m^{-2}$
Electric charge, quantity of electricity	C	coulomb		$A\,s$
Electric conductance	S	siemens	$A\,V^{-1}$	$A^2\,s^3\,kg^{-1}\,m^{-2}$
Electric potential, difference electromotive force	V	volt	$W\,A^{-1}$	$kg\,m^2\,s^{-3}\,A^{-1}$
Electric resistance	Ω	ohm	Ω	$V\,A^{-1} = kg\,m^2\,s^{-3}\,A^2$
Electromotive force, emf	e	volt	$W\,A^{-1}$	$kg\,m^2\,s^{-3}\,A^{-1}$
Energy, work, quantity of heat	J	joule	Nm	$kg\,m^2\,s^{-2}$
Force	N	newton		$kg\,m\,s^{-2}$
Frequency	ν, f	hertz	Hz	s^{-1}
Illuminance	lux	lux	$lm\,m^{-2}$	$cd\,m^{-2}\,m^{-4} = cd\,m^{-2}$
Inductance	H	henry	$W\,A^{-1}$	$m^2\,kg\,A^{-2}$
Luminous flux	Cd, sr	lumen	lm	$m^2\,m^{-2}\,cd\,cd$
Magnetic flux	Wb	weber	$V\,s$	$m^2\,kg\,s^{-2}\,A^{-1}$
Magnetic flux density	T	telsa	$Wb\,m^{-1}$	$kg\,s^{-2}\,A^{-1}$
Plane angle	rd	radian	1	$m\,m^{-1} = 1$
Quantity of heat	Q	joule	J	$N\,m = kg\,m^3\,s^{-2}$
Power, radian flux	P	watt	$J\,s^{-1}$	$W = J\,s^{-1} = kg\,m^2\,s^{-3}$
Pressure, stress	p	pascal	Pa	$Nm^{-2} = kg\,m^{-1}\,s^{-2}$
Solid angle	sr	steradian	1	$m^2\,m^{-2} = 1$

Table A.3 Common Physical Constants (Mohr et al., 2012)

Physical Quantity /Constant	Symbol	Value	SI Unit
Atomic mass constant $(1/12)$ m $(^{12}C) = 1$ unit energy equivalent	m_u	$1.660538921 \times 10^{-27}$	kg
Atomic mass unit u = 1/12 of (^{12}C) mass	u	$1.660538921 \times 10^{-27}$	kg
Avogadro number	N_A, L	$6.02214129 \times 10^{23}$	mol^{-1}
Boltzmann constant, R/NA	k	$1.38061188 \times 10^{-23}$	$J\,K^{-1}$
Electric constant, $1/\mu_o c^2$	ε_o	$8.854187817... \times 10^{-12}$	$F\,m^{-1}$
Electron mass	m_e	$9.10938291 \times 10^{-31}$	kg
Electron volt, e/C	eV	$1.602176565 \times 10^{-19}$	J
Faraday's constant, $N_A e$	F	96485.3365	$C\,mol^{-1}$
Freezing temperature of water	T	273.15	K
Gravity acceleration	g_n	9.80665	$m\,s^{-2}$
Ideal gas volume	V	22.4146	mol^{-1}
Loschmidt constant, N_A/V_m	n_0	2.6516462×10^{25}	m^3
Loschmidt constant, N_A/V_m	n_0	2.6867805×10^{25}	m^3
Magnetic constant	μ_0	$4\pi 10^{-7} = 12.566370614... \times 10^{-7}$	$N\,A^{-2}$
Molar dry air mass	M_a	28.97	$g\,mol^{-1}$
Molar water mass	M_w	18.01599	$g\,mol^{-1}$
Molar vapor mass	M_v	18.01599	$g\,mol^{-1}$
Molar volume of ideal gas, RT/p $T = 273.15$, $p = 100$ kPa	V_m	22.710953×10^{-3}	$m^3\,mol^{-1}$
Molar volume of ideal gas, RT/p $T = 273.15$, $p = 101.325$ kPa	V_m	22.413968×10^{-3}	$m^3\,mol^{-1}$
Planck constant	h	$6.62606957 \times 10^{-34}$	J s
Proton charge	e	$1.60176565 \times 10^{-19}$	C
Proton mass	m_p	$1.6726121777 \times 10^{-27}$	kg
Quantum magnetic flux, $h/2e$	Φ_o	$2.067833758 \times 10^{-15}$	Wb
Radius of electron	r	2.817939×10^{-15}	m
Speed of light (in vacuum)	c, c_o	299 792 458	$m\,s^{-1}$
Standard atmosphere, atm	p	101 325	Pa
Stefan–Boltzmann constant	σ	5.670373×10^{-8}	$W\,m^{-2}\,K^{-4}$
Triple water point	T	273.16	K
Universal gas constant	R	8.3144621	$J\,mol^{-1}\,K^{-1}$
Universal constant for air	R_a	287.045	$J\,kg^{-1}\,K^{-1}$
Universal constant for vapor	R_v	461.51	$J\,kg^{-1}\,K^{-1}$

Table A.4 Basic Units and Symbols Recommended by International Solar Energy Society (ISES) (Beckman et al., 1978)

Quantity	Symbol	Unit
Absorptance	α	—
Conductivity, thermal	k	$\mathrm{W\,m^{-1}\,K^{-1}}$
Density	ρ	$\mathrm{kg\,m^{-3}}$
Diffusivity, thermal	a	$\mathrm{m^2\,s^{-1}}$
Emittance	ε	—
Extinction coefficient	K	$\mathrm{m^{-1}}$
Index of refraction	n	—
Reflectance	ρ	—
Specific heat	c	$\mathrm{J\,kg^{-1}\,K^{-1}}$
Transmittance, transmission coefficient	τ	—

Table A.5 Symbols for Radiation Quantities Recommended by International Solar Energy Society (ISES) (Beckman et al., 1978)

Preferred Name	Symbol	Unit
Beam irradiance	G_b	$\mathrm{W\,m^{-2}}$
Beam irradiation	H_b	$\mathrm{J\,m^{-2}}$
Diffuse irradiance	G_d	$\mathrm{W\,m^{-2}}$
Diffuse irradiation	H_d	$\mathrm{J\,m^{-2}}$
Global irradiance, or solar flux density	G	$\mathrm{W\,m^{-2}}$
Global irradiation	H	$\mathrm{J\,m^{-2}}$
Irradiance	E, H	$\mathrm{W\,m^{-2}}$
Irradiation of radian exposure	H	$\mathrm{J\,m^{-2}}$

Table A.6 Recommended Subscripts by International Solar Energy Society (ISES) (Beckman et al., 1978)

Subject	Subscript	Subject	Subscript
ambient	a	reflected	r
black body	b	solar	s
beam, direct	b	solar constant	sc
diffuse, scattered	d	sunrise (sunset)	sr (ss)
horizontal	h	total	t*
incident	i	thermal	th
normal	n	useful	u
outside temperature	o	spectral	λ

Table A.7 Recommended Symbols for Miscellaneous Quantities by International Solar Energy Society (ISES) (Beckman et al., 1978)[a]

Quantity	Symbol	Unit
Area	A	m^2
Air mass (or air mass factor)	M	—
Efficiency	η	—
Frequency	ν	s^{-1}
Heat	Q	J
Heat flow rate	$\dot{Q}$	W
Heat flux	$\dot{q}$	$W\,m^{-2}$
Heat transfer coefficient	h	$W\,m^{-2}\,K^{-1}$
Mass flow rate	$\dot{m}$	$kg\,s^{-1}$
Overall heat transfer coefficient	U	$W\,m^{-2}\,K^{-1}$
Stefan–Boltzmann constant	σ	$W\,m^{-2}\,K^{-4}$
System mass	m	kg
Temperature	T	K
Time	t, τ	s
Wavelength	λ	m

[a]The above recommended by ISES symbols are SI symbols and units and are valid for any thermal calculations in desalination and/or solar energy systems.

Table A.8 Recommended Abbreviations for Desalination Methods

a.c.	Alternative current
b.p.	Boiling point
d.c.	Direct current
DIN	States for German norms
f.p.	Freezing point
PP	Polypropelene
ppb	Parts per billion
ppm	Parts per million = $g\,m^{-3}$
ppt	Parts per trillion
rpm	Rotations per minute

A.2 GENERAL UNITS AND SYMBOLS BASED IN SI

Table A.9 Symbols and Units for Material Properties, Physical and Thermal and Related SI Coherent Derived Units (Mohr et al., 2012)

Quantity	Symbol	Unit Symbol
Activity	α	—
Activity coefficient	γ	—
Absorptance	α	—
Alkalinity	A	$eq\,m^{-3}$
Chlorinity, applied only to seawater	Cl	‰
Coefficient of convective heat transfer	a, h	$W\,m^{-2}\,K^{-1}$
Coefficient of conductive heat transfer	λ, k	$W\,m^{-1}\,K^{-1}$
Coefficient of heat transfer	k	$W\,m^{-2}\,K^{-1}$
Coefficient of diffusion (diffusivity)	D	$m^2\,s^{-1}$
Coefficient of expansion	β	K^{-1}
Coefficients of empirical equations	$a, b, c, \ldots$	—
Coefficient of mass transfer	b, k	$m^2\,s^{-1}$
Coefficient of radiation	C	$W\,m^{-2}\,K^{-4}$
Concentration of solute in liquid phase	c	$kg\,m^{-3}$
Concentration ratio: brine S/Seawater S	S_b/S_{in}	—
Density of material, of fluids	ρ	$kg\,m^{-3}$
Diffusivity, molar	D_{mol}	$m^2\,s^{-1}$
Heat capacity	C	$J\,K^{-1}$
Heat loss coefficient	U_L	$W\,m^{-2}\,K^{-1}$
Overall heat transfer coefficient	U	$W\,m^{-2}\,K^{-1}$
Porosity	ε, ψ	—
Reflectance	ρ	—
Salinity	S	$g\,kg^{-1}\,(S_A\,kg\,kg^{-1})$
Specific molar enthalpy	$\tilde{h}$	$J\,mol^{-1}$
Specific gas constant for air = 287.1	R_a	$J\,kg^{-1}\,K^{-1}$
Specific gas constant for vapor = 461.5	R_v	$J\,kg^{-1}\,K^{-1}$
Specific enthalpy of evaporation	$h_e, \Delta h_v, h_{fg}$	$J\,kg^{-1}$
Specific enthalpy of melting (or freezing)	$h_m, \Delta h_m$	$J\,kg^{-1}$
Specific heat capacity at constant pressure	c_p	$J\,kg^{-1}\,K^{-1}$
Specific heat capacity at constant volume	c_v	$J\,kg^{-1}\,K^{-1}$
Specific heat ratio	γ	—
Surface tension	σ	$N\,m^{-1}$
Thermal conductivity	λ	$W\,m^{-1}\,K^{-1}$
Thermal resistance	R	$K\,W^{-1}$
Viscosity, dynamic	η	$Pa\,s$
Viscosity, kinematic	ν	$m^2\,s^{-1}$

Table A.10 Symbols and Units for Coherent Derived Units in the SI Expressed in Terms of Base Units (Taylor and Thompson, 2008a,b)

Quantity	Symbol	Unit
Amount of concentration, concentration	c	mol m^{-3}
Mass flux	$\dot{m}$	$\text{kg m}^{-2}\,\text{s}^{-1}$
Mass density	ρ	kg m^{-3}
Mole	N	mol
Mole flow rate	$\dot{N}$	mol s^{-1}
Mole flux	$\dot{n}$	$\text{mol s}^{-1}\,\text{m}^{-2}$
Mole density	n	mol m^{-3}
Molecular mass	M	kg kmol^{-1}
Molar volume	V_m	$\text{m}^3\,\text{mol}^{-1}$
Specific volume	v	$\text{m}^3\,\text{kg}$
Speed, velocity	v	ms^{-1}
Surface cross-sectional area	f	m^2
Surface density	ρ_A	kg m^{-1}
Volume	V	m^3
Volume flow rate	$\dot{m}_V$	$\text{m}^3\,\text{s}^{-1}$
Volume flux	$\dot{v},\ \dot{q}_V$	$\text{m}^3\,\text{m}^{-2}\,\text{s}^{-1}$
Volume, specific	υ	$\text{m}^3\,\text{kg}^{-1}$

Table A.11 Symbols and Units for Energy, and Thermodynamic Quantities

Quantity	Symbol	Unit
Energy	E	J, Wh
Energy of mass	e	J kg^{-1}
Energy, internal	U	J
Energy, specific internal	u	J kg^{-1}
Energy, internal molar	$\bar{u}$	J mol^{-1}
Energy Gibbs, free	G	J
Energy Gibbs, specific	g	J kg^{-1}
Enthalpy	h	J kg^{-1}
Entropy	S	J K^{-1}
Force	F	N
Force, gravitational (Newtonian constant of gravitation)	G, f	$\text{m}^3\,\text{kg}^{-1}\,\text{s}^{-2}$

Table A.12 Symbols and Units for Pressure

Quantity	Symbol	Unit
Pressure, total	p	Pa, bar
Pressure difference	Δp	Pa, bar
Pressure at critical point	p_c	Pa, bar
Pressure, p/p_c	$p^\star$	Pa, bar
Partial vapor pressure	p_v	Pa, bar
Partial vapor pressure of pure substance	p_{ov}	Pa, bar
Partial vapor pressure at saturation	p_{sat}	Pa, bar
Osmotic pressure	Π	Pa, bar
Fugacity	$\tilde{p}$	Pa, bar
Fugacity of A (in gaseous mixtures)	p_A	Pa, bar

Table A.13 Symbols and Units for Geometric Parameters

Quantity	Symbol	Unit
Circumference	u	m
Diameter	d, D	m
Height	h	m
Length	l, L	m
Radius	r, R	m
Thickness	s	m
Width	s	m

REFERENCES

Beckman, W.A., Bugler, J.W., Coopers, P.I., Duffie, J.A., Dunkle, R.V., Glaser, P.E., et al., 1978. Units and symbols in solar energy. Solar Energ. 21 (1), 65–68, Reprints in Solar Energ. 57 (1), XVII–XVIII, 1996, Solar Energ. 70 (1), III–V, 2001).

Mohr, P.J., Taylor, B.N., Newell, D.B., 2012. CODATA (2010) recommended values of fundamental physical constants. Rev. Mod. Phys. 84, 1527.

NIST Special Publication 330. Taylor, B.N., Thompson, A. (Eds.), 2008a. SI, The International System of Units. National Institute of Standards and Technology, Washington, DC.

Taylor, B.N., Thompson, A. (Eds.), 2008b. Guide for the Use of International System of Units. NIST Special Publication 811. US Department of Commerce, Washington, DC.

ABBREVIATIONS

AGMD	Air gap membrane distillation
ASME	American Society of Mechanical Engineers
ASTM	American Society for Testing Materials
CODATA	Committee on Data for Science and Technology of the International Council for Science
CPC	Compound parabolic collectors

DCMD	Direct contact membrane distillation
ED	Electrodialysis
EDR	Electrodialysis reversal
EOLSS	Encyclopedia of Life Support Systems
EIA	Environmental impact assessment
ETC	Evacuated tube collectors
FESR	Fuel energy saving ratio
FPC	Flat-plate collectors

GOR

$$\text{Gain output ratio} = \frac{\text{Amount of distilate produced, kg}}{\text{Amount of heating steam applied to the 1st effect, kg}}$$

It is slightly less than the number of effects in a distillation unit.

H/D	Humidification—dehumidification
HDD	Humidification—dehumidification desalination
IAPWS	International Association of the Properties of Water and Steam
IAPSO	International Association of Physical Sciences of Oceans
IOC	International Oceanographic Commission
ITS-90	International Temperature Scale of 1990
IUPAC	Commission on Atomic Weights and Isotopic Abundances
LSI	Langelier saturation index
LT-MED	Low Temperature MED
MD	Membrane distillation
MED	Multiple effect distillation
MES	Multiple effect stack
MSF	Multistage flash
MVC	Mechanical vapor compression
NHR	Net heat rate
PHI	Power hydro-impurities

PR   Performance ratio, which is the economy of the steam economy and was established initially in pounds of distillate per 1000 BTU:

$$PR = \frac{\text{pounds of distillate produced}}{1000 \text{ BTU of heat applied}} = \frac{\text{kg of distilled water produced}}{2330 \text{ kJ applied}}$$

PTC	Parabolic trough collectors
SCOR	Scientific Committee on Oceanic Research
SGMD	Sweep gas membrane distillation
SI	Système International d' unites (International System of Units)
TDS	Total dissolved solids
TIC	Total inorganic carbon
TOC	Total dissolved organic carbon
TPC	Temperature polarization coefficient
TVC	Thermal vapor compression
VMD	Vacuum membrane distillation
WHO	World Health Organization

APPENDIX B

B.1 SEAWATER CONSTITUENTS

The ratio of most seawater constituents remains almost stable but the concentration may flocculate according to various parameters, such as heavy rains, flutes, and intensive evaporation rate.

Table B.1 Composition of Ocean Standard Seawater[a]

Element	Average Concentration	Range	Amount per kg of Seawater
Aluminum	540	< 10−1200	ng
Arsenic	1.7	1.1−1.9	μg
Barium	14	4−20	mg
Boron	4.5	−	mg
Bromine	67	−	mg
Cadmium	80	0.1−120	ng
Cesium	0.29	−	mg
Calcium	−0.412	−	g
Carbon	27.6	24−30	mg
Chlorine	19.354	−	g
Copper	0.25	0.03−0.4	mg
Fluorine	1.3	−	mg
Iodine	50	0.1−65	μg
Iron	55	5−140	ng
Lead	2	1.35	ng
Lithium	174	−	mg
Magnesium	1.29	−	g
Manganese	14	5−200	ng
Mercury	1	0.4−0.2	ng
Nickel	0.5	0.1−0.7	μg
Nitrogen	420	> 1−630	μg
Phosphorus	70	< 0.1−110	mg
Potassium	0.399	−	g
Rubidium	120	−	mg
Silicon	2.8	< 0.2−5	mg
Sodium	10.77	−	g
Strontium	7.9	−	mg
Sulfur	0.904	−	g
Zink	0.4	<0.01−0.6	mg
Uranium	3.3	−	mg

[a]Composition of the most important chemical species of some dissolved elements in seawater at salinity 35 (available in OSIL, http://www.osil.co.uk/AboutOSIL/Resources/SeawaterTechnicalPapers/tabid/104/articleType/ArticleView/articleId/220/Composition-of-Ocean-Standard-Seawater.aspx).

Table B.2 Ratio of Chemical Compounds in Normal and Natural Seawater

Molar Ratio	Natural Seawater	Normal Seawater IP/ASTM	Normal Seawater Free of Calcium
Na/Cl	0.5556	0.5559	0.5573
Ca/Cl	0.0	0.0	—
Mg/Cl	0.1	0.1	0.1
K/Cl	0.0	0.0	0.0
SO_4/Cl	0.1395	0.1396	0.0925
Br/Cl	0.0034	0.0035	0.0037
Sr/Cl	0.0007	0.0005	—
F/Cl	7×10^{-5}	4×10^{-5}	4×10^{-5}
H_3BO_3/Cl	0.0014	0.0015	0.0

Table B.3 Mean Ionic Composition of Seawater, $g\ kg^{-1}$

Ions	Ions	Normal Seawater	Natural Seawater
Chlorides	Cl^-	19.3605	18.9799
Sodium	Na^+	10.7678	10.5561
Sulfates	SO_4^{2-}	2.7017	2.6486
Magnesium	Mg^{2+}	1.2975	1.2720
Calcium	Ca^{2+}	0.4081	0.4001
Potassium	K^+	0.3876	0.3800
Bicarbonates	HCO_3	0.1425	0.1397
Bromides	Br^-	0.0659	0.0646
Boric acid	H_3BO_3	0.0265	0.0260
Strontium	Sr^{2+}	0.0136	0.0133
Fluorides	F^-	0.0013	0.0013
Iodides	I^-	0.00005	
Silicon	Si^{++++}	0.00002–0.0004	
Various		0.00013	
TDS		35.1745	34.4816
Water	H_2O	964.8255	965.5184
Water Characteristics			
Salinity	$S\ (g\ kg^{-1})$	36.01	34.3
Chlorinity	$Cl\ (g\ kg^{-1})$	19.4	19.0

The values of Tables B.2 and B.3 are indicative of mean values. In practice, they fluctuate according to the region and may be from site to site in the same region.

Table B.4 Complete Composition of Seawater at Salinity 35 (Turekian, 1968)

Element	Atomic Mass[a]	Amount (g kg^{-1})	Element	Atomic Mass[a]	Amount (g kg^{-1})
Hydrogen	1.008	110	Molybdenum	95.96	1×10^{-5}
Helium	4.002602	7.2×10^{-9}	Ruthenium	98	7×10^{-10}
Lithium	6.94	0.00017	Podium	101.07	–
Beryllium	9.012182	0.6×10^{-9}	Palladium	106.42	–
Boron	10.81	4.45×10^{-3}	Argentum (silver)	107.86	2.8×10^{-7}
Carbon	12.011	0.028	Cadmium	112.411	1.1×10^{-7}
Nitrogen ion	14.007	0.0155	Indium	114.818	–
Oxygen	15.999	883	Stannum (tin)	118.710	8.1×10^{-7}
Fluorine	18.998403	0.013	Antimony	121.760	3.3×10^{-7}
Neon	20.1797	12×10^{-6}	Tellurium	127.60	–
Sodium	22.989769	10.8	Iodine	126.904	6.4×10^{-4}
Magnesium	24.3050	19.4	Xenon	131.293	4.7×10^{-7}
Aluminum	26.98153	1×10^{-6}	Cesium	132.905	3×10^{-6}
Silicon	28.085	0.0029	Barium	137.327	2.1×10^{-5}
Phosphorus	30.973762	0.88×10^{-4}	Lanthanum	138.9054	2.9×10^{-9}
Sulfur	32.06	0.904	Cerium	140.116	1.2×10^{-9}
Chlorine	35.45	19.4	Praseodymium	140.907	6.4×10^{-10}
Argon	39.948	0.45×10^{-3}	Neodymium	144.242	2.8×10^{-9}
Potassium	39.0983	0.392	Samarium	150.36	4.5×10^{-10}
Calcium	40.078	0.411	Europium	151.25	1.3×10^{-9}
Scandium	44.955912	$<4 \times 10^{-8}$	Gadolinium	157.25	7×10^{-9}
Titanium	47.867	1×10^{-6}	Terbium	158.925	1.4×10^{-9}
Vanadium	50.9415	1.9×10^{-6}	Dysprosium	162.500	9.1×10^{-9}
Chromium	51.9961	2×10^{-6}	Holmium	164.930	2.2×10^{-9}
Manganese	54.93804	4×10^{-6}	Erbium	167.259	8.7×10^{-9}
Ferrum (iron)	55.845	3.4×10^{-5}	Thulium	168.934	1.7×10^{-9}
Cobalt	58.993	3.9×10^{-5}	Ytterbium	173.054	8.2×10^{-9}
Nickel	58.710	6.6×10^{4}	Lutetium	174.966	1.5×10^{-9}
Copper	63.54	9×10^{-7}	Hafnium	178.49	8×10^{-8}
Zinc	65.37	5×10^{-6}	Tantalum	180.947	$<2.5 \times 10^{-8}$
Gallium	69.72	3×10^{-8}	Tungsten	183.84	$<1 \times 10^{-8}$
Germanium	72.59	6×10^{-8}	Rhenium	186.207	8.4×10^{-8}
Arsenic	74.922	2.6×10^{-6}	Osmium	190.23	–
Selenium	78.96	0.9×10^{6}	Iridium	192.217	–
Bromine	79.909	0.0673	Platinum	195.084	–
Krypton	83.80	0.21×10^{-6}	Aurum (gold)	196.966	1.1×10^{-7}
Rubidium	85.47	0.12×10^{-3}	Mercury	200.59	1.5×10^{-7}
Strontium	87.62	0.0081	Thallium	204.38	–
Yttrium	88.905	1.3×10^{-8}	Plumbum (lead)	207.2	3×10^{-7}
Zirconium	91.22	2.6×10^{-8}	Bismuth	209.980	3×10^{-7}
Niobium	92.906	1.5×10^{-8}	Uranium	238.0289	3.3×10^{-5}

List of elements in atomic number order.

[a]According to IUPAC, 2009 (Pure Appl. Chem. 83, 2011, 1485–1498). K.K. Turekian, The Oceans, 1968, Second Edition 1976, 146 pp, Prentice Hall Upper Saddle River, New Jersey, 146 pp.

Table B.5 The Relative Contribution of All Major Components for Seawater With Salinity 35°C and 25°C[a]

Element	g kg^{-1} of Solution/Cl	Atomic Mass (g mol^{-1})	mol kg^{-1} of Chlorine Equivalent	10^7 Times the Mole Fractions of the Reference Composition
Na^+	0.5564924	22.989766928	24.2060889	4188071
Mg^{2+}	0.0662600	24.30500000	2.7261880	471678
Ca^{2+}	0.0212700	40.07800000	0.5307151	91823
K^+	0.0206000	39.09830000	0.5268771	91159
Sr^{2+}	0.0004100	87.62000000	0.0046793	810
Cl^-	0.9989041	35.45300000	28.1754464	4874839
SO_4^{2-}	0.1400000	96.06260000	1.4573830	252152
HCO_3^-	0.0054100	61.01684000	0.0886640	15340
Br^-	0.0034730	79.90400000	0.0434647	7520
CO_3^{2-}	0.0007400	60.00890000	0.0123315	2134
$B(OH)_4^-$	0.0004100	78.84036000	0.0052004	900
F^-	0.0000670	18.99840320	0.0035266	610
OH^-	0.0000070	17.00734000	0.0004116	71
$B(OH)_3$	0.0010030	61.83302000	0.0162211	2801
CO_2	0.0000220	44.00950000	0.0004999	86
Sum	1.8150685	—	57.7976977	10 000 000

[a]Millero, F.J., Feistel, R., Wright, D.G., McDougall, T.J., 2008. The composition of standard seawater and the definition of reference-composition salinity scale. Deep-Sea-Res. 55 (1), 50−72.

Table B.6 Henry's Constant for Various Gases Dissolved in Seawater, $K_H \times 10^5$ atm^{-1}

°C	Air	CO$_2$	H$_2$	N$_2$	O$_2$	CH$_4$
0	2.31	137.4	1.73	1.89	3.92	4.46
10	1.82	96.7	1.57	1.50	303	3.37
20	1.51	70.1	1.46	1.24	2.44	2.66
30	1.30	53.8	1.37	1.08	2.11	2.23
40	1.15	41.3	1.33	0.962	188	1.92
50	1.06	35.3	1.34	0.885	1.70	1.73
60	0.99	28.6	1.34	0.883	1.59	1.60

Table B.7 Indicative Concentration of Main Elements in Seawater for Various Salinities[a]

S ‰	Na$^+$	Mg^{2+}	Ca^{2+}	K$^+$	Sr^{2+}	B	Cl$^-$	SO$_4^{2+}$	Br$^-$	F$^-$	HCO$_3^-$
5.0	1.539	0.185	0.058	0.057	0.001	0.001	2.763	0.387	0.010	0.0002	0.020
10.0	3.078	0.370	0.118	0.114	0.002	0.001	5.527	0.0775	0.019	0.0004	0.041
15.0	4.616	0.555	0.177	0.171	0.003	0.002	8.290	1.162	0.029	0.0005	0.061
20.0	6.156	0.739	0.235	0.228	0.005	0.003	11.054	1.550	0.038	0.0007	0.081
25.0	7.696	0.924	0.294	0.285	0.006	0.003	13.817	1.937	0.048	0.0009	0.101
30.0	9.234	1.109	0.353	0.342	0.007	0.004	16.581	2.325	0.058	0.0011	0.122
31.0	9.542	1.146	0.365	0.353	0.007	0.004	17.133	2.402	0.059	0.0011	0.126
32.0	9.850	1.183	0.377	0.365	0.007	0.004	17.685	2.480	0.062	0.0012	0.130
33.0	10.156	1.220	0.388	0.376	0.008	0.004	18.239	2.570	0.063	0.0012	0.134
34.0	10.465	1.257	0.400	0.288	0.008	0.004	18.791	2.635	0.065	0.0012	1.137
35.0	10.773	1.294	0.412	0.399	0.008	0.004	19.344	2.712	0.067	0.0013	0.142
36.0	11.081	1.331	0.424	0.410	0.008	0.005	19.897	2.789	0.069	0.0013	0.146
37.0	11.389	1.368	0.435	0.422	0.009	0.005	20.449	2.857	0.071	0.0013	0.150
38.0	11.696	1.405	0.447	0.433	0.009	0.005	21.002	2.944	0.073	0.0014	0.154
39.0	12.004	1.492	0.459	0.445	0.009	0.005	21.555	3.022	0.075	0.0014	0.158
40.0	12.312	1.497	0.471	0.456	0.009	0.005	22.107	3.099	0.077	0.0015	0.162
41.0	12.620	1.516	0.482	0.467	0.009	0.005	22.660	3.177	0.079	0.0015	0.166
42.0	12.929	1.553	0.294	0.476	0.009	0.005	23.213	3.254	0.081	0.0015	0.170

[a]Hill, M.N., 1963. The Sea, vol. 2, Comparison of Seawater, Interscience Publ.

B.2 PROPERTIES OF PURE AND NATURAL WATERS

Table B.8 Main Constituents of Natural Waters

Major Elements (1.0–10^3 ppm)	Secondary Elements (0.1–10 ppm)	Minor Elements (0.0001–0.1 ppm)	Trace Elements (<0.001 ppm)
Sodium	Iron	Antimony, aluminum	Beryllium, bismuth
Calcium	Potassium	Arsenic, barium	Cerium, cesium
Magnesium	Carbonates	Bromide/cadmium	Gallium, gold
Bicarbonate	Nitrates	Chromium, cobalt	Indium, lanthanum
Sulfates	Fluorides	Copper, germanium	Niobium, platinum
Chlorides	Boron	Iodide, lead	Radium, ruthenium
Silica		Lithium, manganese	Scandium, silver
		Molybdenum, nickel	Thallium, thorium
		Phosphate, rubidium	Tin, tungsten
		Selenium, thallium	Ytterbium, yttrium
		Uranium, vanadium	Zirconium
		Zink	

Table B.9 Physical Properties of Natural Water

T (°C)	Vapor Pressure p_v (kPa)	Density (kg m^{-3})	Expansion Coefficient b (K$^{-1}\times 10^{-3}$)	Compressibility (atm $\times 10^{-8}$)	Surface Tension σ (10^{-2} Nm)
0	0.6102	999.84	~0.070	50.6	76.62
10	1.2259	999.70	0.088	48.6	74.20
20	2.3349	998.20	0.207	47.0	72.75
25	3.1634	997.05	0.255	46.5	71.96
30	4.2370	995.65	0.303	46.0	71.15
40	7.3685	992.22	0.385	45.3	69.55
50	12.3234	988.05	0467	45.0	67.90
60	19.8984	983.21	0.523	45.0	66.17
70	31.1282	977.79	0.585	45.2	64.41
80	47.3117	971.83	0.643	45.7	62.60
90	70.0485	965.32	0.698	46.5	60.74
100	101.2300	958.35	0.752	48.0	58.84

Table B.10 Physical Properties of Natural Water (continuation)

T, °C	Dielectric Constant	Electric Conductance, 10^{-8} (Sc m^{-1})	Refraction Index, n	Gibbs Free Energy, G (kJ kg^{-1})	Ionization Enthalpy (kJ mol^{-1})
0	87.69	1.61	1.33464	0.00	62.81
10	83.82	2.83	1.33389	42.03	59.64
20	80.08	4.94	1.33299	83.86	57.00
25	78.25	6.34	1.33287	104.74	55.84
30	76.49	8.04	1.33192	146.40	54.75
40	73.02	12.53	1.33051	167.33	52.75
50	69.70	18.90	1.32894	209.10	50.90
60	66.51	27.58	1.32718	250.10	49.13
70	63.45	38.91	1.32511	292.75	47.39
80	60.54	53.03	1.32287	334.07	45.64
90	57.77	69.65	1.32050	376.68	43.86
100	55.15	88.10	1.31783	418.77	42.05

Table B.11 Physical Properties of Pure Water

Temperature	Density, ρ (kg m^{-3})	Viscosity, Dynamic, η (10^3 Pas)	Viscosity, Kinematic, ν (10^6 m^2 s^{-1})	Surface[a] Tension, σ (N m^{-1})	Vapor Pressure, p_v (Pa)
0.0	998	1.781	1.785	0.0765	0.61
5.0	1000	1.518	1.519	0.0749	0.87
1.0	999.7	1.307	1.306	0.0742	1.23
15.0	999.1	1.139	1.139	0.0735	1.70
20.0	998.1	1.001	1.003	0.0728	2.34
25.0	997.0	0.890	0.893	0.0720	3.17
30.0	996.7	0.789	0.800	0.0712	4.24
40.0	992.2	0.653	0.658	0.0696	7.38
50.0	988.0	0.547	0.553	0.0679	12.33
60.0	983.2	0.466	0.474	0.0662	19.92
70.0	077.8	0.404	0.413	0.0644	31.16
80.0	971.8	0.354	0.364	0.0625	47.34
90.0	965.3	0.315	0.326	0.0608	70.10
120.0	958.4	0.282	0.294	0.0598	101.33

[a]Surface tension is calculated in contact of water with surrounding air.

Table B.12 Dynamic Viscosity of Natural Waters, η (Pa s)

T (°C)	Viscosity, (10^{-3} Pa s)	T (°C)	Viscosity (10^{-3} Pa s)	T (°C)	Viscosity (10^{-3} Pa s)
0	1.7921	32	0.7679	66	0.4293
2	1.6728	34	0.7371	68	0.4174
4	1.5674	36	0.6814	70	0.4061
6	1.4728	38	0.6560	72	0.3952
8	1.3800	40	0.6321	74	0.3849
10	1.3077	42	0.6097	76	0.3750
12	1.2363	44	0.5883	78	0.3655
14	1.1709	46	0.65683	80	0.3565
16	1.1110	48	0.5494	82	0.3478
18	1.9559	50	0l.5315	84	0.3395
20	1.0050	52	0.5146	86	0.3315
20.20	1.0000	54	0.4980	88	0.3029
22	1.9579	56	0.4832	90	0.3165
24	1.9142	58	0.4688	92	0.3095
26	1.6737	60	0.4550	94	0.3027
28	1.6360	62	0.4418	96	0.2962
30	1.8007	64	0.4293	100	0.2838

Table B.13 Water and Air Constants in Normal Atmospheric Conditions

Constant	Value	SI Unit
Water density	$\sim 10^3$	$(kg\ m^{-3})$
Viscosity at 18°C	$\sim 10^3$	$(N\ s\ m^{-2})$
Specific heat capacity	4.2	$(kg\ J\ kg^{-1}\ K^{-1})$
Thermal conductivity of water	0.6	$(W\ m^{-1}\ K^{-1})$
Enthalpy	2.3	$(MJ\ kg^{-1})$
Melting enthalpy of water at 0°C, 101.3 kPa	333.6	$(kJ\ kg^{-1})$
Evaporation enthalpy of water (100°C)	2258	$(kJ\ kg^{-1})$
Molecular mass of water	18.01534	$(kg\ mol^{-1})$
Air density at 273.15 K, ρ_{air}	1.3	$(kg\ m^{-3})$
Dynamic viscosity of air, η_{air}	1.7×10^{-3}	$(N\ s\ m^{-2} = Pas)$
Specific thermal capacity of air, c_p	~ 1.0	$(kJ\ kg^{-1}\ K^{-1})$
Thermal conductivity of air, λ_{air}	~ 0.024	$(W\ m^{-1}\ K^{-1})$
Molecular mass of air, M_{air}	28.97	$kg\ mol^{-1}$
Ideal volume of air (0°C)	22.4146	$[m^3 (kg\ mol)^{-1}\ K^{-1}$
Universal gas constant, R	1.9871	$[kcal (kg\ mol)^{-1}\ K^{-1}]$
	8.3144	$(J\ mol^{-1}\ K^{-1})$
	0.082057	$m^3\ atm (kg\ mol)^{-1}\ K^{-1}]$

Table B.14 Some Common Properties of Water Ice and Water Vapor

Property	H_2O, liquid	H_2O, solid	H_2O, gas
Boiling point at 101.325 kPa	100°C, 373.1243 K	—	—
Chemical potential, $d\mu/dT$ Negative molar entropy ($-S$)	-69.9 J mol⁻¹ K⁻¹ at 25°C	-44.8 J mol⁻¹ K⁻¹ at 25°C	-188.7 J mol⁻¹ K⁻¹ at 25°C
Thermal conductivity, l, k	0.610 W m⁻¹ K⁻¹ at 25°C	Ice Ih = 2.4 W m⁻¹ K⁻¹ at 20°C Ih 99 at -20°C, 171 at 120°C	0.025 W m⁻¹ K⁻¹ at 100°C 104.3 at 240 K
Dielectric constant	0.1°C = 87.9 e eq. 25°C = 78.4 100°C = 55.6		
Internal energy, U	1.8883 kJ mol⁻¹ at 25°C 101.0325 kPa	Ih ice 6.007 J mol⁻¹, 101.0325 kPa	45.15 kJ mol⁻¹, at 100°C, 101.0325 kPa
Enthalpy ($H = U + PV$)	1.891 kJ mol⁻¹ at 25°C	Ih ice 6.005 J mol⁻¹, 101.0325 kPa	48.20 kJ mol⁻¹, at 100°C, 101.0325 kPa
Enthalpy of vaporization (liquid), h_{fg}, h_{ev}	45.051 kJ mol⁻¹ at 0°C	Ih ice, 46.567 J mol⁻¹, 240 K	40.657 kJ mol⁻¹ at 100°C
Enthalpy of fusion, h_{fis}	—	9.00678 kJ mol⁻¹ at 0°C 101.0325 kPa	—
Enthalpy of sublimation	—	51.059 kJ mol⁻¹ at 0°C, 51.139 kJ mol⁻¹, 240 K	—
Gibbs energy of formation (chemical potential), g	-237.18 kJ mol⁻¹ at 25°C	-236.59 kJ mol⁻¹ at 25°C	-228.59 kJ mol⁻¹ at 25°C
Heat capacity ratio c_p/c_v	—	—	1.3368, 100°C, 101.0325 kPa
Molality, m	55.508472		
Molecular mass	$2.9915051 \times 10^{-23}$ g mol⁻¹		
pH	6.9976 at 25°C		
Prandtl number, Pr	6.1 at 25°C		
Refractive index, n	$n = 1.33286$ at 25°C	$n = 1.3091$	
Surface tension, σ	0.07198 J m⁻² at 25°C		
Viscosity, dynamic, η	0.8909 mPa s at 25°C and 101.0325 kPa		0.0123 mPa–s at 25°C, 101.0325 kPa
Viscosity, kinematic, ν	0.8935×10^{-6} m² s⁻¹		

Table B.15 Natural Water Constants at Environmental Pressure ($p = 1$ bar)

T (°C)	ρ (kg m^{-3})	c_p (kJ kg^1 K^{-1})	λ (W m^{-1} K^{-1})	η (10^{-4} Pa s)	ν (10^{-6} m^2 s^{-1})	a (10^{-6} m^2 s^{-1})	Pr (−)
−10	998.14	4.277	0.5423	26.452	2.650	0.1270	20.86
−8	998.67	4.261	0.5460	24.292	2.432	0.1283	18.96
−6	999.09	4.248	0.5497	22.397	2.242	0.1295	17.31
−4	999.42	4.236	0.5535	20.725	2.074	0.1307	15.86
−2	999.67	4.227	0.5573	19.243	1.925	0.1319	14.60
0	999.84	4.218	0.5610	17.923	1.793	0.1330	13.48
2	999.94	4.211	0.5648	16.741	1.674	0.1341	12.48
4	999.97	4.205	0.5686	15.679	1.568	0.1352	11.60
6	999.90	4.200	0.5724	14.720	1.472	0.1363	10.80
8	999.85	4.196	0.5762	13.853	1.385	0.1373	10.09
10	999.70	4.292	0.5800	13.064	1.307	0.1384	9.443
15	999.10	4.185	0.5893	11.380	1.139	0.1434	8.082
20	998.21	4.181	0.5984	10.020	1.004	0.1489	7.001
25	997.05	4.179	0.6072	8.9045	0.893	0.1457	6.128
30	995.65	4.177	0.6151	7.9768	0.801	0.1480	5.414
35	994.03	4.177	0.6233	7.1962	0.724	0.1501	4.828
40	992.22	4.177	0.6306	6.5325	0.658	0.1521	3.909
45	990.21	4.178	0.6373	5.9632	0.602	0.1540	3.553
50	988.04	4.180	0.6376	5.4708	0.554	0.1558	3.248
55	985.69	4.182	0.6492	5.0419	0.512	0.1575	3.248
60	983.20	4.184	0.6544	4.6659	0.475	0.1591	2.983
65	960.55	4.187	0.6590	4.3344	0.442	0.1605	2.754
70	977.77	4.190	0.6631	4.0406	0.413	0.1619	2.553
75	974.84	4.193	0.6668	3.7790	0.388	0.1631	2.376
80	971.79	4.197	0.6700	3.5449	0.365	0.1643	2.221
65	968.61	4.201	0.6728	3.3348	0.344	0.1653	2.082
90	963.31	4.206	0.6752	3.1453	0.326	0.1663	1.959
95	961.89	4.211	0.6773	2.9740	0.309	0.1672	1.849
99.63[a]	958.61	4.216	0.7689	2.8295	0.295	0.1680	1.757

[a]99.63°C is saturation condition.

Molecular water mass $M = 18.0152$ kg mol^{-1}.
Critical temperature $T_c = 647.12$ K $+ \Delta T$, where -0.10 K $\leq \Delta T \leq 0.10$ K.
Critical pressure $p_c = (220.64 \pm 0.05)$ bar $+ \alpha \Delta T$, $\alpha = 2.4$ bar K^{-1}.
Critical density ρ_c (322 ± 3.0)kg m^{-3}.

B.3 PROPERTIES OF SATURATED VAPOR IN AIR

Table B.16 Constants of Saturated Vapor in Air at Pressure 1.0 bar and Ratio x

T (K)	t (°C)	p (mbar)	ρ (kg m^{-3})	h_{ev} (kJ kg^{-1})	x (kg kg^{-1})	h_{vs} (kJ kg^{-1})
253	−20	1.03	0.000879	2839.6	0.000641	−18.50
257	−16	1.30	0.001209	2838.8	0.000937	−13.78
261	−12	2.17	0.001801	2817.5	0.001352	−8.71
265	−8	3.19	0.002527	2836.3	0.001931	−3.22
269	−4	4.37	0.003519	2835.4	0.002729	+2.76
273.15	0	6.11	0.004846	2834.6	0.003821	9.55
273.16	0	6.11	0.004846	2834.6	0.003821	9.55
277	4	8.13	0.006358	2491.3	0.005100	16.80
281	8	10.72	0.008267	2481.6	0.006749	24.99
285	12	14.02	0.001066	2472.4	0.008849	34.38
289	16	18.17	0.01363	2463.3	0.011513	45.17
293	20	23.38	0.01729	2453.6	0.014895	57.86
297	24	29.83	0.02177	2444.4	0.019135	72.60
301	28	37.80	0.02723	2434.7	0.024435	90.48
305	32	47.55	0.03380	2425.5	0.033050	111.58
309	36	59.42	0.04171	2415.9	0.039289	137.0
313	40	73.77	0.05114	2406.3	0.049532	167.65
317	44	91.02	0.06233	2396.6	0.062278	204.95
321	48	111.66	0.07553	2387.0	0.078146	250.46
329	52	136.20	0.09103	2377.4	0.098018	306.66
331	56	165.15	0.1091	2368.2	0.12297	375.33
335	60	199.24	0.1302	2358.5	0.15472	464..13
339	64	239.17	0.1545	2348.5	0.19541	575.80
341	68	285.67	0.1825	2338.4	0.24866	721.04
345	72	339.7	0.2146	2328.8	0.31966	913.9
349	76	402.0	0.2514	2318.8	0.41790	1179.5
353	80	473.7	0.2933	2308.3	0.55931	1560.9
357	84	555.9	0.3406	2297.8	0.77781	2149.0
361	88	649.7	0.3942	2287.8	1.15244	3136.3
365	92	756.4	0.4545	2277.3	1.92718	5236.3
369	96	877.2	0.5221	2266.8	14.42670	11944.5
372	99	978.0	0.5780	2258.9	27.14538	72905.3
373	100	1014.0	0.5977	2256.4		

B.4 ELECTRICAL PROPERTIES OF CONSTITUENTS IN SEAWATER

Table B.17 Coefficient of Electrical Conductivity of Ions, F_{ai}, in $\mu S\ cm^{-1}$

Ion		$meq\ l^{-1}$	$mg\ l^{-1}$
Calcium	Ca^{2+}	52.0	2.60
Magnesium	Mg^{2+}	46.6	3.82
Potassium	K^{+}	72.0	1.84
Sodium	Na^{+}	48.9	2.13
Hydrogen carbonate	HCO^{3}	43.6	0.715
Carbonates	CO_3^{2-}	84.6	2.82
Chlorides	Cl^{-}	75.9	2.14
Sulfates	SO_4^{2-}	73.9	1.54
Nitrates	NO_3^{-}	71.0	1.15

Table B.18 Specific Electrical Conductance of Seawater in $\mu S\ cm^{-1}$

Cl (‰)	S (‰)	0°C	5°C	10°C	15°C	20°C	25°C
1	1.81	1839	2134	2439	2763	3091	3431
2	3.63	3556	4125	4714	5338	5871	6628
3	5.44	5187	6116	6872	7778	8702	9658
4	7.26	6758	7845	8958	10133	11337	12583
5	9.10	8327	9653	11019	12459	13939	15471
6	10.89	9878	11444	13063	14758	16512	18324
7	12.78	11404	13203	15069	17015	19035	21121
8	14.52	12905	14934	17042	19235	21514	23868
9	16.33	14388	16641	18986	21423	23957	26573
10	18.15	15852	18329	20906	23584	26367	29242
11	19.96	17304	20000	22804	25722	28749	31879
12	21.77	18741	21655	24684	27841	31109	43489
13	23.59	20167	23297	26548	29940	33447	37075
14	25.04	21585	24929	28397	32024	35765	39638
15	27.22	22993	26548	30231	34090	40345	44701
16	29.03	24393	28156	32050	36138	42606	47201
17	30.85	25783	29753	33855	38168	44844	49677
18	32.66	27162	31336	35644	40176	47058	52127
19	34.45	28530	32903	37415	42158	49248	54551
20	36.29	29885	34454	39167	44414	51414	56949
21	38.09	31277	35989	40900	46044	53556	59321
22	39.92	32556	37508	42614	47948	53556	59321

B.5 SOLUBILITY OF GASES IN SEAWATER

Table B.19 Oxygen Concentration in Seawater, in g m^{-3} (Salinity, S g kg^{-1})

°C	0, S	5, S	10, S	15, S	20, S	25, S	30, S	35, S	40, S	45, S
0	14.60	14.11	13.04	13.18	12.74	12.31	11.90	11.50	11.11	10.74
1	14.20	13.73	13.27	12.83	12.40	11.98	11.58	11.20	10.83	10.46
2	13.81	13.36	12.91	12.49	12.07	11.67	11.29	10.91	10.55	10.20
3	13.45	13.00	12.58	12.16	11.76	11.38	11.00	10.64	10.29	9.95
4	13.09	12.67	12.25	11.85	11.47	11.09	10.73	10.38	10.04	9.71
5	12.76	12.34	11.94	11.56	11.18	10.82	10.47	10.13	9.80	9.48
6	12.44	12.04	11.65.	11.27	10.91	10.56	10.22	9.89	9.57	9.27
7	12.13	11.74	11.37	11.00	19.65	10.31	9.98	9.66	9.35	9.06
8	11.83	11.46	11.09	10.74	10.40	10.07	9.75	9.44	9.14	8.85
9	11.55	11.19	10.83	10.49	10.16	9.84	9.53	9.23	8.94	8.66
10	11.28	10.92	10.58	10.25	9.93	9.62	9.32	9.03	8.75	8.47
11	11.02	10.67	10.34	10.02	9.71	9.41	9.12	8.83	8.56	8.30
12	10.77	10.43	10.11	9.80	9.50	9.21	8.92	8.65	8.38	8.12
13	10.53	10.20	9.89	9.59	9.30	9.01	8.74	8.47	8.21	7.96
14	10.29	9.98	9.68	9.38	9.10	8.82	8.55	8.30	8.04	7.80
15	10.07	9.77	9.47	9.19	8.91	8.64	8.38	8.13	7.88	7.65
16	9.86	9.36	9.28	8.82	8.73	8.47	8.21	7.97	7.73	7.50
17	9.65	9.17	9.09	8.64	8.55	8.30	8.05	7.81	7.58	7.36
18	9.45	8.99	8.90	8.47	8.39	8.14	7.90	7.66	7.44	7.22
19	9.26	8.81	8.73	8.31	8.22	7.98	7.75	7.52	7.30	7.09
20	9.08	8.64	8.56	8.15	8.07	7.83	7.60	7.38	7.17	6.96
21	8.90	8.48	8.39	8.00	7.91	7.69	7.46	7.25	7.04	6.84
22	8.73	8.32	8.23	7.85	7.77	7.54	7.33	7.12	6.91	6.72
23	8.56	8.16	8.08	7.71	7.63	7.41	7.20	6.99	6.79	6.60
24	8.40	8.01	7.93	7.57	7.49	7.28	7.07	6.87	6.68	6.49
25	8.24	7.87	7.79	7.44	7.36	7.15	6.95	6.75	6.56	6.38
26	8.09	7.73	7.65	7.31	7.23	7.03	6.83	6.64	6.46	6.28
27	7.95	7.59	7.51	7.18	7.10	6.91	6.72	6.53	6.35	6.17
28	7.81	7.59	7.38	7.18	6.98	6.79	6.61	6.42	6.25	6.08
29	7.67	7.46	7.38	7.06	6.87	6.68	6.50	6.32	6.15	5.98
30	7.54	7.33	7.14	6.94	6.75	6.57	6.39	6.22	6.05	5.89
31	7.41	7.21	7.14	6.83	6.65	6.47	6.29	6.12	5.96	5.80
32	7.29	7.09	6.90	6.72	6.54	6.36	6.19	6.03	5.87	5.71
33	7.17	6.98	6.79	6.61	6.44	6.26	6.10	5.94	5.78	5.63
34	7.05	6.86	6.68	6.51	6.33	6.17	6.01	5.85	5.69	5.54
35	6.93	6.75	6.58	6.40	6.24	6.07	5.92	5.76	5.61	5.46
36	6.82	6.65	6.47	6.31	6.14	5.98	5.83	5.68	5.53	5.19
37	6.72	6.54	6.37	6.21	6.05	5.89	5.74	5.59	5.54	5.31
38	6.61	6.44	6.28	6.12	5.96	5.81	5.66	5.51	5.37	5.24
39	6.51	6.34	6.18	6.03	5.87	5.72	5.58	5.54	5.30	5.16
40	6.41	6.24	6.09	5.94	5.79	5.64	5.50	5.36	5.22	5.09

Table B.20 Solubility of Nitrogen in Seawater (Salinity ‰)[a]

°C	0, S	10, S	20, S	30, S	34, S	35, S	36, S	38, S	40, S
−1	–	–	16.28	15.10	14.65	14.54	14.44	14.22	14.01
0	18.42	17.10	15.87	14.73	14.30	14.19	14.09	13.88	13.67
1	17.95	16.67	15.48	14.38	13.86	13.86	13.75	13.55	13.35
2	17.50	16.26	15.11	14.04	13.64	13.54	13.44	13.24	13.05
3	17.07	15.87	14.75	13.72	13.32	13.23	13.13	12.94	12.76
4	16.65	15.49	ʻ14.41	13.41	13.03	12.93	12.84	12.66	12.47
5	16.26	15.13	14.09	13.11	12.74	12.65	12.56	12.38	12.21
6	15.88	14.79	13.77	12.83	12.47	12.38	12.29	12.12	11.95
8	15.16	14.14	13.18	12.29	11.95	11.87	11.79	11.62	11.46
10	14.51	13.54	12.64	11.80	11.48	11.40	11.32	11.17	11.01
12	13.90	12.99	12.14	11.34	11.04	10.96	10.86	10.74	10.60
14	13.34	12.48	11.67	10.92	10.63	10.56	10.49	10.35	10.21
16	12.83	12.01	11.24	10.53	10.25	10.19	10.12	9.99	9.86
18	12.35	11.57	10.84	10.16	9.90	9.84	9.77	9.65	9.52
20	11.90	11.16	10.47	9.82	9.57	9.51	9.45	9.33	9.21
22	11.48	10.76	10.12	9.50	9.26	9.21	9.15	9.03	8.92
24	11.09	10.42	9.79	9.20	8.98	8.92	8.87	8.76	8.65
26	10.73	10.09	9.49	8.92	8.71	8.65	8.60	8.50	8.39
28	10.38	9.77	9.20	8.66	8.45	8.40	8.35	8.25	8.15
30	10.06	9.48	8.93	8.41	8.21	8.16	8.12	8.02	7.92
32	9.76	9.20	8.67	8.18	7.99	7.94	7.89	7.80	7.71
34	9.48	8.94	8.43	7.96	7.77	7.74	7.68	7.59	7.51
36	9.21	8.69	8.20	7.75	757	7.53	7.48	7.40	7.31
38	8.95	9.46	7.99	7.55	7.38	7.33	7.29	7.21	7.13
40	8.71	8.43	7.7	7.36	7.19	7.15	7.11	7.03	6.95

[a]*Nitrogen* solubility is expressed in $cm^3 \, dm^{-3}$ at 1 atm and relative humidity is 100%. Concentration of nitrogen in air is 78.084%.

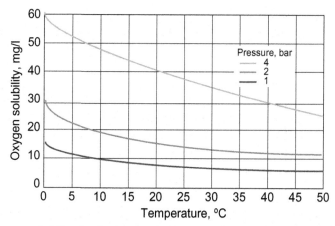

Figure B.1 Oxygen solubility in freshwater (salinity ‰) for pressures 1, 2, and 4 bar (www. EngineeringToolBox.com).

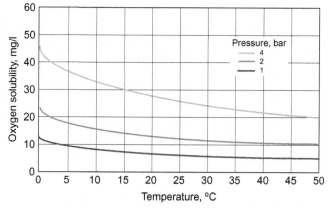

Figure B.2 Oxygen solubility in seawater (salinity 35‰) for pressures 1, 2, and 4 bar (www. EngineeringToolBox.com).

Table B.21 Solubility of Carbon Dioxide in Seawater (mol l^{-1}) of Various Salinities (Salinity, g $kg^{-1} \times 10^{-4}$)

°C	0, S	15, S	16, S	17, S	18, S	19, S	20, S	21, S
0	770	674	667	660	653	646	640	633
2	712	623	617	611	605	599	593	587
4	662	578	573	567	562	557	551	587
6	619	538	533	528	524	519	514	509
8	576	504	499	495	490	486	482	477
10	536	472	468	464	460	456	452	448
12	502	442	438	434	431	428	424	421
14	472	316	413	510	406	403	400	396
16	442	393	390	387	384	381	377	374
18	417	371	368	365	362	359	356	354
20	394	351	348	346	343	340	337	335
22	372	331	329	327	324	321	319	317
26	332	299	297	294	292	298	287	285
28	314	284	281	279	277	275	273	271
30	299	270	268	266	264	262	260	258

Table B.22 Solubility of Oxygen in Seawater

					$\rightarrow$°C				
S (‰)	0°C	3	6	9	12	15	18	21	23
0	10.2	9.41	8.70	8.08	7.53	7.05	6.61	6.23	5.99
5	9.87	9.10	8.70	7.83	7.30	6.84	6.42	6.05	5.82
10	9.54	8.80	8.42	7.58	7.08	6.63	6.23	5.87	5.65
15	9.22	8.51	8.15	7.34	6.86	6.43	6.05	5.70	5.49
20	8.91	8.23	7.89	7.11	6.65	6.24	5.87	5.54	5.34
25	9.61	7.96	7.64	6.89	6.44	6.05	5.69	5.38	5.18
30	8.32	7.70	7.39	6.67	6.24	5.87	5.53	5.22	5.04
31	8.27	7.65	7.15	6.63	6.21	5.83	5.49	5.19	5.01
32	8.21	7.60	7.11	6.59	6.17	5.79	5.46	5.16	4.98
33	8.16	7.55	7.06	6.54	6.13	5.76	5.43	5.13	4.95
34	8.10	7.50	7.01	6.50	6.09	5.72	5.40	5.10	4.92
35	8.05	7.45	6.97	6.46	6.05	5.6	5.36	5.07	4.89
36	7.99	7.40	6.92	6.42	6.01	5.65	5.33	5.04	4.87
37	7.94	7.35	6.88	6.38	5.98	5.62	5.30	5.01	4.84
38	7.88	7.30	6.83	6.34	5.94	5.58	5.27	4.98	4.81

B.6 VARIOUS CHARACTERISTICS

Table B.23 Ice Vapor Pressure at Equilibrium Point and the Corresponding Specific Vapor Density

t (°C)	$p_{v\text{-ice}}$ (μbar)	$\rho_{v\text{-ice}}$ (10^{-3} g m^{-3})	t (°C)	$p_{v\text{-ice}}$ (μbar)	$\rho_{v\text{-ice}}$ (10^{-3} g m^{-3})
−100	0.01403	0.001756	−40	0.1283	0.1192
−98	0.02101	0.02599	−38	0.1606	0.1480
−96	0.03117	0.03812	−36	0.2002	0.1480
−94	0.04584	0.05544	−34	0.2488	0.2254
−92	0.06685	0.07996	−32	0.3079	0.2767
−90	0.09672	0.1144	−30	0.3798	0.3385
−88	0.1388	0.1624	−28	0.4669	0.4127
−86	0.1977	0.2289	−26	0.5720	0.5015
−84	0.2796	0.3203	−24	0.6985	0.6075
−82	0.3925	0.4449	−22	0.8502	0.7336
−80	0.5473	0.6138	−21	0.9370	0.8053
−78	0.7377	0.8413	−20	1.032	0.8815
−76	1.042	1.145	−19	1.135	0.9678
−74	1.425	1.550	−18	1.248	1.060
−72	1.936	2.085	−17	1.371	1.160
−70	2.615	2.789	−16	1.506	1.269
−68	3.511	3.708	−15	1.652	1.387
−66	4.688	4.903	−14	1.811	1.515
−64	6.225	6.449	−13	1.984	1.653
−62	8.223	8.428	−12	2.172	1.803
−60	10.80	10.98	−11	2.376	1.964
−58	14.13	14.23	−10	2.597	2.139
−56	18.38	18.34	−9	2.837	2.328
−54	23.80	23.53	−8	3.097	2.532
−52	30.67	30.5	−7	3.379	2.752
−50	39.35	38.21	−6	3.685	2.990
−48	50.26	48.37	−5	4.015	3.246
−46	63.93	60.98	−4	4.372	3.521
−44	80.97	76.56	−3	4.757	3.817
−42	102.1	95.70	−2	5.173	4.136
			−1.0	5.623	4.479

Table B.24 Conversion Values of g m^{-3} to Equivalents

Component	Chemical Presentation	eq m^{-3} to g m^{-3}	g m^{-3} to eq m^{-3}	eq m^{-3} to eq CaCO$_3$
Calcium	Ca	20.04	0.04991	2.4970
Magnesium	Mg	12.16	0.08224	4.1151
Sodium	Na	23.00	0.04348	2.1756
Potassium	K	39.10	0.02458	1.2798
Chlorides	Cl	35.46	0.02820	1.4112
Carbonates	CO$_3$	30.00	0.03330	1.6680
Bicarbonates	HCO$_3$	61.01	0.01639	0.8202
Sulfates	SO$_4$	48.04	0.02082	1.0416
Nitrates	NO$_3$	62.01	0.01623	0.8070
Hydroxide	OH	17.01	0.05879	2.9263
Phosphates	PO$_4$	31.67	0.03158	1.5800
Calcium bicarbonate	Ca(HCO$_3$)	81.05	0.01234	0.6174
Calcium carbonate	CaCO$_3$	50.04	0.01998	1.0000
Calcium chloride	CaCl$_2$	55.50	0.01802	0.9016
Calcium hydroxide	Ca(OH)$_2$	37.05	0.02699	1.3506
Calcium sulfate	CaSO$_4$	68.07	0.01469	0.7351
Magnesium bicarbonate	Mg(HCO$_3$)	73.17	0.01367	0.6839
Magnesium carbonate	MgCO$_3$	42.16	0.02100	1.1869
Magnesium chloride	MgCl$_2$	47.62	0.03428	1.0508
Magnesium hydroxide	Mg(OH)$_2$	29.17	0.01661	1.7154
Magnesium sulfate	MgSO$_4$	60.20	0.01341	0.8312
Potassium chloride	KCl	74.56	0.01341	0.6711
Sodium bicarbonate	NaHCO$_3$	84.01	0.01190	0.5956
Sodium carbonate	Na$_2$CO$_3$	53.00	0.01887	0.9442
Sodium chloride	NaCl	58.46	0.01711	0.8560
Sodium hydroxide	NaOH	40.01	0.02449	1.2507
Sodium sulfate	Na$_2$SO$_4$	71.04	0.01418	0.7044
Sodium nitrate	NaNO$_3$	85.01	0.01776	0.5886
Sodium phosphate	Na$_2$PO$_4$	54.71	0.01829	0.9153
Iron	Fe	27.92	0.03582	1.7923
Iron carbonate II	FeCO$_3$	57.92	0.01727	0.8640
Iron sulfate II	FeSO$_4$	75.96	0.01316	0.6635

APPENDIX C

C.1 DIAGRAMS OF SEAWATER PROPERTIES

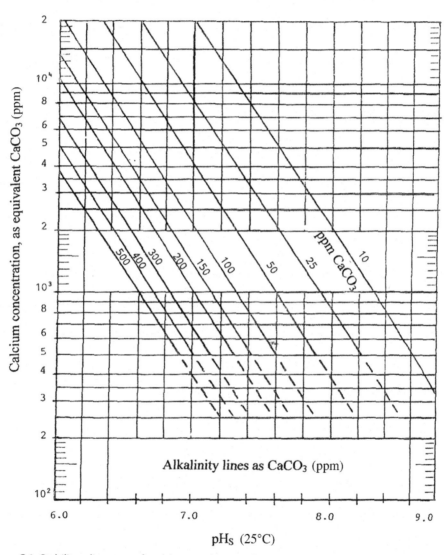

Figure C.1 Stability diagram of calcium carbonate for various solution alkalinities at 100°C. *McKetta, J.J., Executive Editor, 1994. Encyclopedia of Chemical Processing and Design, vols. 44 and 66, Marcel Dekker Inc., New York.*

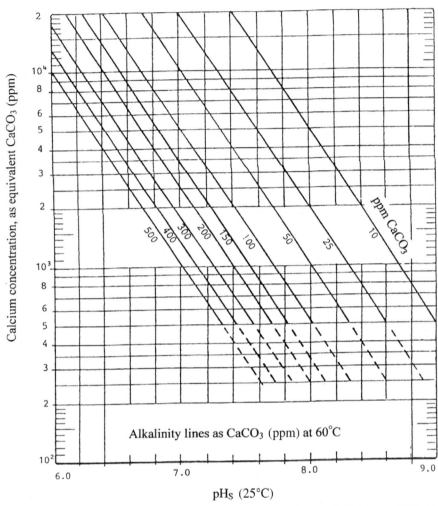

Figure C.2 Stability diagram of calcium carbonate for various solution alkalinities at 60°C. *McKetta, J.J., Executive Editor, 1994. Encyclopedia of Chemical Processing and Design, vols. 44 and 66, Marcel Dekker Inc., New York.*

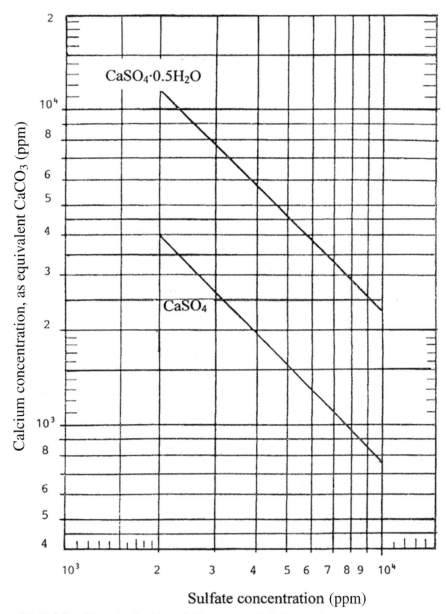

Figure C.3 Stability diagram of calcium sulfate at 100°C (CaSO₄ and CaSO₄ · 0.5H₂O). *McKetta, J.J., Executive Editor, 1994. Encyclopedia of Chemical Processing and Design, vols. 44 and 66, Marcel Dekker Inc., New York.*

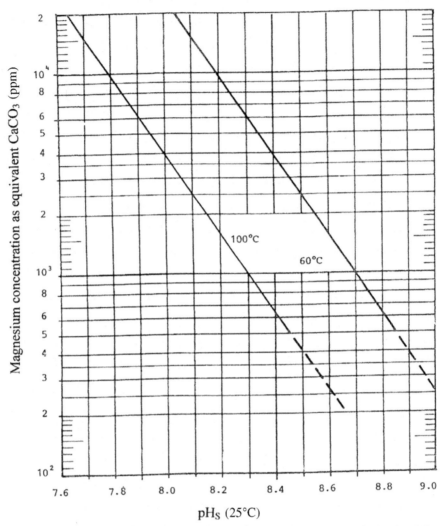

Figure C.4 Stability diagram of magnesium hydroxide (MgOH)$_2$ at 60°C and 100°C. *McKetta, J.J., Executive Editor, 1994. Encyclopedia of Chemical Processing and Design, vols. 44 and 66, Marcel Dekker Inc., New York.*

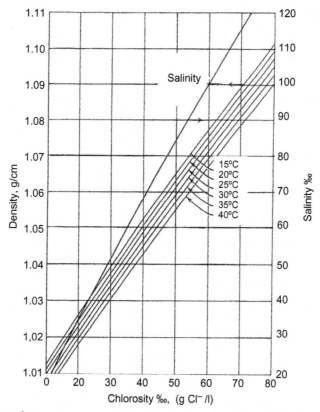

Figure C.5 Density of seawater concentrates at various temperatures. *Spiegler, K.S., El-Sayed, Y.M., 1994. A Desalination Primer, Balaban Desalination Publications, 215 pp.*

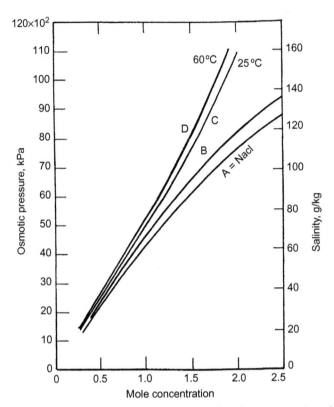

Figure C.6 Curves of osmotic pressure (Π) as a function of mole concentration. A is pure sodium chloride solutions. B is seawater as function of salinity. Both A and B are at ambient temperatures. D and C are seawaters at temperatures of 25°C and 60°C, respectively, as function of mole concentration. *McKetta, J.J., Executive Editor, 1994. Encyclopedia of Chemical Processing and Design, vols. 44 and 66, Marcel Dekker Inc., New York.*

INDEX

Note: Page numbers followed by "*f*" and "*t*" refer to figures and tables, respectively.

Printed in the United States
By Bookmasters